Intelligent Control Systems Using Soft Computing Methodologies

Intelligent Control Systems Using Soft Computing Methodologies

Edited by
Ali Zilouchian
Mo Jamshidi

CRC Press
Boca Raton London New York Washington, D.C.

Library of Congress Cataloging-in-Publication Data

Intelligent control systems using soft computing methodologies / edited by Ali Zilouchian and Mohammad Jamshidi.
    p. cm.
    Includes bibliographical references and index.
    ISBN 0-8493-1875-0
    1. Intelligent control systems—Data processing. 2. Soft computing. I. Zilouchian, Ali.
II. Jamshidi, Mohammad.

TJ217.5 .I5435 2001
629.89′0285′63—dc21   2001016189

Visit the CRC Press Web site at www.crcpress.com

To my late grandfather, Gholam-Reza for his devotion to science and humanitarian causes

A. Zilouchian

To my family, Jila, Ava and Nima for their love and patience

M. Jamshidi

PREFACE

Since the early 1960s, artificial intelligence (AI) has found its way into industrial applications – mostly in the area of expert knowledge-based decision making for the design and monitoring of industrial products or processes. That fact has been enhanced with advances in computer technology and the advent of personal computers, and many applications of intelligence have been realized. With the invention of fuzzy chips in the1980s, fuzzy logic received a high boost in industry, especially in Japan. In this country, neural networks and evolutionary computations were also receiving unprecedented attention in both academia and industry. As a result of these events, "soft computing" was born.

Now at the dawn of the 21st century, soft computing continues to play a major role in modeling, system identification, and control of systems – simple or complex. The significant industrial uses of these new paradigms have been found in the U.S.A and Europe, in addition to Japan. However, to be able to design systems having high MIQ® (machine intelligence quotient, a concept first introduced by Lotfi Zadeh), a profound change in the orientation of control theory may be required.

The principal constituents of soft computing are fuzzy logic, neurocomputing, genetic algorithms, genetic programming, chaos theory, and probabilistic reasoning. One of the principal components of soft computing is fuzzy logic. The role model for fuzzy logic is the human mind. From a control theoretical point of view, fuzzy logic has been intermixed with all the important aspects of systems theory: modeling, identification, analysis, stability, synthesis, filtering, and estimation. Interest in stability criteria for fuzzy control systems has grown in recent years. One of the most important difficulties with the creation of new stability criteria for any fuzzy control system has been the analytical interpretation of the linguistic part of fuzzy controller IF-THEN rules. Often fuzzy control systems are designed with very modest or no prior knowledge of a solid mathematical model, which, in turn, makes it relatively difficult to tap into many tools for the stability of conventional control systems. With the help of Takagi-Sugeno fuzzy IF-THEN rules in which the consequences are analytically derived, sufficient conditions to check the stability of fuzzy control systems are now available. These schemes are based on the stability theory of interval matrices and those of the Lyapunov approach. Frequency-domain methods such as describing functions are also being employed for this purpose.

This volume constitutes a report on the principal elements and important applications of soft computing as reported from some of the active members of this community. In its chapters, the book gives a prime introduction to soft computing with its principal components of fuzzy logic, neural networks, genetic algorithms, and genetic programming with some textbook-type problems given. There are also many industrial and development efforts in the applications of intelligent systems through soft computing given to guide the interested readers on their research interest track.

This book provides a general foundation of soft computing methodologies as well as their applications, recognizing the multidisciplinary nature of the subject. The book consists of 21 chapters, organized as follows:

In *Chapter 1*, an overview of intelligent control methodologies is presented. Various design and implementation issues related to controller design for industrial applications using soft computing techniques are briefly discussed in this chapter. Furthermore, an overall evaluation of the intelligent systems is presented therein.

The next two chapters of the book focus on the fundamentals of neural networks (NN). Theoretical as well as various design issues related to NN are discussed. In general, NN are composed of many simple elements emulating various brain activities. They exploit massive parallel local processing and distributed representation properties that are believed to exist in the brain. The primary purpose of NN is to explore and produce human information processing tasks such as speech, vision, knowledge processing, and motor control. The attempt of organizing human information processing tasks highlights the classical comparison between information processing capabilities of the human and so called hard computing. The computer can multiply large numbers at fast speed, yet it may not be capable to understand an unconstrained pattern such as speech. On the other hand, though humans understand speech, they lack the ability to compute the square root of a prime number without the aid of pencil and paper or a calculator. The difference between these two opposing capabilities can be traced to the processing methods which each employs. Digital computers rely upon algorithm-based programs that operate serially, are controlled by CPU, and store the information at a particular location in memory. On the other hand, the brain relies on highly distributed representations and transformations that operate in parallel, have distributed control through billions of highly interconnected neurons or processing elements, and store information in various straight connections called synapses. *Chapter 2* is devoted to the fundamental issues above. In *Chapter 3*, supervised learning with emphasis on back propagation and radial basis neural functions algorithms is presented. This chapter also addresses unsupervised learning (Kohonen self-organization) and recurrent networks (Hopfield).

In *Chapters 4 – 7*, several applications of neural networks are presented in order to familiarize the reader with design and implementation issues as well as applicability of NN to science and engineering. These applications areas include medicine and biology (*Chapter 4*), digital signal processing (*Chapter 5*), computer networking (*Chapter 6*), and oil refinery (*Chapter 7*).

Chapters 8, 9 and 10 of the book are devoted to the theoretical aspect of fuzzy set and fuzzy logic (FL). The main objective of these three chapters is to provide the reader with sufficient background related to implementation issues in the following chapters. In these chapters, we cover the fundamental concepts of fuzzy sets, fuzzy relation, fuzzy logic, fuzzy control, fuzzification, defuzification, and stability of fuzzy systems.

As is well known, the first implementation of Professor Zadeh's idea pertaining to fuzzy sets and fuzzy logic was accomplished in 1975 by Mamedani, who demonstrated the viability of fuzzy logic control (FLC) for a small model steam engine. After this pioneer work, many consumer products as well as other high tech applications using fuzzy technology have been developed and are currently available on the market. In *Chapters 11 – 16*, several recent industrial applications of fuzzy logic are presented. These applications include navigation of autonomous planetary rover (*Chapter 11*), autonomous underwater vehicle (*Chapter 12*), management of air conditioning, heating and cooling systems (*Chapter 13*), robot manipulators (*Chapter 14*), desalination of seawater (*Chapter 15*), and object recognition (*Chapter 16*).

Chapter 17 presents a brief introduction to evolutionary computations. In *Chapters (18 – 20),* several applications of evolutionary computations are explored. The integration of these methodologies with fuzzy logic is also presented in these chapters. Finally, some examples and exercises are provided in *Chapter 21*. MATLAB neural network and fuzzy logic toolboxes have been utilized to solve several problems.

The editors would like to take this opportunity to thank all the authors for their contributions to this volume and to the soft computing area. We would like to thank Professor Lotfi A. Zadeh for his usual visionary ideas and support. The encouragement and patience of CRC Press Editor Nora Konopka is very much appreciated. Without her continuous help and assistance during the entire course of this project, we could not have accomplished the task of integrating various chapters into this volume. The editors are also indebted to many who helped us realize this volume. Hooman Yousefizadeh, a Ph.D. student at FAU, has modified several versions of various chapters of the book and organized them in camera-ready format. Without his dedicated help and commitment, the production of the book would have taken a great deal longer. We sincerely thank Robert Caltagirone, Helena Redshaw, and Shayna Murry from CRC Press for their assistance. We would like to also thank the project editor, Judith Simon Kamin from CRC Press for her commitment and skillful effort of editing and processing several iterations of the manuscript. Finally, we are indebted to our families for their constant support and encouragement throughout the course of this project.

Ali Zilouchian   *Mo Jamshidi*
Boca Raton, FL   Albuquerque, NM

ABOUT THE EDITORS

Ali Zilouchian is currently a professor and the director of the Intelligent Control laboratory funded by the National Science Foundation (NSF) in the department of electrical engineering at Florida Atlantic University, Boca Raton, FL. His recent works involve the applications of soft computing methodologies to industrial processes including oil refineries, desalination processes, fuzzy control of jet engines, fuzzy controllers for car engines, kinematics and dynamics of serial and parallel robot manipulators. Dr. Zilouchian's research interests include the industrial applications of intelligent controls using neural network, fuzzy logic, genetic algorithms, data clustering, multidimensional signal processing, digital filtering, and model reduction of large scale systems. His recent projects have been funded by NSF and Motorola Inc. as well as several other sources.

He has taught more than 22 different courses in the areas of intelligent systems, controls, robotics, computer vision, digital signal processing, and electronic circuits at Florida Atlantic University and George Washington University. He has supervised 13 Ph.D. and M.S. students during the last 15 years. In addition, he has served as a committee member on more than 25 MS theses and Ph.D. dissertations. He has published over 100 book chapters, textbooks, scholarly journal papers, and refereed conference proceedings. In 1996, Dr. Zilouchian was honored with a Florida Atlantic University Award for Excellence in Undergraduate Teaching.

Dr. Zilouchian is a senior member of IEEE, member of Sigma Xi and New York Academy of Science and Tau Beta Pi. He received the outstanding leadership award for IEEE branch membership development activities for Region III in 1988. He has served as session chair and organizer of nine different sessions in the international conferences within the last five years. He was a keynote speaker at the International Conference on Seawater Desalination Technologies in November 2000. Dr. Zilouchian is currently an associate editor of the *International Journal of Electrical and Computer Engineering* out of Oxford, UK. He is also the local chairman of the next WAC 2002 to be held in June 2002 in Orlando, Florida.

Mohammad (Mo) Jamshidi (Fellow IEEE, Fellow ASME, Fellow AAAS) earned a Ph.D. degree in electrical engineering from the University of Illinois at Urbana-Champaign in February 1971. He holds an honorary doctorate degree from Azerbaijan National University, Baku, Azerbaijan, 1999. Currently, he is the Regents professor of electrical and computer engineering, the AT&T professor of manufacturing engineering, professor of mechanical engineering and founding director of the NASA Center for Autonomous

Control Engineering (ACE) at the University of New Mexico, Albuquerque. He was on the advisory board of NASA JPL's Pathfinder Project mission, which landed on Mars on July 4, 1997. He is currently a member of the NASA Minority Businesses Resource Advisory Committee and a member of the NASA JPL Surface Systems Track Review Board. He was on the USA National Academy of Sciences NRC's Integrated Manufacturing Review Board. Previously he spent 6 years at U.S. Air Force Phillips (formerly Weapons) Laboratory working on large scale systems, control of optical systems, and adaptive optics. He has been a consultant with the Department of Energy's Los Alamos National Laboratory and Oak Ridge National Laboratory. He has worked in various academic and industrial positions at various national and international locations including with IBM and GM Corporations.

He has contributed to over 475 technical publications including 45 books and edited volumes. Six of his books have been translated into at least one foreign language. He is the founding editor, co-founding editor, or editor-in-chief of five journals (including Elsevier's *International Journal of Computers and Electrical Engineering*) and one magazine (*IEEE Control Systems Magazine*). He has been on the executive editorial boards of a number of journals and two encyclopedias. He was the series editor for ASME Press Series on Robotics and Manufacturing from 1988 to 1996 and Prentice Hall Series on Environmental and Intelligent Manufacturing Systems from 1991 to 1998. In 1986 he helped launch a specialized symposium on robotics which was expanded to International Symposium on Robotics and Manufacturing (ISRAM) in 1988, and since 1994, it has been expanded into the World Automation Congress (WAC) where it now encompasses six main symposia and forums on robotics, manufacturing, automation, control, soft computing, and multimedia and image processing. He has been the general chairman of WAC from its inception.

Dr. Jamshidi is a fellow of the IEEE for contributions to "large-scale systems theory and applications and engineering education," a fellow of the ASME for contributions to "control of robotic and manufacturing systems," a fellow of the AAAS – the American Association for the Advancement of Science – for contributions to "complex large-scale systems and their applications to controls and optimization". He is also an associate fellow of Third World Academy of Sciences (Trieste, Italy), member of Russian Academy of Nonlinear Sciences, associate fellow, Hungarian Academy of Engineering, corresponding member of the Persian Academies of Science and Engineering, a member of the New York Academy of Sciences and recipient of the IEEE Centennial Medal and IEEE Control Systems Society Distinguished Member Award and the IEEE CSS Millennium Award. He is an honorary professor at three Chinese universities. He is on the board of Nobel Laureate Glenn T. Seaborg Hall of Science for Native American Youth.

CONTRIBUTORS

Akbarzadeh-T, Mohammad
Department of EECE
Ferdowsi University
Mashad, Iran

Battle, Darryl
Department of Electrical
    Engineering
North Carolina A&T University
Greensboro, NC

Bawazir, Khalid
Aramco
Dhahran, Saudi Arabia

Chen, Tan Kay
The National
    University of Singapore
Singapore

Dozier, Gerry
Computer Science and
    Software Engineering
Auburn University
Auburn, AL

El-Osery, Aly
Department of Electrical and
    Computer Engineering
University of New Mexico
Albuquerque, NM

Fathi, Madjid
Department of Electrical and
    Computer Engineering
University of New Mexico
Albuquerque, NM

Hildebrand, Lars
University of Dortmund
Dortmund, Germany

Homaifar, Abdollah
Department of Electrical
    Engineering
North Carolina A&T University
Greensboro, NC

Howard, Ayanna
Jet Propulsion Laboratory
Pasadena, CA

Howard, David
Department of Electrical
    Engineering
Florida Atlantic University
Boca Raton, FL

Jafar, Mutaz
Kuwait Institute of
    Scientific Research
Kuwait City, Kuwait

Jamshidi, Mo
Department of Electrical and
    Computer Engineering
University of New Mexico
Albuquerque, NM

Lee, T.H.
The National
    University of Singapore
Singapore

Meghdadi, A. H.
Department of Electrical
  Engineering
Ferdowsi University
Mashad, Iran

Ross, Timothy J.
Department of Civil Engineering
University of New Mexico
Albuquerque, NM

Seraji, Homayoun
Jet Propulsion Laboratory
Pasadena, CA

Smith, Samuel M.
Institute for Ocean and
  Systems Engineering
Florida Atlantic University
Dania, FL

Song, Feijun
Institute for Ocean and
  Systems Engineering
Florida Atlantic University,
Dania, FL

Talebi-Daryani, Reza
Department of Control Engineering
University of Applied Sciences
Cologne, Germany

Tan, K. C.
The National
  University of Singapore
Singapore

Tunstel, Edward
Jet Propulsion Laboratory
Pasadena, CA

Valafar, Faramarz
Department of Cognitive and
  Neural Systems
Boston University
Boston, MA

Wang, Dali
STM Wireless, Inc.
Irvine, CA

Wang, M. L.
The National
  University of Singapore
Singapore

Yousefizadeh, Homayoun
Procom Technology, Inc.
Santa Ana, CA

Yousefizadeh, Hooman
Department of Electrical
  Engineering
Florida Atlantic University
Boca Raton, FL

Zilouchian, Ali
Department of Electrical
  Engineering
Florida Atlantic University
Boca Raton, FL

ABBREVIATIONS

1D	One Dimension
2D	Two Dimension
A/C	Air Conditioning
ACS	Average Changes in Slope
ADALINE	ADAptive LINear Element
AI	Artificial Intelligence
ANFIS	Adaptive Neuro-Fuzzy Inference System
ANN	Artificial Neural Network
AUV	Autonomous Underwater Vehicle
BP	Back Propagation
BPA	Back Propagation Algorithm
CBR	Constant Bit Rate
CCSN	Common Channel Signaling Network
CP	Complete Partitioning
CP	Candidate Path
CRDF	Causal Recursive Digital Filters
CS	Complete Sharing
CT	Cellulose Triacetate
CV	Containment Value
D	Derivative
DCS	Distributed Control Systems
DDC	Distributed Digital Control
DNS	Dynamic Neural Sharing
DOF	Degree Of Freedom
EA	Evolutionary Algorithm
EAL	Estimated Average Latency
EC	Evolutionary Computation
ED	Electrodialysis
FAM	Fuzzy Associate Memory
FGN	Fractal Gaussian Noise
FIR	Finite Impulse Response
FIS	Fuzzy Inference System
FL	Fuzzy Logic
FLC	Fuzzy Logic Controller
FNF	False Negative Fraction
FOV	Field of View
FPF	False Positive Fraction
FRBS	Fuzzy Rule Based System
FTDM	Fixed Time Division Multiplexing
FTSA	Fuzzy Tournament Selection Algorithm
GA	Genetic Algorithm
GC-EIMS	Gas Chromatography-Electron Impact Mass Spectroscopy

GEPOA	Global Evolutionary Planning and Obstacle Avoidance
GP	Genetic Programming
GPD	Gallon Per Day
GPM	Gallon Per Minute
HFF	Hollow Fine Fiber
HIS	Health and Safety Indicators
I	Integral
IE	Ion Exchange
IIR	Infinite Impulse Response
LMS	Least Mean Square
LSS	Local State Space
MADALINE	Multiple ADALINE
MAL	Measured Average Latency
MCV	Mean Cell Volume
ME	Multi- Effect
MF	Membership Function
MFC	Membership Function Chromosome
MIMO	Multi Input Multi Output
MISO	Multi Input Single Output
MLE	Maximum Likelihood Types Estimates
MSF	Multi- Stage Flash
NB	Negative Big
NL	Negative Large
NM	Negative Medium
NMR	Nuclear Magnetic Resonance
NN	Neural Network
NS	Negative Small
OAM	Optimal Associative Memory
OEX	Ocean Explorer
OR	Operations Research
P	Proportional
PA	Predictive Accuracy
PB	Positive Big
PCP	Piecewise Continuous Polynomial
PD	Proportional Derivative
PE	Processing Element
PI	Proportional Integral
PID	Proportional Integral-Derivative
PL	Positive Large
PLC	Programmable Logic Controller
PM	Positive Medium
PS	Positive Small
PSI	Pressure Per Square Inch
PV	Predictive Value
RBFN	Radial Basis Function Network

RI	Radius of Influence
RMS	Recursive Mean Square
RO	Reverse Osmosis
ROC	Receiver Operating Characteristic
RVP	Read Vapor Pressure
SCADA	Supervisory Control and Data Acquisition
SCS	Sum of the Changes in Slope
SDF	Separable-in-Denominator Digital Filters
SDI	Silt Density Index
SGA	Simple Genetic Algorithm
SMC	Sliding Mode Controller
SMFC	Sliding Mode Fuzzy Controller
SPS	Static Partial Sharing
STDM	Statistical Time Division Multiplexing
SW	Spiral Wound
TC	Time Control
TCF	Temperature Correction Factor
TDS	Total Dissolved Solid
TNF	True Negative Function
TPF	True Positive Function
TS	Takagi-Sugeno
VBR	Variable Bit Rate
VBR*	Visibility Base Repair
VC	Vapor Compressions
VSC	Variable Structure Controller
XOR	Exclusive Or

TABLE OF CONTENTS

**Chapter 16   COMPUTATIONAL INTELLIGENCE
APPROACH TO OBJECT RECOGNITION   351**
K.C. Tan, T.H. Lee, and M.L. Wang

Chapter 19   EVOLUTIONARY FUZZY SYSTEMS   409
Mohammad.R. Akbarzadeh-T. and A.H. Meghdadi

1 INTRODUCTION

Ali Zilouchian and Mo Jamshidi

1.1   MOTIVATION

With the increasing complexity of various industrial processes, as well as household appliances, the link among ambiguity, robustness and performance of these systems has become increasingly evident. This may explain the dominant role of emerging "intelligent systems" in recent years [1]. However, the definition of intelligent systems is a function of expectations and the status of the present knowledge: perhaps the "intelligent systems" of today are the "classical systems" of tomorrow.

The concept of intelligent control was first introduced nearly two decades ago by Fu and G. Saridis [2]. Despite its significance and applicability to various processes, the control community has not paid substantial attention to such an approach. In recent years, intelligent control has emerged as one of the most active and fruitful areas of research and development (R&D) within the spectrum of engineering disciplines with a variety of industrial applications.

During the last four decades, researchers have proposed many model-based control strategies. In general, these design approaches involve various phases such as modeling, analysis, simulation, implementation and verification. Many of these conventional and model-based methods have found their way into practice and provided satisfactory solutions to the spectrum of complex systems under various uncertainties [3]. However, as Zadeh articulated as early as 1962 [4] "often the solution of real life problems in system analysis and control has been subordinated to the development of mathematical theories that dealt with over-idealized problems bearing little relation to theory".

In one of his latest articles [5] related to the historical perspective of system analysis and control, Zadeh has considered this decade as the era of intelligent systems and urges for some tuning: "I believe the system analysis and controls should embrace soft computing and assign a higher priority to the development of methods that can cope with imprecision, uncertainties and partial truth."

Perhaps the truth is complex and ambiguous enough to accept contributions from various viewpoints while denying absolute validity to any particular viewpoint in isolation. The exploitation of the partial truth and tolerance for imprecision underlie the remarkable human ability to understand distortions and make rational decisions in an environment of uncertainty and imprecision. Such

1

modern relativism, as well as utilization of the human brain as a role model on the decision making processes, can be regarded as the foundation of intelligent systems design methodology.

In a broad perspective, intelligent systems underlie what is called "soft computing." In traditional hard computing, the prime objectives of the computations are precision and certainty. However, in soft computing, the precision and certainty carry a cost. Therefore, it is realistic to consider the integration of computation, reasoning, and decision making as various partners in a consortium in order to provide a framework for the trade off between precision and uncertainty. This integration of methodologies provides a foundation for the conceptual design and deployment of intelligent systems. The principal partners in such a consortium are fuzzy logic, neural network computing, generic algorithms and probabilistic reasoning. Furthermore, these methodologies, in most part, are complementary rather than competitive [5], [6]. Increasingly, these approaches are also utilized in combination, referred to as "hybrid." Presently, the most well-known systems of this type are neuro-fuzzy systems. Hybrid intelligent systems are likely to play a critical role for many years to come.

Soft computing paradigms and their hybrids are commonly used to enhance artificial intelligence (AI) and incorporate human expert knowledge in computing processes. Their applications include the design of intelligent autonomous systems/controllers and handling of complex systems with unknown parameters such as prediction of world economy, industrial process control and prediction of geological changes within the earth ecosystems. These paradigms have shown an ability to process information, adapt to changing environmental conditions, and learn from the environment.

In contrast to analytical methods, soft computing methodologies mimic consciousness and cognition in several important respects: they can learn from experience; they can universalize into domains where direct experience is absent; and, through parallel computer architectures that simulate biological processes, they can perform mapping from inputs to the outputs faster than inherently serial analytical representations. The trade off, however, is a decrease in accuracy. If a tendency towards imprecision could be tolerated, then it should be possible to extend the scope of the applications even to those problems where the analytical and mathematical representations are readily available. The motivation for such an extension is the expected decrease in computational load and consequent increase of computation speeds that permit more robust control. For instance, while the direct kinematics mapping of a parallel manipulator's leg lengths to pose (position and orientation of its end effector) is analytically possible, the algorithm is typically long and slow for real-time control of the manipulator. In contrast, a parallel architecture of synchronously firing fuzzy rules could render a more robust control [7].

There is an extensive literature in soft computing from theoretical as well as applied viewpoints. The scope of this introductory chapter is to provide an overview of various members of these consortiums in soft computing, namely fuzzy logic (FL), neural networks (NN), evolutionary algorithms (EA) as well as their integration. In section 1.2, justification as well as rationale for the utilization of NN in various industrial applications is presented. Section 1.3, introduces the concept of FL as well as its applicability to various industrial processes. The evolutionary computation is presented in section 1.4. Section 1.5 is devoted to the integration of soft-computing methodologies commonly called hybrid systems. Finally the organization of the book is presented in section 1.6 of this chapter.

1. 2   NEURAL NETWORKS

For many decades, it has been a goal of engineers and scientists to develop a machine with simple elements similar to one found in the human brain. References to this subject can be found even in 19th century scientific literature. During the 1940s, researchers desiring to duplicate the human brain, developed simple hardware (and later software) models of biological neurons and their interconnection systems. McCulloch and Pitts in 1943[8] published the first systematic study on biological neural networks. Four years later the same authors explored the network paradigms for pattern recognition using a single-layer perceptron. Along with the progress, psychologists were developing models of human learning. One such model, that has proved most fruitful, was due to D. O. Hebb, who, in 1949, proposed a learning law that became the starting point for artificial neural networks training algorithm [9]. Augmented by many other methods, it is now well recognized by scientists as indicative of how a network of artificial neurons could exhibit learning behavior. In the 1950s and 1960s, a group of researchers combined these biological and psychological insights to produce the first artificial neural network [9], [10]. Initially implemented as electronic circuits, they were later converted into a more flexible medium of computer simulation. However, from 1960 to 1980, due to certain severe limitations on what a NN could perform, as pointed out by Minsky [11], neural network research went into near eclipse. The discovery of training methods for a multi-layer network of the 1980s has, more than any other factor, been responsible for the recent resurgence of NN.

1.2.1   Rationale for Using NN in Engineering

In general, artificial neural networks (ANNs) are composed of many simple elements emulating various brain activities. They exploit massively parallel local processing and distributed representation properties that are believed to exist in the brain. A major motivation to introduce ANN among many researchers has been the exploration and reproduction of human information processing tasks such as speech, vision, and knowledge processing and motor control. The attempt of organizing such information processing tasks highlights the classical comparison between information processing capabilities of the human and so called hard computing. The computer can multiply large numbers at fast speed, yet it may not be capable of understanding an unconstrained pattern such as speech. On the other hand, though a human being understands speech, he lacks the ability to compute the square root of a prime number without the aid of pencil and paper or a calculator. The difference between these two opposing capabilities can be traced to different processing methods which each employs. Digital computers rely upon algorithm-based programs that operate serially, controlled by CPU, and store the information at a particular location in memory. On the other hand, the brain relies on highly distributed representations and transformations that operate in parallel, distribute control through billions of highly interconnected neurons or processing elements, and store information in various straight connections called synapses.

During the last decade, various NN structures have been proposed by researchers in order to take advantage of such human brain capabilities. In general, neural networks are composed of many simple elements operating in parallel. The network function is determined largely by the connections between these elements. Neural networks can be trained to perform complex functions due to the nature of their nonlinear mappings of input to output data set.

In recent years, the NN has been applied successfully to many fields of engineering such as aerospace, digital signal processing, electronics, robotics, machine vision, speech, manufacturing, transportation, controls and medical engineering [12]-[60]. A partial list of NN industrial applications includes temperature control [20], [21]; inverted pendulum controller [22], [23]; robotics manipulators [24]-[30] servo motor control [31]-[34]; chemical processes [35]-[37]; oil refinery quality control [38]; aircraft controls and touchdown [12], [39]; character recognition [16], [40]-[42]; process identification [43]-[47]; failure detection [48]; speech recognition [40]; DSP architectures [49]; truck backer [50]; autonomous underwater vehicle [51], Communication[52];steel rolling mill [53] and car fuel injection system [54],and medical diagnosis and applications [15], [55]-[60]. Detailed descriptions of the works can be found in relevant references.

1.3   FUZZY LOGIC CONTROL

The fuzzy logic has been an area of heated debate and much controversy during the last three decades. The first paper in fuzzy set theory, which is now considered to be the seminal paper on the subject, was written by Zadeh [61], who is considered the founding father of the field. In that work, Zadeh was implicitly advancing the concept of human approximate reasoning to make effective decisions on the basis of available imprecise linguistic information [62], [63]. The first implementation of Zadeh's idea was accomplished in 1975 by Mamdani [64], and demonstrated the viability of fuzzy logic control (FLC) for a small model steam engine. After this pioneer work, many consumer products as well as other high tech applications using fuzzy technology have been developed and are currently available in Japan, the U.S. and Europe.

1.3.1   Rationale for Using FL in Engineering

During the last four decades, most control system problems have been formulated by the objective knowledge of the given systems (e.g., mathematical model). However, as we have pointed out in section 1.1, there are knowledge-based systems and information, which cannot be described by traditional mathematical representations.   Such relevant subjective knowledge is often ignored by the designer at the front end, but often utilized in the last phase in order to evaluate design. Fuzzy logic provides a framework for both information and knowledge-based systems. So called knowledge-based methodology is much closer to human thinking and natural language than the traditionally classical logic.

Fuzzy logic controller (FLC) utilizes fuzzy logic to convert the linguistic control strategy based on expert knowledge into an automatic control strategy. In order to use fuzzy logic for control purposes, we need to add a front-end "fuzzifier" and a rear-end "defuzzifier" to the usual input-output data set. A simple fuzzy logic controller is shown in Figure 1.1. It contains four components: rules, fuzzifier, inference engine, and defuzzifier. Once the rule has been established, it can be considered as a nonlinear mapping from the input to the output.

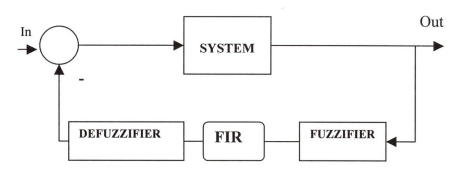

Figure 1.1: A Simple Structure of a Fuzzy Logic Controller.

There are a number of books related to fuzzy logic [65]-[80]. Its applications include automatic train control [6], [67]; robotics [21], [65], [68], [71], [81]-[83]; pattern recognition [2], [7], [67], [71], [75]; servo motor [71], [84], [85], disk drive [86], washing machine [87], [88]; VLSI and fuzzy logic chips [6], [68], [75], [89]; car and helicopter model [6], [65], electronics and home appliances [71], [73], [90]; sensors [71], temperature control [2], [71]; computer vision [71], [73]; aircraft landing systems [71], [73]; navigation and cruise control[71], [91]-[94], inverted pendulum [63],[71],[95]-[97] and cargo ship [98], to name a few. In this book a number of pioneer applications are also presented.

1.4   EVOLUTIONARY COMPUTATION

In recent years, a variety of evolutionary computation methodologies have been proposed to solve problems of common engineering applications. Applications often involve automatic learning of nonlinear mappings that govern the behavior of control systems, as well as parallel search strategies for solving multi-objective optimization problems. These algorithms have been particularly appealing in the scientific communities since they allow autonomous adaptation/optimization without human intervention. These strategies are based on the fact that the biological evolution indeed represents an almost perfect method for adaptation of an individual to the environment according to Darwinian concepts.

There are various approaches to evolutionary optimization algorithms including evolution concept, genetic programming and genetic algorithms. These various algorithms are similar in their basic concepts of evolution and differ mainly in their approach to parameter representation. The evolutionary optimization algorithms operate by representing the optimization parameters via a gene-like structure and subsequently utilizing the basic mechanisms of Darwinian natural selection to find a population of superior parameters. The three basic principles of rules of biological evolution are explained in detail in Chapter 17.

Genetic algorithm (GA), in particular, is an evolutionary algorithm which has performed well in noisy, nonlinear and uncertain processes. Additionally, GAs are also not problem specific, i.e., there is very little, if any, *a priori* knowledge about the system used in design of GAs. Hence, GAs are desirable paradigms for optimizing a wide array of problems with exceeding complexity. The mathematical framework of GA was first developed by Holland [101], and has subsequently been extended [102], [103]. A simple genetic algorithm operates on a finite population of fixed-length binary strings called genes. Genetic algorithms possess three basic operations: reproduction, cross over and mutation. The reproduction is an operation in which the strings are copies based on their fitness. The crossover of genes and mutation of random changes of genes are the other operations in GA. Interested readers are referred to Goldberg [101], Davis [102], Chapter 17 of this book, and the references therein for comprehensive overviews of GA.

Another evolutionary computational approach is genetic programming (GP) which would allow a symbolic-based nonlinear optimization. The GP paradigm [103] also computationally simulates the Darwinian evolution process by applying fitness-based selection and genetic operators to a population of parse trees of a given programming language. It departs from the conventional GA primarily with regard to its representation scheme. Structures undergoing adaptation are executable hierarchical programs of dynamically varying size and structure, rather than numerical strings. Commonly in a hybrid system such as a GP-Fuzzy case, a population comprising fuzzy rule-bases (symbolic structures) that are candidate solutions to the problem, evolves in response to selective pressure induced by their relative success at implementing the desired behavior [103].

1.5   HYBRID SYSTEMS

In many cases, hybrid applications methods have proven to be effective in designing intelligent control systems. As it was shown in recent years, fuzzy logic, neural networks and evolutionary computations are complementary methodologies in the design and implementation of intelligent systems. Each approach has its merits and drawbacks. To take advantage of the merits and eliminate their drawbacks, several integration of these methodologies have been proposed by researchers during the past few years. These techniques include the integration of neural network and fuzzy logic techniques as well as the combination of these two technologies with evolutionary methods.

The merging of the NN and FL can be realized in three different directions, resulting in systems with different characteristics [103]- [108]:
1.   Neuro-fuzzy systems: provide the fuzzy systems with automatic tuning systems using NN as a tool. The adaptive neuro fuzzy inference systems are included in this classification
2.   Fuzzy neural network: retain the functions of NN with fuzzification of some of their elements. For instance, fuzzy logic can be used to determine the learning steps of NN structure.
3.   Fuzzy-neural hybrid systems: utilize both fuzzy logic and neural networks in a system to perform separate tasks for decouple subsystems. The architecture of the systems depends on a particular application. For instance, the NN can be utilized for the prediction where the fuzzy logic addresses the control of the system.

The applications of these hybrid methods to several industrial processes including robot manipulators, desalination plants, and underwater autonomous vehicles will be presented in this book.

On the other hand, the NN, FL and evolutionary computations can be integrated [103], [109]-[123]. For example, the structure and parameter learning problems of neural network can be coded as genes in order to search for optimal structures and parameters of neural network. In addition, the inherent flexibility of the evolutionary computation and fuzzy systems has created a large diversity and variety in how these two complementary approaches can be combined to solve many engineering problems. Some of their applications include control of pH in chemical processes [110], inverted pendulum [111]-[113], cart and poles problem [114], robot trajectory [115], truck-backing problem [116]; automotive active suspension control [117]; temperature control of brine heater [119]; hepatitis diagnosis problem [120]; classification of flowers. [121]and position control of servo systems [122].

In Chapter 18, evolutionary concept and fuzzy logic will be combined for image processing applications. In Chapter 19, the application of GA-fuzzy systems as the most common evolution-based fuzzy system will be presented.

Genetic programming is employed to learn the rules and membership functions of the fuzzy logic controller, and also to handle selection of fuzzy set intersection operators. Finally, Chapter 20 presents a methodology for applying GP to design a fuzzy logic steering controller for a mobile robot.

1.6   ORGANIZATION OF THE BOOK

This book covers basic concepts and applications of intelligent systems using soft computing methodologies and their integration. It is divided into six major parts.

Part I (Chapters 2 – 3) covers the fundamental concepts of neural networks. Single-layer as well as multilayer networks are briefly reviewed. Supervised and unsupervised learning are discussed. Four different NN architectures including back propagation, radial basis functions, Hopfield and Kohonen self-organization are presented.

Part II (Chapters 4 – 7) addresses several applications of NN in science and engineering. The areas of the NN applications include medicine and biology, signal processing, computer networking, chemical process and oil refinery.

Part III (Chapters 8 – 10) of the book covers the fuzzy set theory, fuzzy logic and fuzzy control and stability. In these three chapters, we cover the fundamental concepts of fuzzy sets, fuzzy relation, fuzzy logic, fuzzy control, fuzzification, defuzification and stability of fuzzy systems.

Part IV (Chapters 11 – 16) covers various applications of fuzzy logic control including navigation of autonomous planetary rover, autonomous underwater vehicle, heating and cooling systems, robot manipulators, desalination and object recognition.

Part V (Chapters 17 – 20) covers the concepts of evolutionary computations and their applications to several engineering problems. Chapter 17 presents a brief introduction of evolutionary computations. In the following chapters (18 – 20) several applications of evolutionary computations are explored. Furthermore the integration of these methodologies with the fuzzy logic is presented. Finally, some examples and exercises are provided in Chapter 21. MATLAB neural network and fuzzy logic toolboxes can be used to solve some of these problems.

REFERENCES

1.   Wright, R., Can Machines Think? *Time,* Vol. 147, No. 13, March 1996.
2.   Gupta, M., Saridis, G., and Gaines, B, *Fuzzy Automatica and Decision Processes*, North-Holland, NY, 1977.
3.   Antsaklis, P.J. and Passino, K.M., (eds.), *An Introduction to Intelligent and Autonomous Control*, Kluwer Academic Publishers, Norwell, MA, 1993.
4.   Zadeh, L.A., A Critical View of Our Research in Automatic Control, *IRE Trans. on Automatic Controls*, AC-7, 74, 1962.

5. Zadeh, L.A., The Evolution of Systems Analysis and Control: A Personal Perspective, *IEEE Control Mag.*, Vol. 16, No. 3, 95, 1996.
6. Yager, R. and Zadeh, L.A. (eds.), *An Introduction to Fuzzy Logic Applications in Intelligent Systems,* Kluwer Academic Publishers, Boston, 1992.
7. Diaz-Robainas, R., Zilouchian, A. and Huang, M., Fuzzy Identification on Finite Training-Set with Known Features, *Int. J. Automation Soft Computing.*
8. McCulloch, W.W. and Pitts, W., A Logical Calculus of Ideas Imminent in Nervous Activity, *Bull. Math. Biophy.*, 5, 115, 1943.
9. McClelland, J. L. and Rumelhart, D. E., The PDP Research Group, Parallel Distributed Processing – Explorations in the Microstructure of Cognition, Vol. 2: *Psychological and Biological Models*, MIT Press, MA, 1986.
10. Rosenblatt, F., *Principles of Neurodynamics*, Spartan Press, Washington, DC, 1961.
11. Minsky, M. and Papert, S., *Perceptron: An Introduction to Computational Geometry*, MIT Press, MA, 1969.
12. Miller, W. T., Sutton, R., and Werbos, P., *Neural Networks for Control*, MIT Press, MA, 1990.
13. Zurada, J., *Introduction to Artificial Neural Systems*, West Publishing Co., St. Paul, MN, 1992.
14. Fausett, L, *Fundamentals of Neural Networks*, Prentice-Hall, Englewood Cliffs, NJ. 1994.
15. Croall, I.F. and Mason, J.P. (eds.), *Industrial Applications of Neural Networks*, Springer-Verlag, NY, 1991.
16. Linkens, D.A. (ed.), *Intelligent Control in Biomedicine*, Taylor & Francis, London, 1994.
17. Haykin, S., *Neural Networks: A Comprehensive Foundation*, Prentice-Hall, Upper Saddle River, NJ, 1999.
18. Kosko, B., *Neural Network for Signal Processing*, Prentice-Hall, Englewood Cliffs, NJ, 1992.
19. Kohonen, T., *Self-Organization and Associative Memory*, 3rd ed., Springer-Verlag, NY, 1988.
20. Khalid, M. and Omatu S., A Neural Network Controller for a Temperature Control System, *IEEE Control System Mag.*, 58–64, June 1992.
21. Gupta, M. and Sinha, N. (eds.), *Intelligent Control Systems: Theory and Applications,* IEEE Press, Piscataway, NJ, 1996.
22. Tolat, V., An Adaptive Broom Balancer with Visual Inputs, *Proc. of IEEE Int. Conf. Neural Networks*, 641, 1998.
23. Anderson, C.W., Learning To Control an Inverted Pendulum with Connectionist Networks, *Proc. of Am. Controls Conf.*, 2294, 1988.

24.   Nguyen, L., Patel, R., and Khorasani, K., Neural Network Architectures for the Forward Kinematic Problem in Robotics, *Proc. of IEEE Int. Conf. Neural Networks*, 393, 1990.

25.   Newton, R.T. and Xu, Y., Neural Network Control of a Space Manipulator, *IEEE Control Syst. Mag.*, 14, Dec. 1993.

26.   Miller, W. T., Real-Time Neural Network Control of a Biped Walker Robot, *IEEE Control Syst.*, Vol. 14, No.1, Feb. 1994.

27.   Liu, H., Iberall, T., and Bekey, G., Neural Network Architecture for Robot Hand Control, *IEEE Control Syst.*, Vol. 9, No. 3, 38, April 1989.

28.   Handelman, D., Lane, S., and Gelfand, J., Integrating Neural Networks and Knowledge-Based Systems for Intelligent Robotic Control, *IEEE Control Syst. Mag.*, Vol. 10, No. 3, 77, April 1990.

29.   Rabelo, L.C. and Avula, X., Hierarchical Neuo-controller Architecture for Robotic Manipulation, *IEEE Control Syst. Mag.*, Vol. 12, No.2, 37, April 1992.

30.   Murostsu, Y., Tsujio, S., Sendo, K., and Hayashi, M., Trajectory Control of Flexible Manipulators on a Free-Flying Space Robot, *IEEE Control Syst. Mag.*, Vol. 12, No.3, 51, June 1992.

31.   Zhang, Y., Sen, P., and Hearn, G., An On-Line Trained Adaptive Neural Controller, *IEEE Control Syst.*, Vol. 15, No.5, 67-75, Oct. 1995.

32.   Kupperstien, M. and Rubinstein, J., Implementation of an Adaptive Neural Controller for Sensory-Motor Coordination, *IEEE Control Syst.*, Vol. 9, No. 3, 25, 1989.

33.   Eckmiller, R., Neural Nets for Sensory and Motor Trajectories, *IEEE Control Syst.*, Vol. 9, No. 3, 53, 1989.

34.   Hashimoto, H., Kubota, T., Kudou, M., and Harashima, F., Self-Organization Visual Servo System Based on Neural Networks, *IEEE Control Syst. Mag.*, Vol. 12, No. 2, 31, 1992.

35.   Bhat, N., Minderman, P., McAvoy, T., and Wang, N., Modeling Chemical Process Systems via Neural Network Computation, *IEEE Control Syst. Mag.*, Vol. 10, No.3, 24, 1990.

36.   Borman, S., Neural Network Applications in Chemistry Begin to Appear, *Chemical and Eng. News*, Vol. 67, No. 17, 24, 1989.

37.   Bawazeer, K. H., Prediction of Crude Oil Product Quality Parameters Using Neural Networks, M.S. Thesis, Florida Atlantic University, Boca Raton, FL, 1996.

38.   Draeger, A., Engell, S., and Ranke, H., Model Predictive Control Using Neural Networks, *IEEE Control Syst.*, Vol. 15, No.5, 61, 1995.

39.   Steck, J. E., Rokhsaz, M., and Shue S., Linear and Neural Network Feedback for Flight Control Decoupling, *IEEE Control Syst.*, Vol. 16, No. 4, 22, 1996.

40.   Lippmann, R. P., An Introduction to Computing with Neural Network, *IEEE Acoustic, Speech, and Signal Process. Mag.*, 4, 1987.

41. Guyon, I., Application of Neural Networks to Character Recognition, *Int. J. Pattern Recog. Artif. Intell.,* Vol. 5, Nos. 1 and 2, 353, 1991.
42. LeCun, Y., Jackel, L.D., et al., Handwritten Digital Recognition: Application of Neural Network Chips and Automatic Learning, *IEEE Com. Mag.*, 41,1988.
43. Parlos, A. G., Chong, K.T., and Atiya, A.F., Application of Recurrent Neural Multilayer Perceptron in Modeling Complex Process Dynamic, IEEE *Trans. on Neural Networks*, Vol. 5, No. 2, 255, 1994.
44. Sartori, M. and Antsaklis, P., Implementation of Learning Control Systems Using Neural Networks, *IEEE Control Syst. Mag.*, Vol. 12, No.2, 49-57,1992.
45. Narenda, K. and Parthasarathy, K., Identification and Control of Dynamic Systems Using Neural Networks, *IEEE Trans. on Neural Networks*, Vol.1, 4,1990.
46. Narendra, K.S., Balakrishnan, J., and Cliliz, K., Adaptation and Learning Using Multiple Models, Switching and Tuning, *IEEE Control Syst.*, Vol. 15, No. 3, 37, 1995.
47. Antsaklis, P., Special Issue on Neural Networks for Control System, *IEEE Control Syst. Mag.*, Vol. 10, No.3, 8, 1990.
48. Naida, S., Zafiriou, E., and McAvoy, T., Use of Neural Networks for Sensor Failure Detection in a Control System, *IEEE Control Syst. Mag.*, Vol. 10, No. 3, 49,1990.
49. Radivojevic, I., Herath, J., and Gray, S., High-Performance DSP Architectures for Intelligence and Control Applications, *IEEE Control Syst.,* Vol.11, No. 4, 49, 1991.
50. Nguyen, D. and Widrow, B., Neural Networks for Self-Learning Control Systems, *IEEE Control Syst. Mag.*, 18, 1990.
51. Sanner, R. and Akin, D., Neuromorphic Pitch Attitude Regulation of an Underwater Telerobot, *IEEE Control Syst. Mag.*, Vol. 10, No. 3, 62, 1990.
52. Rauch, H. E. and Winarske, T., Neural Networks for Routing Communication Traffic, IEEE *Control Syst. Mag.*, Vol. 8, No. 2, 26, 1988.
53. Hofer, D. S., Neumerkel, D., and Hunt, K., Neural Control of a Steel Rolling Mill, *IEEE Control Syst.*, Vol. 13, No. 3, 69,1993.
54. Majors, M., Stori, J., and Cho, D., Neural Network Control of Automotive Fuel-Injection Systems, *IEEE Control Syst.*, Vol. 14, No. 3, 31, 1994.
55. Nekovie, R. and Sun, Y., Back-propagation Network and its Configuration for Blood Vessel Detection in Angiograms, *IEEE Trans. on Neural Networks,* Vol. 6, No. 1, 64, 1995.
56. Echauz, J. and Vachtsevanos, G., Neural Network Detection of Antiepileptic Drugs from a Single EEG trace, *Proc. of the IEEE Electro/94 Int. Conf.*, Boston, MA, 346, 1994 .

57.   Charache S., Barton F. B., Moore R. D., et al., *Hydroxyurea and Sickle Cell Anemia*, Vol. 75, No. 6, 300, 1980.

58.   Charache S., Terrin L. M., Moore R. D., et al., Effect of Hydroxyurea on the Frequency of Painful Crises in Sickle Cell Anemia, *N. E. J. Med.*, 332, 1995.

59.   Charache S, Dover G. J., Moore R. D., et al., Hydroxyurea: Effects on Hemoglobin F Production In-Patients With Sickle Cell Anemia, *Blood*, Vol. 79, 10, 1992.

60.   Apolloni, B., Avanzini, G., Cesa-Bianchi, N., and Ronchini, G., Diagnosis of Epilepsy via Backpropagation, *Proc. of the Int. Joint Conf. Neural Networks*, Washington, DC, Vol. 2, 571, 1990.

61.   Zadeh, L.A., Fuzzy Sets, *Information and Control*, 8, 338, 1965.

62.   Zadeh, L.A., A Rationale for Fuzzy Control, *J. Dynamic Syst., Meas. and Control*, Vol. 94, Series G, 3, 1972.

63.   Zadeh, L.A., Making the Computers Think Like People, *IEEE Spectrum*, 1994.

64.   Mamdani, E. H., Application of Fuzzy Algorithms for Control of Simple Dynamic Plant, *Proc. of IEE*, Vol. 121, No. 12, 1974.

65.   Surgeno, M. (ed.*), Industrial Applications of Fuzzy Control*, North-Holland, Amsterdam, 1985.

66.   Yagar, R., Ovchinnikov, S., Tong, R.M., and Nguyen, H.T, *Fuzzy Sets and Applications*, Wiley Interscience, NY, 1987.

67.   Zimmermann, H., *Fuzzy Set Theory and its Applications*, Kluwer Academic Publishers, Boston, 1991.

68.   Ralescu, A. (ed.), *Applied Research in Fuzzy Technology*, Kluwer Academic Publishers, Boston, 1994.

69.   Kaufmann, A. and Gupta, M. (eds.), *Introduction to Fuzzy Arithmetic Theory and Applications*, Van Nostrand Reinhold, NY, 1985.

70.   Bezdek, J., *Pattern Recognition with Fuzzy Objective Function Algorithms*, Plenum Press, NY, 1981.

71.   Marks II,R. (ed.), *Fuzzy Logic Technology and Applications*, IEEE Press, Piscataway, NJ, 1994.

72.   Jamshidi, M., Vadiee N., and Ross, T.J. (eds.), *Fuzzy Logic and Control: Software and Hardware Applications*, Prentice Hall, Englewood Cliffs, NJ, 1993.

73.   Amizadeh, F. and Jamshidi, M., *Soft Computing, Fuzzy Logic, Neural Networks, and Distributed Artificial Intelligence*, Vol. 4, Prentice Hall, Englewood Cliffs, NJ, 1994.

74.   Nguyen, H., Sugeno, M., Tong, R., and Yager, R., *Theoretical Aspects of Fuzzy Control*, John Wiley & Sons, NY, 1995.

75.   Kosko, B., *Fuzzy Engineering*, Prentice Hall, Upper Saddle River, NJ, 1997.

76.   Passino, K., and Yurkovich, S., *Fuzzy Control*, Addison Wesley, Menlo Park, CA, 1998.

77. Cox, E.D., *Fuzzy Logic for Business and Industry*, Charles River Media, Inc., Rockland, MA, 1995.

78. DeSilva, C.W., *Fuzzy Logic and Application*, CRC Press, Boca Raton, FL, 1998.

79. Jamshidi, M., *Large-Scale Systems – Modeling, Control and Fuzzy Logic*, Prentice Hall, Upper Saddle River, NJ, 1996.

80. Langari, G., A Framework for Analysis and Synthesis of Fuzzy Logic, Ph.D. Dissertation, University of California, Berkeley, 1990.

81. Lime, C.M. and Hiyama, T., Application of Fuzzy Control to a Manipulator, *IEEE Trans. on Robotics and Automation*, Vol. 7, 5, 1991.

82. Li, W., Neuro-Fuzzy Systems for Intelligent Robot Navigation and Control Under Uncertainty, *Proc. of IEEE Robotics and Automation Conf.*, 1995.

83. Nedungadi, A., Application of Fuzzy Logic to Solve the Robot Inverse Kinematic Problem, *Proc. of Fourth World Conf. on Robotics Research*, 1, 1991.

84. Li, Y. and Lau, C., Development of Fuzzy Algorithms for Servo System, *IEEE Control Mag.*, 65, 1989.

85. Ready, D. S., Mirkazemi-Moud, M., Green, T., and Williams, B., Switched Reluctance Motor Control Via Fuzzy Adaptive Systems, *IEEE Control Syst.*, Vol. 15, No. 3, 8, 1995.

86. Yoshida, S. and Wakabayashi, N., A Fuzzy Logic Controller for a Rigid Disk Drive, *IEEE Control Syst. Mag.*, Vol. 12, No. 3, 65, 1992.

87. Benison, P., Badami, V., Chiang, K., Khedkar, P., Marcelle, K., and Schutten, M., Industrial Applications of Fuzzy Logic at General Electric, *Proc. of IEEE*, Vol. 83, No, 3, 450, 1995.

88. Schwartz, D., Klir, G., Lewis, H., and Ezawa, Y., Application of Fuzzy Sets and Approximate Reasoning, *Proc. of IEEE*, Vol. 82, No. 4, 482, 1994.

89. Costa, A., DeGloria, A., Faraboschi, P., Pagni, A., and Rizzoto, G., Hardware Solutions for Fuzzy Control, *Proc. of IEEE*, Vol. 83, No. 3, 422, 1995.

90. Takagi, H., *Cooperative System of Neural Network and Fuzzy Logic and its Applications to Consumer Products,* Van Nostrand Reinhold, NY, 1993.

91. Kwong, W.A., Passino, K., Laukonen, E.G., and Yurkovich, S., Expert Supervision of Fuzzy Learning Systems for Fault Tolerant Aircraft Control, *Proc. of IEEE*, Vol. 83, No. 3, 466, 1995.

92. Hessburg, T. and Tomizuka, M., Fuzzy Logic Control for Lateral Vehicle Guidance, *IEEE Control Syst.*, Vol. 14, No. 4, 55, 1994.

93. Chiu, S., Chand, S., Moore, D., and Chaudhary, A., Fuzzy Logic for Control of Roll and Moment for a Flexible Wing Aircraft, *IEEE Control Syst.*, Vol.11, No. 4, 42, 1991.

94.   Vachtesanos, G., Farinwata, S., and Pirovolou, D., Fuzzy Logic control of an Automotive Engine, *IEEE Control System*, Vol. 13, No. 3, 62, 1993.

95.   Takagi, T. and Sugeno, M., Fuzzy Identification of Systems and its Applications to Modeling and Control, *IEEE Trans. on Syst., Man, and Cyb.*, Vol. 15, No.1, 1985.

96.   Lee, C. H., Fuzzy logic in Control Systems: Fuzzy Logic Controller, Part II, *IEEE Trans. on Syst., Man and Cyb.*, Vol. 20, No. 2, 419, 1990.

97.   Berenji, H. and Khedhar, P., Learning and Tuning Logic Controller Through Reinforcements, *IEEE Trans. on Neural Networks*, Vol. 3, No. 5, 724, 1992.

98.   Layne, J. and Passino, K., Fuzzy Model Reference Learning Control for Cargo Ship Steering, *IEEE Control Syst.*, Vol. 13, No. 5, 23, 1993.

99.   Fogel, L. J., *Intelligence Through Simulated Evolution*, John Wiley & Sons, NY, 1999.

100.   Holland, J.H., *Adaptation in Natural and Artificial Systems*, University of Michigan Press, MI, 1975.

101.   Goldberg, D.E., *Genetic Algorithms in Search, Optimization and Machine Learning*, Addison-Wesley, Reading, MA, 1989.

102.   Davis, L. (ed.), *Handbook of Genetic Algorithms*, Van Nostrand Reinhold, N Y, 1991.

103.   Koza, J. R., *Genetic Programming – On the Programming of Computers by Means of Natural Selection*, MIT Press, MA, 1992.

104.   Linkens, D.A. and H.O. Nyongeso, Learning systems in Intelligent Approach of Fuzzy, Neural and Genetic Algorithm Control Application, *IEE Proc. Control Theory and Application*, Vol. 143, No. 4, 367, 1996.

105.   Lin, C.T. and Lee, C.S.G., *Neural Fuzzy Systems*, Prentice Hall, Upper Saddle River, NJ, 1996.

106.   Jang, J., Sun, C., and Mizutani, E., *Neuro Fuzzy and Soft Computing*, Prentice Hall, Upper Saddle River, NJ, 1997.

107.   Jang, J.S. and Sun, C., Neuro-Fuzzy Modeling and Control, *Proc. of IEEE*, Vol. 83, No. 3, 378, 1995.

108.   Mitra, S. and Pal, S.K., Self-Organizing Neural Network as a Fuzzy Classifier, *IEEE Trans. on Syst., Man, and Cyb.*, Vol. 24, No. 3, 1994.

109.   Kim, J., Moor, Y., and Zeigler, B., Designing Fuzzy Net Controllers Using Genetic Algorithms, *IEEE Control Syst.*, Vol. 15, No. 3, 66, 1995.

110.   Karr, C.L., Design of an Adaptive Fuzzy Logic Controller Using a Genetic Algorithm, *Proc. of the Fifth Int. Conf. on Genetic Algorithm*, 450, 1991.

111.   Lee, M. A. and Takagi, H., Integrating Design Stages of Fuzzy Systems using Genetic Algorithms, *Proc. of 2nd IEEE Int. Conf. on Fuzzy Syst.*, 612, San Francisco, 1993.

112. Tan, G.V. and Hu, X., On Designing Fuzzy Controllers Using Genetic Algorithms, *IEEE Int. Conf. on Fuzzy Syst.*, 905, 1996.

113. Kinzel, J., Modification of Genetic Algorithms for Design and Optimizing Fuzzy Controllers, *IEEE Int. Conf. on Fuzzy Syst.*,28, 1994.

114. Cooper, M.G. and Vidal, J., Genetic Design of Fuzzy Controllers: The Cart and Jointed -Pole Problem, *IEEE Int. Conf. on Fuzzy Syst.*, 1332, 1994.

115. Xu, H.Y. and Vukovich, G., Fuzzy Evolutionary Algorithms and Automatic Robot Trajectory Generation, *FUZZ-IEEE'94*, 595, 1994.

116. Homaifar A. and McCormick, E., Simultaneous Design of Membership Functions and Rule Sets for Fuzzy Controllers Using Genetic Algorithms, *IEEE Trans. on Fuzzy Syst.*, Vol. 3, No 2, 129, 1995.

117. Moon, S.Y. and Kwon, W. H., Genetic-Based Fuzzy Control for Automotive Active Suspensions, *FUZZ-IEEE'96*, 923, 1996.

118. Akbarzadeh, M.R., Fuzzy Control and Evolutionary Optimization of Complex Systems, Ph.D. Dissertation, The University of New Mexico, 1998.

119. Wang, C.H., Integrating Fuzzy Knowledge by Genetic Algorithms, *IEEE Trans. on Evolutionary Computations*, Vol. 2, No. 4, 138, 1998.

120. Shi,Y., Implementation of Evolutionary Fuzzy Systems, *IEEE Trans. on Fuzzy Syst.*, Vol. 7, No. 2, 109, 1999.

121. Park, Y.J., Lee, S.Y., and Cho, H.S., A Genetic Algorithm-Based Fuzzy Control of an Electro-Hydraulic Fin Position Servo System, *Proc. IEEE Int. Fuzzy Syst.*, 1999, Seoul, Korea.

122. Kumbla, K.K., Adaptive Neuro-Fuzzy Systems for Passive Systems, Ph.D. Dissertation, University of New Mexico, 1997.

123. Tustel, E., Adaptive Hierarchy of Distributed Fuzzy Control, Application to Behavior Control of Rovers, Ph.D. Dissertation, University of New Mexico, 1996.

2 FUNDAMENTALS OF NEURAL NETWORKS

Ali Zilouchian

2.1   INTRODUCTION

For many decades, it has been a goal of science and engineering to develop intelligent machines with a large number of simple elements. References to this subject can be found in the scientific literature of the 19th century. During the 1940s, researchers desiring to duplicate the function of the human brain, have developed simple hardware (and later software) models of biological neurons and their interaction systems. McCulloch and Pitts [1] published the first systematic study of the artificial neural network. Four years later, the same authors explored network paradigms for pattern recognition using a single layer perceptron [2]. In the 1950s and 1960s, a group of researchers combined these biological and psychological insights to produce the first artificial neural network (ANN) [3,4]. Initially implemented as electronic circuits, they were later converted into a more flexible medium of computer simulation. However, researchers such as Minsky and Papert [5] later challenged these works. They strongly believed that intelligence systems are essentially symbol processing of the kind readily modeled on the Von Neumann computer. For a variety of reasons, the symbolic–processing approach became the dominant method. Moreover, the perceptron as proposed by Rosenblatt turned out to be more limited than first expected. [4]. Although further investigations in ANN continued during the 1970s by several pioneer researchers such as Grossberg, Kohonen, Widrow, and others, their works received relatively less attention. The primary factors for the recent resurgence of interest in the area of neural networks are the extension of Rosenblatt, Widrow and Hoff's works dealing with learning in a complex, multi-layer network, Hopfield mathematical foundation for understanding the dynamics of an important class of networks, as well as much faster computers than those of 50s and 60s.

The interest in neural networks comes from the networks' ability to mimic human brain as well as its ability to learn and respond. As a result, neural networks have been used in a large number of applications and have proven to be effective in performing complex functions in a variety of fields. These include pattern recognition, classification, vision, control systems, and prediction [6], [7]. Adaptation or learning is a major focus of neural net research that provides a degree of robustness to the NN model. In predictive modeling, the goal is to map a set of input patterns onto a set of output patterns. NN accomplishes this task by learning from a series of input/output data sets presented to the network. The trained network is then used to apply what it has learned to approximate or predict the corresponding output [8].

This chapter is organized as follows. In section 2.2, various elements of an artificial neural network are described. The Adaptive Linear Element (ADALINE) and single layer perceptron are discussed in section 2.3 and 2.4 respectively. The multi-layer perceptron is presented in section 2.5. Section 2.6 discusses multi-layer perceptron and section 2.7 concludes this chapter.

2.2   BASIC STRUCTURE OF A NEURON

2.2.1   Model of Biological Neurons

In general, the human nervous system is a very complex neural network. The brain is the central element of the human nervous system, consisting of near 10^{10} biological neurons that are connected to each other through sub-networks. Each neuron in the brain is composed of a body, one axon and multitude of dendrites. The neuron model shown in Figure 2.1 serves as the basis for the artificial neuron. The dendrites receive signals from other neurons. The axon can be considered as a long tube, which divides into branches terminating in little endbulbs. The small gap between an endbulb and a dendrite is called a synapse. The axon of a single neuron forms synaptic connections with many other neurons. Depending upon the type of neuron, the number of synapses connections from other neurons may range from a few hundreds to 10^{4}.

The cell body of a neuron sums the incoming signals from dendrites as well as the signals from numerous synapses on its surface. A particular neuron will send an impulse to its axon if sufficient input signals are received to stimulate the neuron to its threshold level. However, if the inputs do not reach the required threshold, the input will quickly decay and will not generate any action. The biological neuron model is the foundation of an artificial neuron as will be described in detail in the next section.

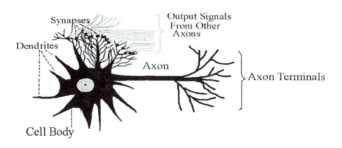

Figure 2.1: A Biological Neuron.

2.2.2 Elements of Neural Networks

An artificial neuron as shown in Figure 2.2, is the basic element of a neural network. It consists of three basic components that include weights, thresholds, and a single activation function.

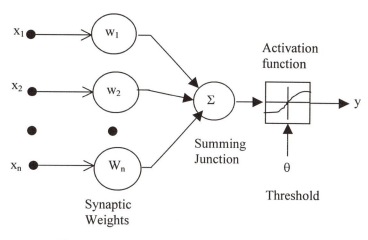

Figure 2.2: Basic Elements of an Artificial Neuron.

2.2.2.1 Weighting Factors

The values $W_1, W_2, W_3, \ldots, W_n$ are weight factors associated with each node to determine the strength of input row vector $X = [x_1\ x_2\ x_3 \ldots, x_n]^T$. Each input is multiplied by the associated weight of the neuron connection $X^T W$. Depending upon the activation function, if the weight is positive, $X^T W$ commonly excites the node output; whereas, for negative weights, $X^T W$ tends to inhibit the node output.

2.2.2.2 Threshold

The node's internal threshold θ is the magnitude offset that affects the activation of the node output y as follows:

$$y = \sum_{i=1}^{n} (X_i W_i) - \theta_k$$

(2.1)

2.2.2.3 Activation Function

In this subsection, five of the most common activation functions are presented. An activation function performs a mathematical operation on the signal output. More sophisticated activation functions can also be utilized depending upon the type of problem to be solved by the network. All the activation functions as described herein are also supported by MATLAB package.

Linear Function
    As is known, a linear function satisfies the superposition concept.   The function is shown in Figure 2.3.

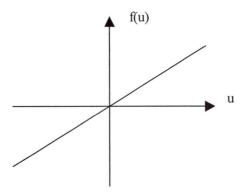

Figure 2.3: Linear Activation Function.

The mathematical equation for the above linear function can be written as

$$y = f(u) = a.u \qquad (2.2)$$

where α is the slope of the linear function 2.2. If the slope α is 1, then the linear activation function is called the identity function. The output (y) of identity function is equal to input function (u). Although this function might appear to be a trivial case, nevertheless it is very useful in some cases such as the last stage of a multilayer neural network.

Threshold Function
    A threshold (hard-limiter) activation function is either a *binary* type or a *bipolar* type as shown in Figures 2.4 and 2.5, respectively. The output of a *binary* threshold function can be written as:

$$y = f(u) = \begin{cases} 0 & if \quad u < 0 \\ \\ 1 & if \quad u \geq 0 \end{cases} \qquad (2.3)$$

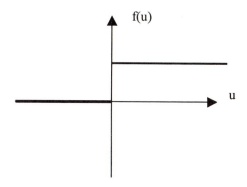

Figure 2.4: Binary Threshold Activation Function.

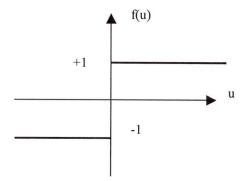

Figure 2.5: Bipolar Threshold Activation Function.

The neuron with the hard limiter activation function is referred to as the McCulloch-Pitts model.

Piecewise Linear Function
This type of activation function is also referred to as saturating linear function and can have either a binary or bipolar range for the saturation limits of the output. The mathematical model for a symmetric saturation function (Figure 2.6) is described as follows:

$$y = f(u) = \begin{cases} -1 & \text{if} & u < -1 \\ u & \text{if} & -1 \geq u \geq 1 \\ 1 & \text{if} & u \geq 1 \end{cases} \qquad (2.4)$$

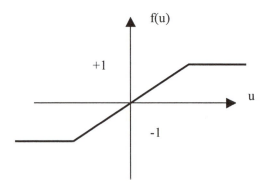

Figure 2.6: Piecewise Linear Activation Function.

Sigmoidal (S shaped) function

This nonlinear function is the most common type of the activation used to construct the neural networks. It is mathematically well behaved, differentiable and strictly increasing function. A sigmoidal transfer function can be written in the following form:

$$f(x) = \frac{1}{1+e^{-\alpha x}} \ , 0 \le f(x) \le 1 \quad (2.5)$$

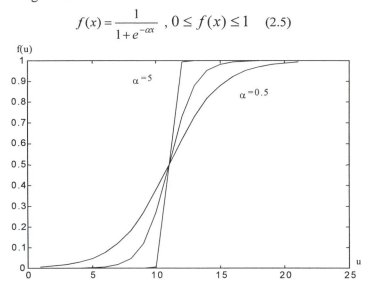

Figure 2.7: A Sigmoid Activation Function.

where α is the shape parameter of the sigmoid function. By varying this parameter, different shapes of the function can be obtained as illustrated in Figure 2.7. This function is continuous and differentiable.

Tangent hyperbolic function
This transfer function is described by the following mathematical form:

$$f(x) = \frac{e^{\alpha x} - e^{-\alpha x}}{e^{\alpha x} + e^{-\alpha x}} \qquad -1 \leq f(x) \leq 1 \tag{2.6}$$

It is interesting to note that the derivatives of Equations 2.5 and 2.6 can be expressed in terms of the individual function itself (please see problems appendix). This is important for the learning development rules to train the networks as shown in the next chapter.

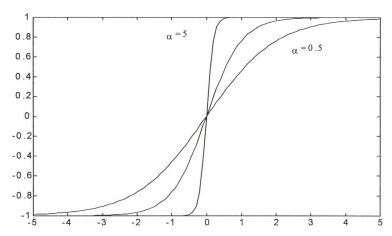

Figure 2.8: A Tangent Hyperbolic Activation Function.

Example 2.1:
Consider the following network consists of four inputs with the weights as shown

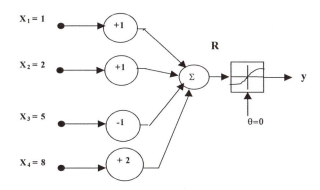

Figure 2.9: Neuron Structure of Example 2.1.

The output R of the network, prior to the activation function stage, is calculated as follows:

$$R = W^T.X = \begin{bmatrix} 1 & 1 & -1 & 2 \end{bmatrix} \begin{bmatrix} 1 \\ 2 \\ 5 \\ 8 \end{bmatrix} = 14 \tag{2.7}$$

With a binary activation function, and a sigmoid function, the outputs of the neuron are respectively as follow:

$$y(Threshold) = 1;$$

$$y(Sigmoid) = 1.5*2^{-8}$$

2.3   ADALINE

An ADAptive LINear Element (ADALINE) consists of a single neuron of the McCulloch-Pitts type, where its weights are determined by the normalized least mean square (LMS) training law. The LMS learning algorithm was originally proposed by Widrow and Hoff [6]. This learning rule is also referred to as delta rule. It is a well-established supervised training method that has been used over a wide range of diverse applications [7]- [11]. Curve fitting approximations can also be used for training a neural network [10]. The learning objective of curve fitting is to find a surface that best fits to the training data. In the next chapter the implementation of LMS algorithms for backpropagation, and curve fitting algorithms for radial basis function network, will be described in detail.

The architecture of a simple ADALINE is shown In Figure 2.10. It is observed that the basic structure of an ADALINE is similar to a linear neuron (Figure 2.2) with the activation function f(.) to be a linear one with an extra feedback loop. Since ADALINE is a linear device, any combination of these units can be accomplished with the use of a single unit.

During the training phase of ADALINE, the input vector $X \in R^n$: $X = \begin{bmatrix} x_1 & x_2 & x_3 & \cdots & x_n \end{bmatrix}^T$ as well as desired output are presented to the network. The weights are adaptively adjusted based on delta rule. After the ADALINE is trained, an input vector presented to the network with fixed weights will result in a scalar output. Therefore, the network performs a mapping of an *n* dimensional mapping to a scalar value. The activation function is not used during the training phase. Once the weights are properly adjusted, the response of the trained unit can be tested by applying various inputs, which are not in the training set. If the network produces consistent responses to a high degree with the test inputs, it said that the network could *generalize*. Therefore, the process of training and generalization are two important attributes of the network.

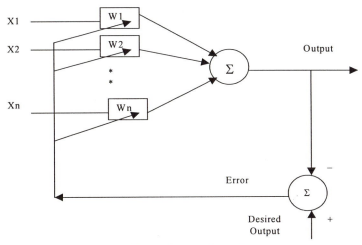

Figure 2.10: ADALINE.

In practice, an ADALINE is usually used to make binary decisions. Therefore, the output is sent through a binary threshold as shown in Figure 2.4. Realizations of several logic gates such as AND, NOT and OR are common applications of ADALINE. Only those logic functions that are linearly separable can be realized by the ADALINE, as is explained in the next section.

2.4   LINEAR SEPARABLE PATTERNS

For a single ADALINE to function properly as a classifier, the input pattern must be linearly separable. This implies that the patterns to be classified must be sufficiently apart from each other to ensure the decision surface consists of a single hyperplane such as a single straight line in two-dimensional space. This concept is illustrated in Figure 2.11 for a two-dimensional pattern.

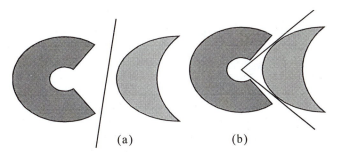

(a)   (b)

Figure 2.11: A Pair of Linearly Separable (a), and Non-Linearly Separable Patterns (b).

A classic example of a mapping that is not separable is XOR (the exclusive or) gate function. Table 2.1 shows the input-output pattern of this problem. Figure 2.12 shows the locations of the symbolic outputs of XOR function corresponding to four input patterns in X1-X2 plane. There is no way to draw a single straight line so that the circles are on one side of the line and the triangular sign on the other side. Therefore, an ADALINE cannot realize this function.

Table 2.1: Inputs/Outputs Relationship for XOR.

X1	*X2*	*Output*
0	0	0
0	1	1
1	0	1
1	1	0

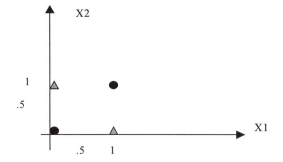

Figure 2.12: The Output of XOR in X1-X2 Plane.

One approach to solve this nonlinear separation problem is to use MADALINE (Multiple ADALINE) networks. The basic structure of a MADALINE network consists of combining several ADALINE with their correspondence activation functions into a single forward structure. When suitable weights are chosen, the network is capable of implementing complicated and nonlinear separable mapping such as XOR gate problems. We will address this issue later in this chapter.

2.5   SINGLE LAYER PERCEPTRON

2.5.1   General Architecture

The original idea of the perceptron was developed by Rosenblatt in the late 1950s along with a convergence procedure to adjust the weights. In Rosenblatt's perceptron, the inputs were binary and no bias was included. It was based on the McCulloch-Pitts model of the neuron with the hard limitation activation function. The single layer perceptron as shown in Figure 2.13 is very similar to ADALINE except for the addition of an activation function.

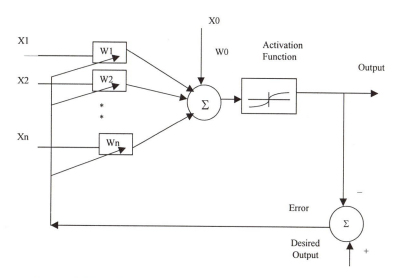

Figure 2.13: A Perceptron with a Sigmoid Activation Function.

Connection weights and threshold in a perceptron can be fixed or adapted using a number of different algorithms. Here the original perceptron convergence procedure as developed by Minsky and Papert[5] is described. First, connection weights W_1, W_2,...,W_n and the threshold value W_0 are initialized to small non-zero values. Then, a new input set with N values received through sensory units (measurement devices) and the input is computed. Connection weights are only adapted when an error occurs. This procedure is repeated until the classification of all inputs is completed.

2.5.2   Linear Classification

For clarification of the above concept, consider two input patterns classes C1 and C2. The weight adaptation at the kth training phase can be formulated as follow:

1.   If k member of the training vector x(k) is correctly classified, no correction action is needed for the weight vector. Since the activation function is selected as a hard limiter, the following conditions will be valid:

$W(k+1) = W(k)$ *if output>0 and x (k)* $\in C1$ *, and*

$W(k+1) = W(k)$ *if output<0 and x(k)* $\in C2$.

2.   Otherwise, the weight should be updated in accordance with the following rule:

$W(k+1) = W(k) + \eta x(k)$ *if output$\geq$0 and x(k)ε C1*

$W(k+1) = W(k) - \eta x(k)$ *if output$\leq$0 and x(k)ε C2*

Where η is the learning rate parameter, which should be selected between 0 and 1.

Example 2.2:

Let us consider pattern classes C1 and C2, where C1: {(0,2), (0,1)} and C2: {(1,0), (1,1)}. The objective is to obtain a decision surface based on perceptron learning. The 2-D graph for the above data is shown in Figure 2.14

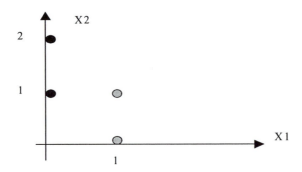

Figure 2.14: 2-D Plot of Input Data Sets for Example 2.2.

Since, the input vectors consist of two elements, the perceptron structure is simply as follows:

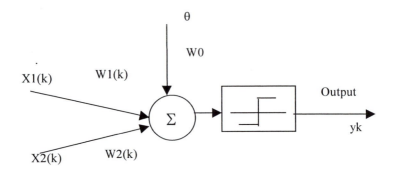

Figure 2.15: Perceptron Structure for Example 2.2.

For simplicity, let us assume $\eta=1$ and initial weight vector W(1)=[0 0]. The iteration weights are as follow:

Iteration 1: $\qquad W^{T}(1).x(1) = \begin{bmatrix} 0 & 0 \end{bmatrix}\begin{bmatrix} 0 \\ 2 \end{bmatrix} = 0$

Weight Update: $\quad W(2) = W(1) + x(1) = \begin{bmatrix} 0 \\ 0 \end{bmatrix} + \begin{bmatrix} 0 \\ 2 \end{bmatrix} = \begin{bmatrix} 0 \\ 2 \end{bmatrix}$

Iteration 2: $\quad W^T(2).x(2) = \begin{bmatrix} 0 & 2 \end{bmatrix} \begin{bmatrix} 0 \\ 1 \end{bmatrix} = 2 > 0$

Weight Update: $\quad W(3) = W(2)$

Iteration 3: $\quad W^T(3).x(3) = \begin{bmatrix} 0 & 2 \end{bmatrix} \begin{bmatrix} 1 \\ 0 \end{bmatrix} = 0$

Weight Update: $\quad W(4) = W(3) - x(3) = \begin{bmatrix} 0 \\ 2 \end{bmatrix} - \begin{bmatrix} 1 \\ 0 \end{bmatrix} = \begin{bmatrix} -1 \\ 2 \end{bmatrix}$

Iteration 4: $\quad W^T(4).x(4) = \begin{bmatrix} -1 & 2 \end{bmatrix} \begin{bmatrix} 1 \\ 1 \end{bmatrix} = 1$

Weight Update: $\quad W(5) = W(4) - x(4) = \begin{bmatrix} -1 \\ 2 \end{bmatrix} - \begin{bmatrix} 1 \\ 1 \end{bmatrix} = \begin{bmatrix} -2 \\ 1 \end{bmatrix}$

Now if we continue the procedure, the perceptron classifies the two classes correctly at each instance. For example for the fifth and sixth iterations:

Iteration 5: $\quad W^T(5).x(5) = \begin{bmatrix} -2 & 1 \end{bmatrix} \begin{bmatrix} 0 \\ 2 \end{bmatrix} = 2 > 0 : Correct\ Classification$

Iteration 6: $\quad W^T(6).x(6) = \begin{bmatrix} -2 & 1 \end{bmatrix} \begin{bmatrix} 0 \\ 1 \end{bmatrix} = 1 > 0 : Correct\ Classification$

In a similar fashion for the seventh and eighth iterations, the classification results are indeed correct.

Iteration 7: $\quad W^T(7).x(7) = \begin{bmatrix} -2 & 1 \end{bmatrix} \begin{bmatrix} 1 \\ 0 \end{bmatrix} = -2 < 0 : Correct\ Classification$

Iteration 8: $\quad W^T(8).x(8) = \begin{bmatrix} -2 & 1 \end{bmatrix} \begin{bmatrix} 1 \\ 1 \end{bmatrix} = -1 < 0 : Correct\ Classification$

Therefore, the algorithm converges and the decision surface for the above perceptron is as follows:

$$d(x) = -2X_1 + X_2 = 0 \qquad (2.8)$$

Now, let us consider the input data {1,2}, which is not in the training set. If we calculate the output:

$$Y = W^T.X = \begin{bmatrix} -2 & 1 \end{bmatrix} \begin{bmatrix} 2 \\ 1 \end{bmatrix} = -3 < 0 \qquad (2.9)$$

The output Y belongs to the class C2 as is expected.

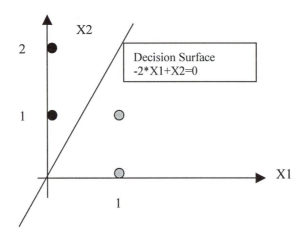

Figure 2.16: Decision Surface for Example 2.2.

2.5.3   Perceptron Algorithm

The perceptron learning algorithm (Delta rule) can be summarized as follows:

Step 1: Initialize the weights $W_1, W_2...W_n$ and threshold θ to small random values.
Step 2: Present new input X1, X2,..Xn and desired output d_k.
Step 3: Calculate the actual output based on the following formula:

$$y_k = f_h (\sum_{i=1}^{n} (X_i Wi) - \theta_k) \qquad (2.10)$$

Step 4: Adapt the weights according to the following equation:

$$W_i(new) = W_i \ (old) + \eta(d_k - y_k)x_i, 0 \le i \le N \qquad (2.11)$$

Where η is a positive gain fraction less than 1 and d_k is the desired output. Note that the weights remain the same if the network makes the correct decision.
Step 5: Repeat the procedures in steps 2–4 until the classification task is completed.

Similar to ADALINE, if the presented inputs pattern is linearly separable, then the above perceptron algorithm converges and positions the decision hyperplane between two separate classes. On the other hand, if the inputs are not separable and their distribution overlaps, then the decision boundary may oscillate continuously. A modification to the perceptron convergence procedure is the utilization of Least Mean Square (LMS) in this case. The algorithm that forms the LMS solution is also called the Widrow-Hoff. The LMS algorithm is similar to the procedure above except a threshold logic nonlinearity, replaces the hard limited non-linearity. Weights are thus corrected on every trail by an amount that depends on the difference between the desired and actual values. Unlike the learning in the ADALINE, the perceptron learning rule has been shown to be capable of separating any linear separable set of the training patterns.

2.6   MULTI-LAYER PERCEPTRON

2.6.1   General Architecture

Multi-layer perceptrons represent a generalization of the single-layer perceptron as described in the previous section. A single layer perceptron forms a half–plane decision region. On the other hand multi-layer perceptrons can form arbitrarily complex decision regions and can separate various input patterns. The capability of multi-layer perceptron stems from the non-linearities used within the nodes. If the nodes were linear elements, then a single-layer network with appropriate weight could be used instead of two- or three-layer perceptrons. Figure 2.17 shows a typical multi-layer perceptron neural network structure. As observed it consists of the following layers:

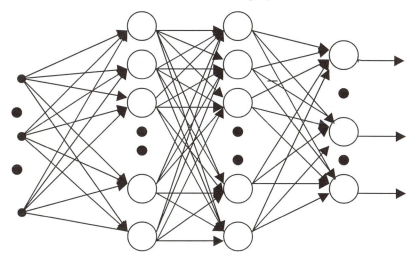

Figure 2.17: Multi-layer Perceptron.

Input Layer: A layer of neurons that receives information from external sources, and passes this information to the network for processing. These may be either sensory inputs or signals from other systems outside the one being modeled.

Hidden Layer: A layer of neurons that receives information from the input layer and processes them in a hidden way. It has no direct connections to the outside world (inputs or outputs). All connections from the hidden layer are to other layers within the system.

Output Layer: A layer of neurons that receives processed information and sends output signals out of the system.

Bias: Acts on a neuron like an offset. The function of the bias is to provide a threshold for the activation of neurons. The bias input is connected to each of the hidden and output neurons in a network.

2.6.2   Input-Output Mapping

The input/output mapping of a network is established according to the weights and the activation functions of their neurons in input, hidden and output layers. The number of input neurons corresponds to the number of input variables in the neural network, and the number of output neurons is the same as the number of desired output variables. The number of neurons in the hidden layer(s) depends upon the particular NN application. For example, consider the following two-layer feed-forward network with three neurons in the hidden layer and two neurons in the second layer:

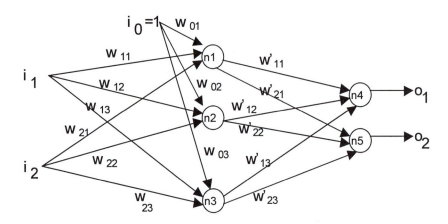

Figure 2.18: An Example of Multi-layer Perceptron.

As is shown, the inputs are connected to each neuron in hidden layer via their corresponding weights. A zero weight indicates no connection. For example, if $W_{23} = 0$, it is implied that no connection exists between the second input (i_2) and the third neuron (n_3). Outputs of the last layer are considered as the outputs of the network.

The structure of each neuron within a layer is similar to the architecture as described in section 2.5. Although the activation function for one neuron could be different from other neurons within a layer, for structural simplicity, similar neurons are commonly chosen within a layer. The input data sets (or sensory information) are presented to the input layer. This layer is connected to the first hidden layer. If there is more than one hidden layer, the last hidden layer should be connected to the output layer of the network. At the first phase, we will have the following linear relationship for each layer:

$$A_1 = W_1 X \tag{2.12}$$

where A_1 is a column vector consisting of m elements, W_1 is an m×n weight matrix and X is a column input vector of dimension n. For the above example, the linear activity level of the hidden layer (neurons n_1 to n_3) can be calculated as follows:

$$\begin{cases} a_{11} = w_{11}i_1 + w_{21}i_2 \\ a_{12} = w_{12}i_1 + w_{22}i_2 \\ a_{13} = w_{13}i_1 + w_{23}i_2 \end{cases} \tag{2.13}$$

The output vector for the hidden layer can be calculated by the following formula:

$$O_1 = F . A_1 \tag{2.14}$$

where A_1 is defined in Equation 2.12, and O_1 is the output column vector of the hidden layer with m element. F is a diagonal matrix comprising the non-linear activation functions of the first hidden layer:

$$F = \begin{bmatrix} f_1(.) & 0 & 0 & ... & 0 \\ 0 & f_2(.) & & & 0 \\ . & & .. & & .. \\ . & & & .. & 0 \\ 0 & 0 & ... & 0 & f_m(.) \end{bmatrix} \tag{2.15}$$

For example, if all activation functions for the neurons in the hidden layer of Figure 2.18 are chosen similarly, then the output of the neurons n_1 to n_3 can be calculated as follows:

$$\begin{cases} O_{11} = f(a_{11}) \\ O_{12} = f(a_{12}) \\ O_{13} = f(a_{13}) \end{cases} \tag{2.16}$$

In a similar manner, the output of other hidden layers can be computed. The output of a network with only one hidden layer according to Equation 2.14 is as follows:

$$A_2 = W_2 . O_1 \tag{2.17}$$

$$O_2 = G . A_2 \tag{2.18}$$

Where A_2 is the vector of activity levels of output layer and O_2 is the q output of the network. G is a diagonal matrix consisting of nonlinear activation functions of the output layer:

$$G = \begin{bmatrix} g_1(.) & 0 & 0 & ... & 0 \\ 0 & g_2(.) & & & 0 \\ . & & .. & & .. \\ . & & & .. & 0 \\ 0 & 0 & ... & 0 & g_q(.) \end{bmatrix} \tag{2.19}$$

For Figure 2.18, the activity level of output neurons n_4 and n_5 can be calculated as follows:

$$\begin{cases} a_{21} = W'_{11}O_{11} + W'_{12}O_{21} + W'_{13}O_{31} \\ a_{22} = W'_{21}O_{11} + W'_{22}O_{21} + W'_{23}O_{31} \end{cases} \tag{2.20}$$

The two outputs of the network with the similar activation functions can be calculated as follows:

$$\begin{cases} O_1 = g(a_{21}) \\ O_2 = g(a_{22}) \end{cases} \tag{2.21}$$

Therefore, the input-output mapping of a multi-layer perceptron is established according to relationships 2.12–2.22. In sequel, the output of the network can be calculated using such nonlinear mapping and the input data sets.

2.6.3   XOR Realization

As it was shown in section 2.4, a single-layer perceptron cannot classify the input patterns that are not linearly separable such as an Exclusive OR (XOR) gate. This problem may be considered as a special case of a more general non-linear mapping problem. In the XOR problem, we need to consider the four corners of the unit square that correspond to the input pattern. We may solve the problem with a multi-layer perceptron with one hidden layer as shown in Figure 2.19.

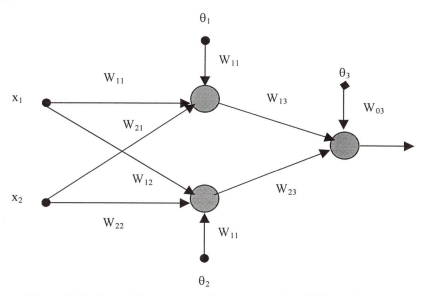

Figure 2.19: Neural Network Architecture to Solve XOR Problem.

In the above configuration, a McCulloh-Pitts model represents each neuron, which uses a hard limit activation function. By appropriate selections of the network weights, the XOR could be implemented using decision surfaces as shown in Figure 2.20.

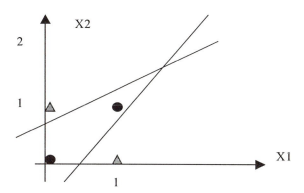

Figure 2.20: Decision Surfaces to Solve XOR Problem.

Example 2.3:
   Suppose weights and biases are selected as shown in Figure 2.21. The McCulloh-Pitts model represents each neuron (binary hard limit activation function). Show that the network solves XOR problem. In addition, draw the decision boundaries constructed by the network.

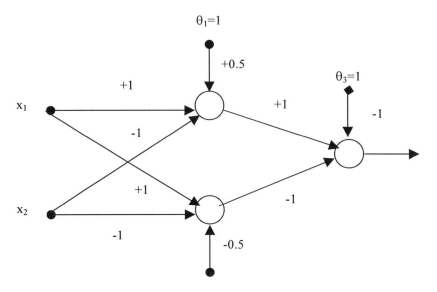

Figure 2.21: Neural Network Architecture for Example 2.3.

In Figure 2.21, suppose the outputs of neurons (before activation function) denote as O_1, O_2, and O_3. The outputs of the summing points at the first layer are as follow:

$$O_1 = x_1 - x_2 + 0.5 \tag{2.22}$$
$$O_2 = x_1 - x_2 - 0.5 \tag{2.23}$$

With the binary hard limited functions, the output y_1 and y_2 are shown in Figures 2.22 and 2.23.

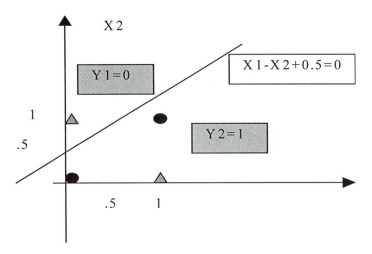

Figure 2.22: Decision Surface for Neuron 1 of Example 2.3.

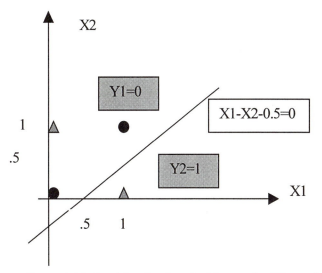

Figure 2.23: Decision Surface for Neuron 2 of Example 2.3.

The outputs of the summing points at the second layer are:

$$O_3 = y_1 - y_2 - 1 \qquad (2.24)$$

The decision boundaries of the network are shown in Figure 2.24. Therefore, XOR realization can be accomplished by selection of appropriate weights using Figure 2.19.

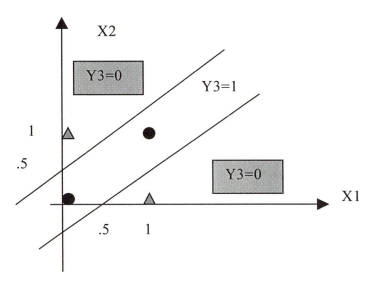

Figure 2.24: Decision Surfaces for Example 2.3.

2.7   CONCLUSION

In this chapter, the fundamentals of neural networks were introduced. The perceptron is the simplest form of neural network used for the classification of linearly separable patterns. Multi-layer perceptron overcome many limitations of single-layer perceptron. They can form arbitrarily complex decision regions in order to separate various nonlinear patterns. The next chapter is devoted to several neural network architectures. Applications of NN will be presented in Chapters 4–7 and Chapter 15 of the book.

REFERENCES

1.   McCulloch, W.W. and Pitts, W., A Logical Calculus of Ideas Imminent in Nervous Activity. *Bull. Math. Biophys.*, 5, 115–133, 1943.
2.   Pitts, W. and McCulloch, W.W., How we Know Universals, Bull. *Math.* 127–147, 1947.
3.   McClelland, J.L. and Rumelhart, D.E., *Parallel Distributed Processing -Explorations in the Microstructure of Cognition*, Vol. 2, *Psychological and Biological Models*, MIT Press, Cambridge, MA, 1986.
4.   Rosenblatt, F., *Principles of Neurodynamics*, Spartan Press, Washington, DC, 1961.
5.   Minsky, M. and Papert, S., *Perceptron: An Introduction to Computational Geometry*, MIT Press, Cambridge, MA, 1969.
6.   Widrow, B. and Hoff, M.E, Adaptive Switching Circuits, IRE WESCON Convention Record, Part 4, NY, IRE, 96–104, 1960.
7.   Fausett, L., *Fundamentals of Neural Networks*, Prentice-Hall, Englewood Cliffs, NJ, 1994.
8.   Haykin, S., *Neural Networks: A Comprehensive Foundation*, Prentice Hall, Upper Saddle River, NJ, 1999.
9.   Kosko, B., *Neural Network for Signal Processing*, Prentice Hall, Englewood Cliffs, NJ, 1992.
10.   Ham, F. and Kostanic, I., *Principles of Neurocomputing for Science and Engineering*, McGraw Hill, New York, NY, 2001.
11.   Lippmann, R.P., An Introduction to Computing with Neural Network, *IEEE Acoustic, Speech, and Sig. Proces. Mag.*, 4, 1987.

3 NEURAL NETWORK ARCHITECTURES

Hooman Yousefizadeh and Ali Zilouchian

3.1   INTRODUCTION

Interest in the study of neural networks has grown remarkably in the last two decades. This is due to the conceptual viewpoint regarding the human brain as a model of a parallel computation device, a very different one from a traditional serial computer. Neural networks are commonly classified by their network topology, node characteristics, learning, or training algorithms. On the other hand, the potential benefits of neural networks extend beyond the high computation rates provided by massive parallelism of the networks. They typically provide a greater degree of robustness or fault tolerance than Von Neumann sequential computers.   Additionally, adaptation and continuous learning are integrated components of NN. These properties are very beneficial in areas where the training data sets are limited or the processes are highly nonlinear. Furthermore, designing artificial neural networks to solve problems and studying real biological networks (Chapter 4) may also change the way we think about the problems and may lead us to new insights and algorithm improvements.

The main goal of this chapter is to provide the readers with the conceptual overviews of several neural network architectures. The chapter will not delve too deeply into the theoretical considerations of any one network, but will concentrate on the mechanism of their operation.   Examples are provided for each network to clarify the described algorithms and demonstrate the reliability of the network. In the following four chapters various applications pertaining to these networks will be discussed.

This chapter is organized as follows. In section 3.2, various classifications of neural networks according to their operations and/or structures are presented. Feedforward and feedback networks are discussed. Furthermore, two different methods of training, namely supervised and unsupervised learning, are described. Section 3.3 is devoted to error back propagation (BP) algorithm. Various properties of this network are also discussed in this section. Radial basis function network (RBFN) is a feedforward network with supervised learning, which is the subject of the discussion in section 3.4. Kohonen self-organizing as well as Hopfield networks are presented in sections 3.5 and 3.6, respectively. Finally section 3.7 presents the conclusions of this chapter.

3.2   NN CLASSIFICATIONS

3.2.1   Feedforward and Feedback Networks

In a feedforward neural network structure, the only appropriate connections are between the outputs of each layer and the inputs of the next layer. Therefore, no connections exist between the outputs of a layer and the inputs of either the same layer or previous layers.   Figure 3.1 shows a two-layer feedforward network. In this topology, the inputs of each neuron are the weighted sum of the outputs from the previous layer. There are weighted connections between the outputs of each layer and the inputs of the next layer. If the weight of a branch is assigned a zero, it is equivalent to no connection between correspondence nodes. The inputs are connected to each neuron in hidden layer via their correspondence weights. Outputs of the last layer are considered the outputs of the network.

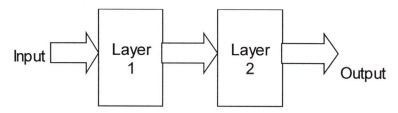

Figure 3.1: General Structure of Two-Layer Feedforward Network.

For feedback networks the inputs of each layer can be affected by the outputs from previous layers. In addition, self feedback is allowed. Figure 3.2 shows a simple single layer feedback neural network.

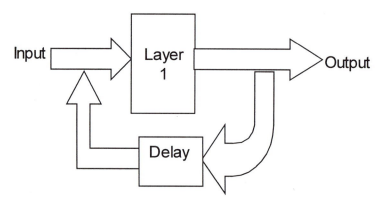

Figure 3.2: General Structure of a Sample Feedback Network.

As observed, the inputs of the network consist of both external inputs and the network output with some delays. Examples of feedback algorithms include the Hopfield network, described in detail in section 3.6, and the Boltzman Machine.

An important issue for feedback networks is the stability and convergence of the network.

3.2.2   Supervised and Unsupervised Learning Networks

There are a number of approaches for training neural networks. Most fall into one of two modes:

- *Supervised Learning*: Supervised learning requires an external teacher to control the learning and incorporates global information. The teacher may be a training set of data or an observer who grades the performance. Examples of supervised learning algorithms are the least mean square (LMS) algorithm and its generalization, known as the back propagation algorithm[1]-[4], and radial basis function network [5]-[8]. They will be described in the following sections of this chapter.

In supervised learning, the purpose of a neural network is to change its weights according to the inputs/outputs samples. After a network has established its input output mapping with a defined minimum error value, the training task has been completed. In sequel, the network can be used in recall phase in order to find the outputs for new inputs. An important factor is that the training set should be comprehensive and cover all the practical areas of applications of the network. Therefore, the proper selection of the training sets is critical to the good performance of the network.

- *Unsupervised Learning*: When there is no external teacher, the system must organize itself by internal criteria and local information designed into the network. Unsupervised learning is sometimes referred to as *self-organizing learning*, i.e., learning to classify without being taught. In this category, only the input samples are available and the network classifies the input patterns into different groups. Kohonen network is an example of unsupervised learning.

3.3   BACK PROPAGATION ALGORITHM

Back propagation algorithm is one of the most popular algorithms for training a network due to its success from both simplicity and applicability viewpoints. The algorithm consists of two phases: *Training phase and recall phase*. In the *training* phase, first, the weights of the network are randomly initialized. Then, the output of the network is calculated and compared to the desired value. In sequel, the error of the network is calculated and used to adjust the weights of the output layer. In a similar fashion, the network error is also propagated backward and used to update the weights of the previous layers. Figure 3.3 shows how the error values are generated and propagated for weights adjustments of the network.

In the *recall* phase, only the feedforward computations using assigned weights from the training phase and input patterns take place. Figure 3.4 shows both the feedforward and back propagation paths. The feedforward process is used in both recall and training phases. On the other hand, as shown in Figure 3.4(b), back propagation of error is only utilized in the training phase.

In the training phase, the weight matrix is first randomly initialized. In sequel, the output of each layer is calculated starting from the input layer and moving forward toward the output layer. Thereafter, the error at the output layer is calculated by comparison of actual output and the desired value to update the weights of the output and hidden layers.

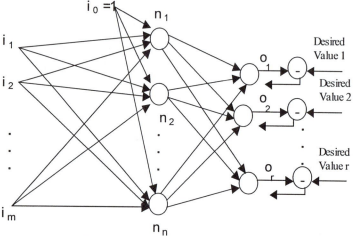

Figure 3.3. Back Propagation of the Error in a Two-Layer Network.

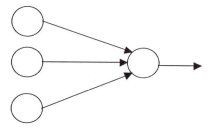

a) Forward propagation (Training and Recall Phase)

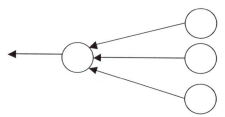

b) Backward propagation (Training Phase)

Figure 3.4: Forward Propagation in Recall and Training Phase and Backward Propagation in Training Phase.

There are two different methods of updating the weights. In the first method, weights are updated for each of the input patterns using an iteration method. In the second method, an overall error for all the input output patterns of training sets is calculated. In other words, either each of the input patterns or all of the patterns together can be used for updating the weights. The training phase   will be terminated when the error value is less than the minimum set value provided by the designer. One of the disadvantages of back propagation algorithm is that the training phase is very time consuming.

During the recall phase, the network with the final weights resulting from the training process is employed. Therefore, for every input pattern in this phase, the output will be calculated using both linear calculation and nonlinear activation functions. The process provides a very fast performance of the network in the recall phase, which is one of its important advantages.

3.3.1   Delta Training Rule

The back propagation algorithm is the extension of the perceptron structure as discussed in the previous chapter with the use of multiple adaptive layers. The training of the network is based on the delta training rule method. Consider a single neuron in Figure 3.5.

The relations among input, activity level and output of the system can be shown as follows:

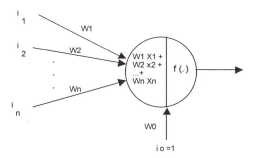

Figure 3.5: A Single Neuron.

$$a = w_0 + w_1 i_1 + w_2 i_2 + \cdots + w_n i_n \tag{3.1}$$

or in the matrix form:

$$a = w_0 + W^T I \tag{3.2}$$

$$o = f(a) \tag{3.3}$$

where W and I are weight and input vectors of the neuron, a is activity level of the neuron and o is the output of the neuron. w_0 is called bias value.

Suppose the desired value of the output is equal to d. Error e can be defined as follows:

$$e = \frac{1}{2}(d - o)^2 \tag{3.4}$$

by substituting Equations 3.2 and 3.3 into Equation 3.4, the following relation holds:

$$e = \frac{1}{2}(d - f(w_0 + W^T I))^2 \tag{3.5}$$

The error gradient vector can be calculated as follows:

$$\nabla e = -(d - o)f'(w_0 + W^T I)I \tag{3.6}$$

The components of gradient vector are equal to:

$$\frac{\partial e}{\partial w_j} = -(d - o)f'(w_0 + W^T I)I_j \tag{3.7}$$

where $f'(.)$ is derivative of activation function. To minimize the error the weight changes should be in negative gradient direction. Therefore we will have

$$\Delta W = -\eta \nabla e \tag{3.8}$$

where η is a positive constant, called learning factor. By Equations (3.6) and 3.7, the ΔW is calculated as follows:

$$\Delta W = -\eta(d - o)f'(a)I \tag{3.9}$$

For each weight j Equation 3.9 can be written as:

$$\Delta w_j = -\eta(d - o)f'(a)I_j \qquad j = 0,1,2,...,n \tag{3.10}$$

Therefore we update the weights of the network as:

$$w_j\,(new) = w_j\,(old) + \Delta w_j \qquad j = 0,1,2,...,n \tag{3.11}$$

For Figure 3.3, the Delta rule can be applied in a similar manner to each neuron. Through generalization of Equation 3.11 for normalized error and using Equation 3.10 for every neuron in output layer we will have:

$$w_j\,(new) = w_j\,(old) + \frac{\eta(d_j - o_j)f'(a_j)x_j}{\|X\|^2} \qquad j = 0,1,2,...,n \tag{3.12}$$

where $X \in R^n$ is the input vector to the last layer, xj is the jth element of X and $\|.\|$ denotes L2-Norm.

The above method can be applied to the hidden layers as well. The only difference is that the o_j will be replaced by y_j in 3.12. y_j is the output of hidden layer neuron, and not the output of network.

One of the drawbacks in the back propagation learning algorithm is the long duration of the training period. In order to improve the learning speed and avoid the local minima, several different methods have been suggested by researchers. These include addition of first and second moments to the learning phase, choosing proper initial conditions, and selection of an adaptive learning rate.

To avoid the local minima, a new term can be added to Equation 3.12. In such an approach, the network memorizes its previous adjustment, and, therefore it will escape the local minima, using previous updates. The new equation can be written as follows:

$$w_j\,(new) = w_j\,(old) + \frac{\eta(d_j - o_j)f'(a_j)x_j}{\|X\|^2} + \alpha[w_j\,(new) - w_j\,(old)] \quad (3.13)$$

where α is a number between 0 and 1, namely the momentum coefficient.

Nguyen and Widrow [9] have proposed a systematic approach for the proper selection of initial conditions in order to decrease the training period of the network. Another approach to improve the convergence of the network and increase the convergence speed is the adaptive learning rate. In this method, the learning rate of the network (η) is adjusted during training. In the first step, the training coefficient is selected as a large number, so the resulting error values are large. However, the error will be decreased as the training progresses, due to the decrease in the learning rate. It is similar to coarse and fine tunings in selection of a radio station.

In addition to the above learning rate and momentum terms, there are other neural network parameters that control the network's performance and prediction capability. These parameters should be chosen very carefully if we are to develop effective neural network models. Two of these parameters are described below.

Selection of Number of Hidden Layers

The number of input and output nodes corresponds to the number of network inputs and desired outputs, respectively. The choice of the number of hidden layers and the nodes in the hidden layer(s) depends on the network application. Selection of the number of hidden layers is a critical part of designing a network and is not as straightforward as input and output layers. There is no mathematical approach to obtain the optimum number of hidden layers, since such selection is generally fall into the application oriented category. However, the number of hidden layers can be chosen based on the training of the network using various configurations, and selection of the configuration with the fewest number of layers and nodes which still yield the minimum root-mean-squares (RMS) error quickly and efficiently. In general, adding a second hidden layer improves the network's prediction capability due to the nonlinear separability property of the network. However, adding an extra hidden layer commonly yields prediction capabilities similar to those of two-hidden layer networks, but requires longer training times due to the more complex structures. Although using a single hidden layer is sufficient for solving many functional approximation problems, some problems may be easier to solve with a two-hidden-layer configuration.

Normalization of Input and Output Data Sets

Neural networks require that their input and output data be normalized to have the same order of magnitude. Normalization is very critical for some applications. If the input and the output variables are not of the same order of magnitude, some variables may appear to have more significance than they actually do. The training algorithm has to compensate for order-of-magnitude differences by adjusting the network weights, which is not very effective in many of the training algorithms such as back propagation algorithm. For example, if input variable i_1 has a value of 50,000 and input variable i_2 has a value of 5, the assigned weight for the second variable entering a node of hidden layer 1 must be much greater than that for the first in order for variable 2 to have any significance. In addition, typical transfer functions, such as a sigmoid function, or a hyperbolic tangent function, cannot distinguish between two values of x_i when both are very large, because both yield identical threshold output values of 1.0.

The input and output data can be normalized in different ways. In Chapters 7 and 15, two of these normalized methods have been selected for the appropriate applications therein.

The training phase of back propagation algorithm can be summarized in the following steps:

1. Initialize the weights of the network.
2. Scale the input/output data.
3. Select the structure of the network (such as the number of hidden layers and number of neurons for each layer).
4. Choose activation functions for the neurons. These activation functions can be uniform or they can be different for different layers.
5. Select the training pair from the training set. Apply the input vector to the network input.
6. Calculate the output of the network based on the initial weights and input set.
7. Calculate the error between network output and the desired output (the target vector from the training pair).
8. Propagate error backward and adjust the weights in such a way that minimizes the error. Start from the output layer and go backward to input layer.
9. Repeat steps 5–8 for each vector in the training set until the error for the set is lower than the required minimum error.

After enough repetitions of these steps, the error between the actual outputs and target outputs should be reduced to an acceptable value, and the network is said to be trained. At this point, the network can be used in the recall or generalization phases where the weights are not changed.

Network Testing

As we mentioned before, an important aspect of developing neural networks is determining how well the network performs once training is complete. Checking the performance of a trained network involves two main criteria: (1) how well the neural network recalls the output vector from data sets used to train the network (called the *verification step*); and (2) how well the network predicts responses from data sets that were not used in the training phase (called the *recall* or *generalization step*).

In the verification step, we evaluate the network's performance in specific initial input used in training. Thus, we introduce a previously used input pattern to the trained network. The network then attempts to predict the corresponding output. If the network has been trained sufficiently, the network output will differ only slightly from the actual output data. Note that in testing the network, the weight factors are not changed: they are frozen at their last values when training ceased.

Recall or generalization testing is conducted in the same manner as verification testing; however, now the network is given input data with which it was not trained. Generalization testing is so named because it measures how well the network can generalize what it has learned, and form rules with which to make decisions about data it has not previously seen. In the generalization step, we feed new input patterns (whose results are known to us, but not to the network) to the trained network. The network generalizes well when it sensibly interpolates these new patterns. The error between the actual and predicted outputs is larger for generalization testing and verification testing. In theory, these two errors converge upon the same point corresponding to the best set of weight factors for the network.

In the following subsection, two examples are presented to clarify various issues related to BP.

Example 3.1:

Consider the network of Figure 3.6 with the initial values as indicated. The desired values of the output are $d_0 = 0$ $d_1 = 1$. We show two iterations of learning of the network using back propagation. Suppose the activation function of the first layer is a sigmoid and activation function of the output is a linear function.

$$f(x) = \frac{1}{1+e^{-x}} \Rightarrow f'(x) = f(x)[1 - f(x)] \qquad (3.14)$$

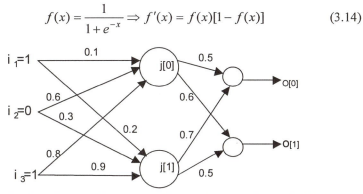

Figure 3.6: Feedforward Network of Example 3.1 with Initial Weights.

Iteration Number 1:

Step 1: Initialization: First the network is initialized with the values as shown in Figure 3.6.

Step 2: Forward calculation, using Equations (3.1–3.3):
$$J[0] = f(W_j[0].I) = f(0.1*1 + 0.6*0 + 0.8*1) = f(0.9) = 0.7109$$
$$J[1] = f(W_j[1].I) = f(1.1) = 0.7503$$

$$O[0] = f(W_k[0].J) = 0.5*0.7109 + 0.7*0.7503 = 0.88066$$
$$O[1] = f(W_k[1].J) = 0.6*0.7109 + 0.5*0.7503 = 0.80169$$

Step 3: According to Equation 3.5 the errors are calculated as follows:

$$\Delta k[0] = d_0 - k[0] = 0 - 0.88066 = -0.88066$$
$$\Delta k[1] = d_1 - k[1] = 1 - 0.80169 = 0.19831$$

Step 4: The updated weights of the network are calculated according to Equations 3.10 and 3.11 as follows:

$$W_{k00\ (new)} = W_{k00\ (old)} + n*\Delta k[0]*f(k[0])*j[0] =$$
$$0.5 + 1*(-0.88066) + 0.2072*0.7109 = 0.3694$$

$W_{k01\ (new)} = 0.56309 \qquad W_{k10\ (new)} = 0.6301 \qquad W_{k11\ (new)} = 0.5138$

$$W_{j00\ (new)} = W_{j00\ (old)} + n*I[0]*\Sigma w\Delta =$$
$$0.1 + 1*1*(0.5* -0.88066 + 0.6*0.19831) = -0.2213$$

$W_{j01\ (new)} = 0.6 \qquad W_{j02\ (new)} = 0.4787 \qquad W_{j10\ (new)} = -0.3173$

$W_{j11\ (new)} = 0.3 \qquad W_{j12\ (new)} = 0.3827$

Iteration Number 2: For this iteration the new weight values in Iteration 1 are utilized. Steps 2–4 of the previous iteration are repeated.

Step 2:
$J[0] = 0.5640 \qquad J[1] = 0.5163 \qquad O[0] = 0.4991 \qquad O[1] = 0.6299$

Step 3:
$\Delta k[0] = -0.4991 \qquad \Delta k[1] = 0.3701$

Step 4:
$W_{k00\ (new)} = 0.3032 \quad W_{k01\ (new)} = 0.5025 \quad W_{k10\ (new)} = 0.6774$

$W_{k11\ (new)} = 0.5751 \quad W_{j00\ (new)} = -0.17248 \quad W_{j01\ (new)} = 0.6$

$W_{j02\ (new)} = 0.5275 \quad W_{j10\ (new)} = -0.4015 \quad W_{j11\ (new)} = 0.3$

$W_{j12\ (new)} = 0.2985$

The weights after the two iterations of training of the network can be calculated as follows:

$J[0] = 0.5878 \quad J[1] = 0.5257 \quad O[0] = 0.4424 \quad O[1] = 0.7005$

Table 3.1 summarizes the results for the training phase. As can be seen, the values of the output are closer to the desired value and the error value has been decreased. Training should be continued until the error values become less than a predetermined value as set by the designer (for example, 0.01). It should be noted that the selection of small values for maximum error level will not necessarily lead to better performance in the recall phase.

Table 3.1: Summary of Outputs and Error Norm after Iterations

Error	Initial	Iteration 1	Iteration 2
Output 1	- 0.8807	- 0.4991	- 0.4424
Output 2	0.1983	0.3701	0.2995
Error Norm	0.9027	0.6213	0.5342

Choosing a very small value for this maximum error level may force the network to learn the inputs very well, but it will not lead to better overall performance.

Example 3.1 is also solved using MATLAB as shown in Chapter 21. Below is the output result of the program.

$$\frac{Final}{Output} = \begin{bmatrix} -0.0088 \\ 1.0370 \end{bmatrix} \qquad \frac{Input\ Layer}{Weight} = \begin{bmatrix} -0.0255 & 0.6 & 0.6475 \\ 0.0763 & 0.3 & 0.7763 \end{bmatrix}$$

$$\frac{Hidden\ Layer}{Weight} = \begin{bmatrix} 0.1170 & 0.2897 \\ 0.4987 & 0.4159 \end{bmatrix} \qquad \frac{Bias}{Weight} = \begin{bmatrix} -0.1255 \\ -0.1237 \end{bmatrix}$$

As observed, only four iterations are needed to complete the training task for this example. (In this case, the training sets include only one input output set, so each epoch is equivalent to an iteration.) The initial weights of the network for the program are selected as indicated in this example. The final values of the outputs are equal to -0.0088 and 1.0370. These values are close enough to the desired values. The training error is less than 0.001, which the network has achieved during the training phase.

Example 3.2: Forward Kinematics of Robot Manipulator

In this example a simple back propagation neural network has been used to solve the forward kinematic of a robot manipulator. Therefore, θ_1 θ_2 are the inputs with X, Y as the outputs of the network. A set of 200 samples is applied to the network in the training phase.

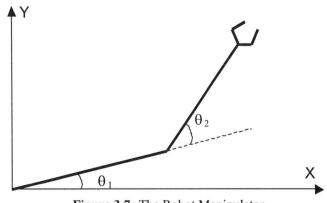

Figure 3.7: The Robot Manipulator.

The relation between (θ_1 and θ_2) and (X and Y) is as follows:

$$X = l_1 \cos\theta_1 + l_2 \cos(\theta_1 + \theta_2)$$
$$Y = l_1 \sin\theta_1 + l_2 \sin(\theta_1 + \theta_2)$$

(3.15)

Figure 3.8 shows how the error of the network changes until the performance goal has been met.

Figure 3.8: The Error of the Network During Training.

After the network has established input and output mapping during the training phase, new inputs are applied to the network to observe its performance in the recall phase. Figure 3.9 shows the simulation result of the network.

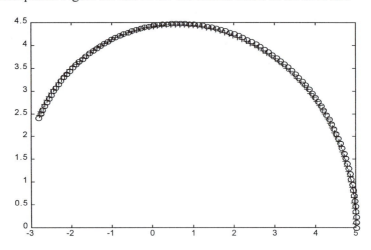

Figure 3.9: The Network Output and Prediction of the Neural Network Using the Back Propagation Algorithm.

3.4   RADIAL BASIS FUNCTION NETWORK (RBFN)

The back propagation method as described in the previous section, has been widely used to solve a number of applications [1],[2]. However, despite the practical success, the back propagation algorithm has serious training problems and suffers from slow convergence [3]. While optimization of learning rate and momentum coefficient parameters yields overall improvements on the networks, it is still inefficient and time consuming for real time applications [4].

Radial Basis Function Networks (RBFN) provide an attractive alternative to BP networks [5]. They perform excellent approximations for curve fitting problems and can be trained easily and quickly. In addition, they exhibit none of the BP's training pathologies such as local minima problems. However, RBFN usually exhibits a slow response in the recall phase due to the large number of neurons associated in the second layer [6],[7]. One of the advantages of RBFN is the fact that linear weights associated with the output layer can be treated separately from the hidden layer neurons. As the hidden layer weights are adjusted through a nonlinear optimization, output layer weights are adjusted through linear optimization.

RBFN approximation accuracy and speed may be further improved with a strategy for selecting appropriate centers and widths of the receptive fields. The redistribution of centers to locations where input training data are meaningful can lead to more efficient RBFN [8].

In this section, the fundamental idea pertaining the RBFN is presented. Furthermore, two examples are provided to clarify the training and recall phases associated with these networks. The network is inspired by Cover's theorem as explained below.

Cover's Theorem[6]: A complex pattern classification problem cast in a high dimensional space nonlinearity is more likely to be linearly separable than in a low dimensional space.

Example 3.3:

Consider the XOR problem as presented previously. As it was shown in chapter 2, an XOR gate cannot be implemented by a single perceptron due to nonlinear separability property of the input pattern. However, suppose, the following pair of Guassian hidden functions are defined:

$$h_1(x) = e^{-\|x - u_1\|^2} \qquad u_1 = \begin{bmatrix} 1 \\ 1 \end{bmatrix}$$

$$h_2(x) = e^{-\|x - u_2\|^2} \qquad u_2 = \begin{bmatrix} 0 \\ 0 \end{bmatrix} \qquad (3.16)$$

If we calculate $h_1(x), h_2(x)$ for the above input patterns we will have the Table 3.2. Figure 3.10 shows the graph of the outputs in the $h_1 - h_2$ space.

Table 3.2: Mapping of XY to $h_1 - h_2$

Input pattern: X	$h_1(x)$	$h_2(x)$
(1, 1)	1	0.1353
(0, 1)	0.3678	0.3678
(0, 0)	0.1353	1
(1, 0)	0.3678	0.3678

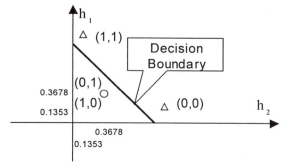

Figure 3.10: XOR Problem in $h_1 - h_2$ Space.

As can be seen, the XOR problem in $h_1 - h_2$ space is mapped to a new problem, which is linearly separable. Therefore, Guassian functions can be used to solve the above interpolation problem with one layer network. The above interpolation problem can be generalized as: Suppose there exist N points $(X_1, ..., X_N)$ and a corresponding set of N real values (d_1, d_2, d_3, ..., d_1); find a function that satisfies the following interpolation condition:

$$F(x_i) = d_i \qquad i = 1, 2, ..., N \qquad (3.17)$$

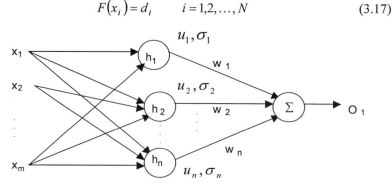

Figure 3.11: A Simple Radial Basis Network.

Figure 3.11 shows a simple radial basis network. This network is a feedforward network similar to back propagation, but it has totally different performance. The first difference is the initial weights. Despite random initial selection of the weights in back propagation, here the initial weights are not chosen randomly. The weights of each hidden layer neuron are set to values that produce a desired response. Such weights are assigned so that the network gives the maximum output for inputs equal to its weights. The activation functions h_i can be defined as follows:

$$h_i = e^{-D_i^2/2\sigma^2} \tag{3.18}$$

where D_i is defined as the distance of the input to the center of the neuron which is identified by the weight vector of hidden layer neuron i. Equation (3.19) shows this relation:

$$\begin{cases} D_i^2 = (x - u_i)^T (x - u_i) \\ x : input \quad vector \\ u_i : Weight \quad vector \quad of \quad hidden \quad layer \quad neuron \quad i \end{cases} \tag{3.19}$$

Therefore, the final contribution of the neuron will decrease for the inputs, which are far from the center of the neuron. With this fact in mind, it is reasonable to give the values of each input of the training set to a neuron, which will result in faster training of the network. The main part of the training of the network is adjusting the weights of the output layer. Figure 3.12 shows a single neuron.

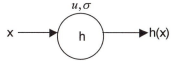

Figure 3.12: A Simple Radial Basis Neuron

Function h(x) as shown in Figure 3.13 can be defined as follows:

$$h(x) = e^{-\frac{(x-u)^2}{2\sigma^2}} \tag{3.20}$$

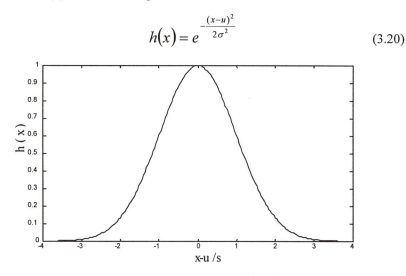

Figure 3.13: The Graph of h(x).

As both graph and formula show:

$$\begin{cases} h(x)=1 & x=u \\ h(x)=0 & |x-u|>3\sigma \\ 0<h(x)<1 & |x-u|<3\sigma \end{cases} \qquad (3.21)$$

The above formula indicates that each neuron only possesses contributions from the inputs that are close to the center of the weight function. For other values of x, the neuron will have zero output value with no contribution in the final output of the network. Figure 3.14 shows a radial basis neuron with two inputs, X_1 and X_2.

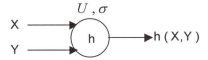

Figure 3.14: A Simple Radial Basis Neuron with Two Inputs.

Figure 3.15 shows the three-dimensional graph of this neuron. As is seen, the fundamental idea is similar. As Figure 3.15 shows, the function is radially symmetric around the center U.

Training of the radial basis network includes two stages. In the first stage, the center U_i and diameter of receptive σ_i of each neuron will be assigned. At the second stage of the training, the weight vector W will be adjusted accordingly. After the training phase is completed, the next step is the recall phase in which the outputs are applied and the actual outputs of the network are produced.

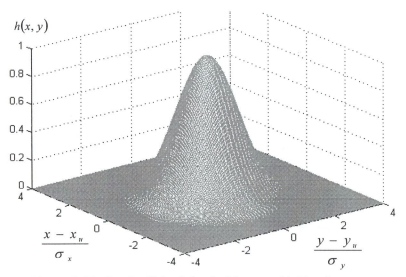

Figure 3.15: Graph of h(x,y) for the Neuron with Two Inputs.

Finding the center U$_i$ of each neuron

One of the most popular approaches to locate the centers U$_i$ is to divide the input vector to some clusters and then find the center of each cluster and locate a hidden layer neuron at that point.

Finding the diameter of the receptive region

The value of σ can have significant effect on the performance of the network. There are different approaches to find this value. One of the popular methods is based on the similarity of the clustering of the input data. For each hidden layer neuron, the RMS distance of each neuron and its first nearest neighbor will be calculated; this value is considered as σ. The training phase of RBFN can be summarized as follows:

1. Apply an input vector X from the training set.
2. Calculate the output of the hidden layer.
3. Compute the output Y and compare it to the desired value. Adjust each weight W accordingly:

$$w_{ij}(n+1) = w_{ij}(n) + \eta.(x_j - y_j)x_i \qquad (3.22)$$

4. Repeat steps 1 to 3 for each vector in the training set.
5. Repeat steps 1 to 4 until the error is smaller than a maximum acceptable amount.

The advantage of radial basis network to back propagation network is faster training. The main problem of back propagation is its lengthy training; therefore radial basis networks have caught a lot of attention lately. The major disadvantage of radial basis network is that it is slow in the recall phase due to its nonlinear functions.

Example 3.4:

This example is the same as Example 3.1, where p and o are input and output consecutively. We try to solve the problem using the radial basis network by MATLAB. The details of the program are provided in Chapter 21. The output of the program is shown below. As is observed, the output is very accurate for the same input values. Also, execution of this simple code shows that the network's training is very fast. The answer can be obtained quickly, with high accuracy. The output of the network to a similar input is also shown. $\tilde{o}$ is the output for the new applied input $\tilde{p}$, which is close to p. It can be seen that this value is close to the output of the training input.

$$p = \begin{bmatrix} 1 \\ 0 \\ 1 \end{bmatrix} \qquad o = \begin{bmatrix} 0 \\ 1 \end{bmatrix} \qquad \tilde{P} = \begin{bmatrix} 1.1 \\ -0.3 \\ 0.9 \end{bmatrix} \qquad \tilde{o} = \begin{bmatrix} 0 \\ 0.9266 \end{bmatrix}$$

Example 3.5:

In this example the inverse kinematics of the robot manipulator of Example 3.2 is solved by RBFN, using MATLAB program. Figure 3.16 compares the actual path and the network prediction of this example. The actual path is shown with circles and the network output with +. As can be seen, the network can predict the path very accurately. In comparison with back propagation, prediction of RBFN is more accurate and the training of this network is much faster. However, due to the number of neurons, the recall phase of the network is usually slower than back propagation.

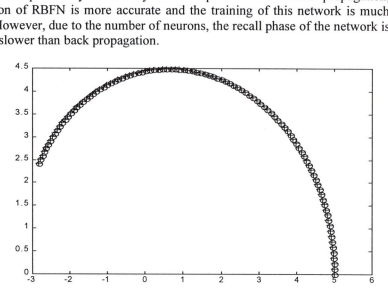

Figure 3.16: Output of the RBFN and Actual Output of the System.

3.5   KOHONEN SELF-ORGANIZATION NETWORK

The Kohonen self-organization network uses unsupervised learning and organizes itself to topological characteristics of the input patterns. The discussion in this section will not seek to explain fully all the intricacies involved in self-organization networks, but rather seek to explain the simple operation of the network with two examples. Interested readers can refer to Kohonen[10], Zurada[11], and Haykin and Simon[12] for more detailed information on unsupervised leaning and self-organization networks.

Learning and brain development phenomena of newborns are very interesting from several viewpoints. As an example, consider how a baby learns to focus its eyes. The skill is not originally present in newborns, but they generally acquire it soon after birth. The parents cannot ask their baby what to do in order to make sense of the visual stimuli impinging on the child's brain. However, it is well known that after a few days, a newborn has learned to associate sets of visual stimuli with objects or shapes. Such remarkable learning occurs naturally with little or no help and intervention from outside. As another example, a baby learns to develop a particular trajectory to move an object or grab a bottle of milk in a special manner. How can these phenomena happen?

One possible answer is provided by a self-learning system, originally proposed by Teuvo Kohonen [10]. His work provides a relatively fast and yet powerful and fascinating model of how neural networks can self-organize. In general, the term *self-organization* refers to the ability of some networks to learn without being given the correct answer for an input pattern. These networks are often closely modeled after neurobiological systems to mimic brain processing and evolution phenomena.

A Kohonen network is not a hierarchical system, but consists of a fully interconnected array of neurons. The output of each neuron is an input to all other inputs in the network including itself. Each neuron has two sets of weights: one set is utilized to calculate the sum of weighted external inputs, and another one to control the interactions between different neurons in the network. The weights on the input pattern are adjustable, while the weights between neurons are fixed.

The other two networks that have been discussed so far in this chapter (BP and RBFN) have neurons that receive input from previous layers and generate output to the next layer or the external world. However, the neurons in the network have neither input nor output to the neurons in the same layer. On the contrary, the Kohonen network receives not only the entire input pattern into the network, but also numerous inputs from the other neurons with the same layer. A block diagram of a simple Kohonen network with N neurons is shown in Figure 3.17.

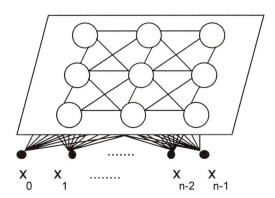

$$X_0 \quad X_1 \quad \text{........} \quad X_{n-2} \quad X_{n-1}$$

Figure 3.17: A Two Dimensional Kohonen Network.

Notice that the input is connected to all the nodes and there are interconnections between the neurons of the same layer. During each presentation, the complete input pattern is presented to each neuron. Each neuron computes its output as a sigmoidal function on the sum of its weighted inputs. The input pattern is then removed and the neurons interact with each other. The neuron with the largest activation output is declared the winner neuron and only that neuron is allowed to provide the output. However, not only the winning neuron's weight is updated, but also all the weights in a neighborhood around the winning neuron. The neighborhood size decreases slowly with each iteration [11].

3.5.1   Training of the Kohonen Network

When we construct a Kohonen network, we must do two things that have not been generally required by the other networks. First, we must properly initialize the weight vectors of the neurons. Second, the weight vectors and the input vectors should be normalized. These two steps are vital to the success of the Kohonen network. The procedure to train a Kohonen self-organization is as follows:

1. Normalize the random selected weights W_i.
2. Present an input pattern vector x to the network. All neurons in the Kohonen layer receive this input vector.
3. Choose the winning neuron as the one with the largest similarity measure between all weight vectors W_i and the input vector x. If the shortest Euclidean distance is selected as similarity measure within a cluster, then the winning unit m satisfies the following equation:

$$\|x - W_m\| = \min_i \{\|x - w_i\|\} \qquad (3.23)$$

where m is referred to as the winning unit.
4. Decrease the radius of Nm region as the training progress, where Nm denotes, as a set of index associated with the winning neighborhood around the winner unit C. The radius of Nm region can be fairly large as the learning starts and slowly reduced to include only the winner and possibly its immediate neighbors.
5. The weight of the winner unit and its neighborhood units are obtained as follows:

$$(W_i)_{new} = (W_i)_{old} + \alpha \left[x - (W_i)_{old} \right] \qquad (3.24)$$

where, Wi is the weight vector, x is the input pattern vector and a is the leaning rate ($0<\alpha<1$) Since a depend on the size of neighborhood function, Equation (3.25) can be rewritten as

$$(W_i)_{new} = (W_i)_{old} + \alpha \, N_{ci} \left[x - (W_i)_{old} \right] \qquad (3.25)$$

where the function N_c can be chosen appropriately such as a Gaussion function or a Mexican hat function.
6. Present the next input vector. Repeat steps 3–5 until the training phase is completed for all inputs.

In order to achieve a good convergence for the above procedure, the learning rate α, as well as the size of neighborhood Nc should be decreased gradually with each iteration. As was mentioned before, at the beginning of the training phase, the selected region around the winner unit might be fairly large. Therefore, a substantial portion of the network can learn each pattern. As the training proceeds, the size of the neighborhood slowly decreases, so fewer and fewer neurons learn with each iteration. Finally the winner itself will adjust its weights. After the completion of this procedure, the network is trained for the next input vector in a similar fashion.

Kohonen self-organization network has some interesting capabilities that can be extremely useful. One possible application is vector quantization. The network can also be used to perform dimension reduction and feature extraction as well as classification.

In MATLAB, the leaning rate, α, and the neighborhood size are altered through two phases: an ordering phase and a tuning phase.

3.5.2 Examples of Self–Organization

In this subsection, two examples of self-organization maps are provided. The detailed description of examples can be found in chapter 21

Example 3.6: 1-D self-organization Mapping
Consider 200 2-Element unit vectors spread uniformly between 0 and 180 as shown in Figure 3.18. We now consider a 1-D self-organization map with 20 neurons.

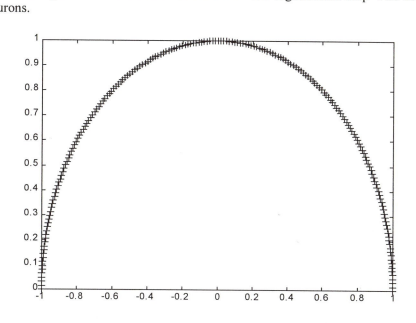

Figure 3.18: Original Distribution of the Input of the Kohonen Network.

Figure 3.19 shows the weights of the Kohonen self-organizing network after training. It can easily be observed that the weights of the network have the pattern of the input. In the other words, the network is being adjusted to the form of the pattern of input of the network.

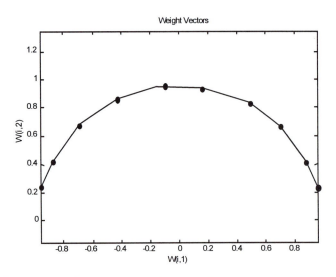

Figure 3.19: Weights of the Kohonen Self-organizing Network.

Example 3.7: 2-D Self-organization Mapping

Suppose we have created 2000 input vectors randomly (Figure 3.20). We will define a two-dimensional map of 35 neurons to classify these input vectors. The two dimensional map is five neurons by seven neurons in horizontal and vertical directions, respectively. The map is then trained for 5,000 presentation cycles in the MATLAB. The results are displayed in Figure 3.22. The details of the program are given in Chapter 21.

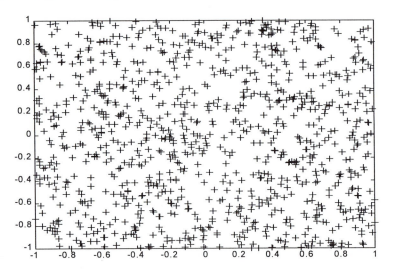

Figure 3.20: Initial Inputs of the Network of Example 3.7.

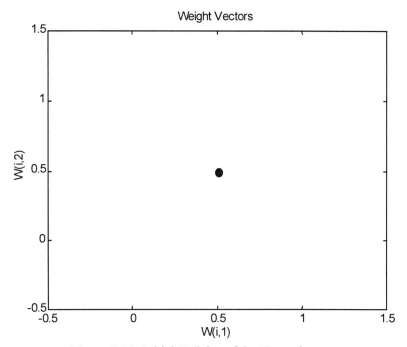

Figure 3.21: Initial Weights of the Network.

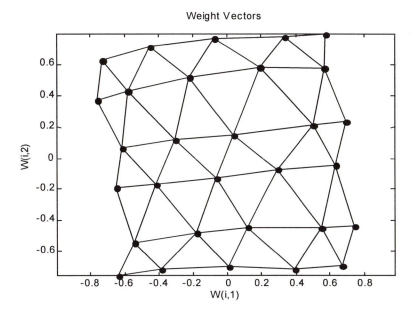

Figure 3.22: Weights of the Kohonen Self-organizing Network after Training (Example 3.7).

3.6 HOPFIELD NETWORK

Hopfield rekindled interest in neural networks by his extensive work on different versions of the Hopfield network [13],[14]. The network can be utilized as an associative memory or to solve optimization problems. One of the original network [13], which can be used as a content addressable memory is described in this chapter. The network is a typical recursive model in which nodes are connected to one another. Figure 3.23 shows a Hopfield network.

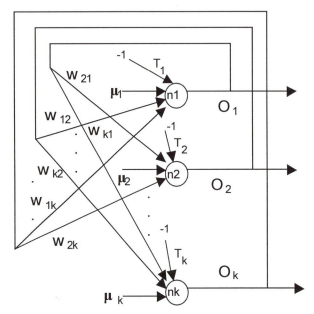

Figure 3.23: Hopfield Network.

As is shown, the output of each neuron consists of the inputs from other neurons, with the exception of itself. Therefore, the activity level of the neurons can be calculated using the following formula:

$$a_i = \sum_{\substack{j=1 \\ j \neq i}}^{n} w_{ij} o_j + \mu_i - T_i \qquad i = 1, \ 2, \ ..., \ n \qquad (3.26)$$

or in the vector form as:

$$a_i = W_i O + \mu_i - T_i \qquad i = 1, \ 2, \ ..., \ n \qquad (3.27)$$

where:

$$W_i = \begin{bmatrix} w_{i1} & w_{i2} & \cdots & w_{in} \end{bmatrix} \qquad i = 1, \ 2, \ ..., \ n \qquad (3.28)$$

Wi is the weight vector for the i-th input of the neural network and the i-th element of this vector is equal to zero. On the other hand,

$$O_i = \begin{bmatrix} o_1 \\ o_2 \\ \vdots \\ o_n \end{bmatrix} \qquad i = 1, \ 2, \ ..., \ n \qquad (3.29)$$

is the output vector of the neural network. Equation 3.27 in the matrix form can be rewritten as follows:

$$A = WO + I - T \qquad i = 1, \ 2, \ ..., \ n \qquad (3.30)$$

The weight matrix W is a symmetric matrix with all diagonal elements equal to zero. If the activation function of the neuron is a sign function, we will have:

$$o_i = \begin{cases} -1 & \text{if} & a_i < 0 \\ +1 & \text{if} & a_i > 0 \end{cases} \qquad (3.31)$$

The output transition between old value and new value will happen at certain times. At that time, if the value of the additive weighted sum of a neuron is greater than threshold of that neuron, the new output of that neuron will remain or change to +1, otherwise it will remain or change to −1.

Considering this fact, we can define the state of the network, which is the value of the outputs at one time. For example, $O = \begin{bmatrix} 1 & -1 & 1 & \cdots & 1 \end{bmatrix}$ is a state of the network. For each neuron we have two values. Therefore 2^n states exist for a network with n neurons.

In a Hopfield network, we apply an input at certain times and then it will be removed. This causes transitions in states of the network. These transitions continue until the network reaches to a stable point, which is called an *attractor.* An important point about this network is that at each time one neuron will calculate its activity level and change its output. In other words, updating of the outputs of the neuron is being done in an asynchronous fashion. Therefore to calculate activity level of the next neuron, and find the output of that neuron, we use some updated value for the output of the other neurons. The updating order of the neurons is random. It depends on random propagation delays and noise. When using the formula in matrix form, we should be careful, because it offers synchronous or parallel updating. If we consider $E = \begin{bmatrix} 1 & -1 \end{bmatrix}$, each state of the system is an edge of the graph in E^n space. After applying an input pattern, the state of the network goes from edge to adjacent edge until it reaches an attractor of 2^n edges. An attractor should satisfy the equation:

$$\text{sgn}[A_a] = O_a \qquad (3.32)$$

Where A_a and O_a are activity level and output at the attractor. Note that if the network satisfies this equation, the next state of the network is equal to its present state and therefore no transition will happen until a new input pattern is applied to the network.

As mentioned earlier, input will be applied momentarily and then will be removed. Considering this fact and using Equation 3.30, Equation 3.32 will change to:

$$O_a = \text{sgn}[WO_a - T] \tag{3.33}$$

If we define the energy function for the system as:

$$E = -\frac{1}{2}O^T WO + \mu \cdot O - T^T O = -\frac{1}{2}\sum_{\substack{i=1 \\ j \neq i}}^{n}\sum_{j=1}^{n} w_{ij}o_i o_j - \sum_{i=1}^{n} i_i o_i + \sum_{i=1}^{n} T_i o_i \tag{3.34}$$

The gradient of the energy can be calculated from Equation 3.34 as:

$$\nabla E = -\frac{1}{2}(W' + W)O - \mu^T + T^T = -WO - \mu^T + T^T \tag{3.35}$$

Here we have used the fact that the weight matrix is symmetric. The energy increment is equal to:

$$\Delta E = (\nabla E)^T \Delta O \tag{3.36}$$

As discussed earlier outputs will be updated one at a time. Therefore only i-th output will be updated,

$$\Delta O = \begin{bmatrix} 0 & \cdots & o_i & \cdots & 0 \end{bmatrix}^T \tag{3.37}$$

The energy increment will be equal to:

$$\Delta E = (-W_i^T O - \mu_i^T + T_i^T)\Delta o_i = -A_i \Delta o_i \tag{3.38}$$

It is obvious that for positive A_i, $\Delta o_i \geq 0$ and for negative A_i, $\Delta o_i \leq 0$. Looking at Equation 3.38 it can be seen that $\Delta E \leq 0$. Therefore it can be concluded that state transitions of the network are in a way that the energy is either decreased or retained. This means that the attractors are the edges with lowest levels of energy. Following is an example to clarify these ideas.

Example 3.8:
Shows the state transitions and attractors in a fourth order Hopfield network. Consider the weight matrix as follows:

$$W = \begin{bmatrix} 0 & -1 & -1 & 2 \\ -1 & 0 & 1 & -1 \\ -1 & 1 & 0 & -1 \\ 2 & -1 & -1 & 0 \end{bmatrix} \tag{3.39}$$

Considering the threshold and external inputs equal to zero, energy level can be calculated as follows:

$$E = \frac{1}{2}O^T WO \tag{3.40}$$

or:

$$E = -\frac{1}{2}\begin{bmatrix} o_1 & o_2 & o_3 & o_4 \end{bmatrix}\begin{bmatrix} 0 & -1 & -1 & 2 \\ -1 & 0 & 1 & -1 \\ -1 & 1 & 0 & -1 \\ 2 & -1 & -1 & 0 \end{bmatrix}\begin{bmatrix} o_1 \\ o_2 \\ o_3 \\ o_4 \end{bmatrix} \tag{3.41}$$

After simplification we will have:

$$E = -o_1(-o_2 - o_3 + 2o_4) - o_2(o_3 - o_4) + o_3 o_4 \qquad (3.42)$$

Now if we consider all the states of the network starting from $\begin{bmatrix} -1 & -1 & -1 & -1 \end{bmatrix}$ to $\begin{bmatrix} 1 & 1 & 1 & 1 \end{bmatrix}$, we can calculate all the energy levels of the network. The result will be the levels 1, 1, -1 3, -1, 3, -7, 1, 1, -7, -1, 3, 3, -1, 3, 1 respectively. Therefore the energy levels are –7, -1, 1, 3. The two states with the lowest energy level -7 are $\begin{bmatrix} -1 & 1 & 1 & -1 \end{bmatrix}$ and $\begin{bmatrix} 1 & -1 & -1 & 1 \end{bmatrix}$. We can see that these states are attractors of the network. In other words, they satisfy Equation 3.33. If we try any other state of the network, we will see that they do not satisfy this equation, which means that they are not attractors of the network.

In other words, the attractors are the states with minimum levels of energy. In fact we can see that the transition in the network will be from an state to another state with a lower or the same level of energy. On the other hand, we know that the transition is asynchronous. Therefore, at each single step we will go from one state to its adjacent state. These transitions are in the direction of reduction of energy level until we reach a state with a minimum level of energy, which is the attractor of the network. Figure 3.24 shows the state transition of the network.

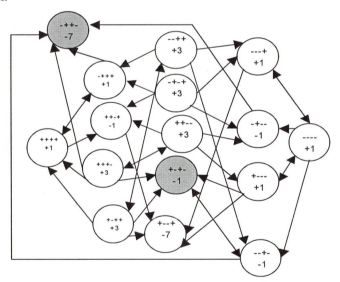

Figure 3.24: State Transition of the Hopfield Network to Reach to a Stable State.

3.7   CONCLUSIONS

In this chapter, four different neural networks were presented. Several numerical examples were provided to demonstrate the effectiveness of these networks. The described networks consist of highly parallel building blocks that illustrate NN design principles. They can be used to construct more complex systems. In general, the NN architectures cannot compete with the conventional techniques at performing precise numerical operations. However, there are large classes of problems that often involve ambiguity, prediction, or classifications that are more amenable to solution by NN than other available techniques. In the following chapters several of these problems will be addressed in detail.

REFERENCES:

1. Widrow, B. and Lehr, M. A., Thirty Years of Adaptive Neural Networks: Perceptron, MADALINE, and back propagation. *Proc. of the IEEE,* Vol. 78, 1415-1442, 1990.
2. Rumelhart, D.E., Hinton G.E. and Williams, R.J., Learning Internal Representations by Error Propagation, *Parallel Data Processing,* Vol. 1, Chap. 8, the MIT Press, Cambridge, MA, 1986.
3. Specht, D.F., A General Regression Neural Network, *IEEE Trans. on Neural Networks,* Vol. 2, 568–576, 1991.
4. Wasserman P.D., *Advanced Methods in Neural Computing*, Van Nostrand Reinhold, New York, 1993.
5. Moody, J. and Darken, C., Fast Learning in Networks of Locally-Tuned Processing Units, *Neural Computation*, Vol. 1, 281–294, 1989.
6. Jang, J. S., Sun, C. T., and Mizutani, E., *Neuro-Fuzzy and Soft Computing*, Prentice Hall, Englewood Cliffs, NJ, 1997.
7. Lowe, D., Adaptive Radial Basis Function Nonlinearities and the Problem of Generalization, *Proc. First IEEE Int. Conf. on Artificial Networks*, London, UK, 1989.
8. Wettschereck D. and Dietterich, T., Improving the Performance of Radial Basis Function Networks by Learning Center Locations, *Advances in Neural Information Processing Systems*, Vol. 4, 1133-1140, Morgan Kaufmann, San Mateo, CA, 1992.
9. Nguyen, D. and Widrow B., The Truck Backer-Upper, *Int. Joint Conf. on Neural Networks*, Washington, DC, Vol. 2, 357–363, 1989.
10. Kohonen, T., *Self-organization and Associative Memory*, 3rd ed., Springer-Verlag, New York, 1988.
11. Zurada, J., *Introduction to Artificial Neural Systems*, West Publishing Co, St. Paul, MN, 1992.
12. Haykin, S., *Neural Networks: A Comprehensive Foundation*, Prentice Hall, Upper Saddle River, NJ, 1999.
13. Hopfield J.J., Neurons with Grades Response Have Collective Computational Properties Like Those of Two State Neurons", *Proc. of National Academic of Science*, Vol. 81, 3088–3092, 1984.
14. Hopfield, J.J. and Tank, D.W., Computing with Neural Circuits: A Model, *Science*, Vol. 233, 625–633, 1986.

4 APPLICATIONS OF NEURAL NETWORKS IN MEDICINE AND BIOLOGICAL SCIENCES

Faramarz Valafar

4.1 INTRODUCTION

In this chapter, we will discuss applications of artificial neural networks (ANNs) in medicine and biological sciences. In particular, we will discuss ANN solutions to classical engineering problems of detection, estimation, extrapolation, interpolation, control, and pattern recognition as it pertains to these sciences. We will discuss some of these applications in detail to introduce the readers to typical problems that researchers face in the area.

Research in ANNs' applications as an alternative to classical engineering and mathematical techniques in medicine and biological sciences has intensified in recent years. Since the early 1990s, many applications of ANNs have replaced classical solutions to the engineering problems mentioned above. This is also true in medicine and biological sciences. [1 – 20] To discuss applications and accomplishments of ANNs in medicine and biological sciences, we will first introduce a few standard measures that will be used throughout this chapter to compare or report various results. These measures have been recommended and used to evaluate physicians and healthcare workers by various organizations, and therefore are good measures for evaluating the performance of any automated system that is designed to assist these healthcare professionals.

4.2. TERMINOLOGY AND STANDARD MEASURES

The American Heart Association (AHA) recommends the use of four measures to evaluate procedures for diagnosing CAD. [21] Since these measures are useful in other areas of diagnosis as well, we will be using them in evaluating most diagnostic systems.

$$sensitivity \equiv TPF = \frac{TP * 100}{TP + FN} \qquad (4.1)$$

$$specificity \equiv TNF = \frac{TN * 100}{TN + FP} \qquad (4.2)$$

$$PA = sensitivity * P(D) + specificity * [1 - P(D)] \qquad (4.3)$$

$$PV = \frac{sensitivity * P(D)}{sensitivity * P(D) + (100 - specificity) * [1 - P(D)]} \tag{4.4}$$

Where *TP* stands for true positive, *FN* stands for false negative, *TN* stands for true negative, and *FP* stands for false positive. *Sensitivity*, or *true-positive fraction (TPF)*, is the probability of a patient who is suffering from a disease to be diagnosed as such. *Specificity*, or *true-negative fraction (TNF)*, is the probability that a healthy individual is diagnosed as such by a diagnosis mechanism for a specific disease. *PA* is the *predictive accuracy*, or the overall percentage of correct diagnosis. *PV* is the *predictive value* of a positive test, or the percentage of those who have the disease and have tested positive for it. *P(D)* is the *a priori* probability of a patient who is referred to the diagnosis procedure actually having cancer.

In addition to *TPF* and *TNF*, we define two other related values. *False-positive fraction (FPF)* is the probability of a healthy patient being incorrectly diagnosed as having a specific disease. And *false-negative fraction (FNF)* is the probability that a patient who is suffering from a disease will be incorrectly diagnosed as healthy. In this way, the following relations can be established:

$$FPF = 1 - TNF \tag{4.5}$$
$$FNF = 1 - TPF \tag{4.6}$$

To clarify the terminology and symbols, let us consider the following example.

Example 4.1:

Let us assume that 100 patients were referred to the mammography department for diagnosis of breast cancer. Let us further assume that of the 100 individuals, 38 actually had a cancerous tumor, and the remaining 62 either did not have any tumor or did not have one that was malignant (cancerous). Let us further assume that a diagnosis procedure (manually conducted by physicians, by an automated system, or by both) correctly diagnosed 32 of the 38 cancer sufferers as having breast cancer. It, however, misdiagnosed six of those as being cancer free. Let us also assume that the procedure correctly classified 58 of the 62 cancer-free patients as such, and misclassified the remaining 4 as having breast cancer. Finally, let us assume that on the average, 35 % of those who are referred to the mammography procedure actually have breast cancer.

In this example *TP = 32, FN = 6, TN = 58, FP = 4, and P(D) = 0.35.* Hence,

$$Sensitivity \equiv TPF = \frac{32 * 100}{32 + 6} = 84.21\% \implies FNF = 1 - 84.21 = 15.79\%,$$

$$Specificity = TNF = \frac{58 * 100}{58 + 4} = 93.55\% \implies FPF = 1 - 93.55 = 6.45\%,$$

$$PA = 84.21 * 0.35 + 93.55 * [1 - 0.35] = 90.28\%,$$

$$PV = \frac{84.21 * 0.35}{84.21 * 0.35 + (100 - 93.55) * [1 - 0.35]} = 87.55\%$$

In this example the overall system accuracy is 90.28 %, while the predictive value of a positive test is at 87.55 %.

Another commonly used measure of ANNs' performance that has found its way into the medical community (among others) is the receiver operating characteristic (ROC) curve. [22,23]. ANNs that perform pattern recognition or detection could be viewed as a receiver system (in the sense of a radar signal receiver) that receives a noisy signal and attempts to identify it. In the radar example, identification of the signal could mean classifying an aircraft as friend or foe. In medical decision-making, it usually means the diagnosis of a patient as healthy or sick. For simplicity, let us assume that the ANN has one output neuron. The following discussion can be expanded to cover multi output ANNs as well.

An important variable in the performance of the ANN is the threshold value θ of the output neuron. If $\theta=1$, all incoming signals in radar technology would be classified as noise. In medical technology, it would translate into having a negative diagnosis for all patients and, thus, categorizing them as healthy. If $\theta=0$, we would be classifying all patients as sick. In the first case, the probability of detection, or *TPF*, would have a value of zero, but so would the probability of false alarm, or *FPF*. In the second case, *TPF* would have a value of one, as would *FPF*. Neither of these receivers (detectors) would be desirable. The ROC curve gives an idea as to how the receiver would perform for all values of threshold (θ) between 0 and 1.

Let us assume that $f(z|H_s)$ is the conditional probability density function of z, the activation level of the output neuron, given that the input to the network contains a signal (patient actually is carrying the disease) (hypothesis H_s). Similarly, $f(z|H_n)$ is the conditional probability density function of z, given that the input to the network does not contain any signal (patient is actually healthy) (noise only, hypothesis H_n). A hypothetical example of these two density functions is shown in Figure 4.1.

With θ_0 being the alarm threshold, the probability of detection, or *TPF*, and the probability of positive error, or *FPF*, can then be calculated as follows:

$$TPF(\theta_0) = \int_{\theta_0}^{1} f(z|H_s)dz \qquad (4.7)$$

$$FPF(\theta_0) = \int_{\theta_0}^{1} f(z|H_n)dz \qquad (4.8)$$

Figure 4.2 shows the *TPF and FPF* graphs for the probability density functions of the hypothetical example shown in Figure 4.1.

Definition 4.1: The ROC curve of a system is the plot of that system's *TPF* curve versus its *FPF* curve. The operating variable of the three curves is θ_0, the alarm threshold.

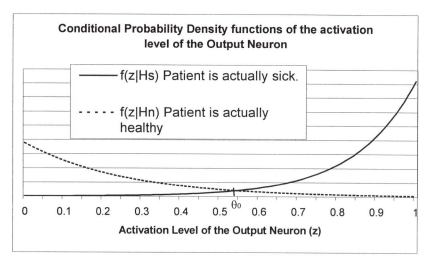

Figure 4.1: Hypothetical Conditional Probability Density Functions of the Activation Level z of the Output Neuron, Given that the Patient is Known to be Sick (H_s) or Healthy (H_n).

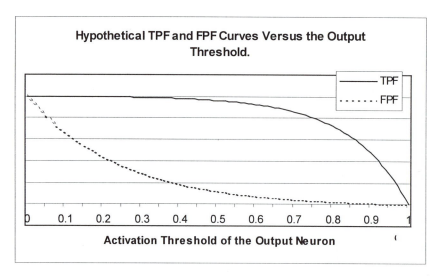

Figure 4.2: The True Positive Fraction and the False Positive Fraction Functions of the Hypothetical Example Shown in Figure 4.1, Plotted vs. the Alarm Threshold of the Output Neuron.

The ROC plot of the hypothetical example can now be plotted according to Definition 4.1. Figure 4.3 shows the ROC plot for the hypothetical example of

Figure 4.1. This figure also shows a worst case classifier (dashed line), and a theoretical best case classifier.

The ROC curve demonstrates an important property of any detection system: namely, that the probability of "true positive" is directly related to the probability of "false positive." They rise and fall together. The ideal classifier is one whose TPF is one for all values of FPF, including when $\theta_0=1$ and $FPF=0$ (the red curve in Figure 4.3). The worst classifier is one that has no discrimination. A positive detection always has equal probability of being true or false. In other words, $TPF = FPF$ for all values of θ_0. This, in turn, would produce the dashed line ROC curve shown in Figure 4.3.

A consolidated measure that is a good representation of the overall quality of the receiver, and of the model used to build the receiver, is the area under the ROC curve. This area is commonly referred to as A_z. [23] This area varies between 0.5 (worst receiver) and 1 (best receiver). The area under the hypothetical ROC curve of the example in Figure 4.1 is 97.49 %. Furthermore, the best operating point of a receiver can be determined from the ROC curve by determining the point with a maximum distance from the diagonal line of the worst case classifier. In Figure 4.3, the label "Ideal Operating Point" shows this point of the hypothetical receiver. As can be seen from the figure, the best operating point of the hypothetical receiver has a *TPF* value of about 84 %, and FPF value of about 3.5 %.

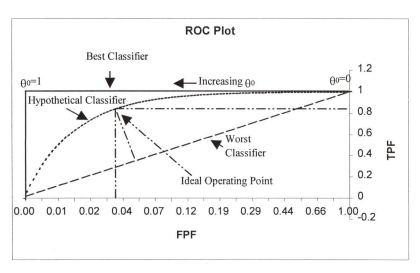

Figure 4.3: ROC Plots of a Hypothetical Receiver, a Theoretical Best Case Receiver, and a Worst Case Receiver.

The measures introduced in this section are particularly helpful in comparing the performance of most diagnosis procedures (systems) and therefore will be used in several parts of this chapter.

4.3   RECENT NEURAL NETWORK RESEARCH ACTIVITY IN MEDICINE AND BIOLOGICAL SCIENCES

ANNs have enjoyed success in various areas of medicine and biological sciences. ANNs have been successfully applied to areas such as radiology [16], cancer research, [12,14,24–29] biochemical spectrum analysis, [30] sleep disorder, [31] cardiac disease, [1,2,15,18,19] biochemistry of a disease, [4,5] HIV and AIDS, [5,29] epilepsy, [6,20] vision, [7] motor control, [8] lunge disease, [10,11] pathology and laboratory data analysis, [13,14] diagnosis decision support, [17,18,32] and many more.

In the following we present a brief summary of three research projects as examples of some of the most active research areas in ANNs' applications in medicine and biological sciences. These three applications are only meant to give an indication as to the breadth of the activity areas, and to demonstrate the typical problems (and give some ideas as to possible solutions) that researchers often face when dealing with real world data in the areas of medicine and biological sciences. It is also our hope that through these examples we can indicate the level of achievement of the ANN research community in various fields.

4.3.1   ANNs in Cancer Research

Pattern recognition using ANNs in cancer research is likely to be the most active area in terms of application of ANNs in medicine. ANNs have been used extensively in various roles in cancer research anywhere from tumor detection and analysis, [24,25,26] to the detection of biochemical changes in the body due to cancer, [29] to analysis of follow-up data in primary breast cancer, [27] to visualizing anticancer drug databases. [28] Among various types of cancer and detection methods, breast cancer diagnosis by the means of ANN classification of mammography images has been one of the most widely studied.

T.C.S.S. André and A. C. Roque [24] offer one of the most recent studies in this area. The authors have developed a medical decision support system using neural networks to aid in the diagnosis of breast cancer. This system uses digital mammogram images to classify a case as having one of three possible outcomes: suspicion of malignant breast cancer, suspicion of benign breast cancer, or no suspicion of breast cancer. André and Roque [33] used a staged (layered) neural network with a set of identical single layer networks as the input layer. These input layer networks used localized receptive fields without overlapping in the mammogram image. The hidden and the output layers of the network were each a single layer of perceptrons. The input layer was first trained, with regions taken from several mammograms, to become a feature extractor using the competitive learning algorithm. [33] The perceptron layers were then trained with the backpropagation learning algorithm.

The authors report a *TPF* of 0.75, and an *FPF* value of 0.06 for the optimum operating point of the ANN system described above. Furthermore, they report an A_z value of 0.84 for their system. To put this value in perspective, it should be mentioned that A_z values typically fall in the 0.80 to 0.90 range for mammography analysis. In a similar study, Wu et al. [34] report an A_z value of 0.84 for a group of attending radiologists, and an A_z value of 0.80 for a group of resident radiologists.

Wu et al. [34] also conducted a similar study using a neural network trained with the backpropagation algorithm. They used a set of features of mammogram images that were selected by experienced radiologists as the input signal to the neural networks. In this case, they report an A_z value of 0.95 for textbook cases, and an A_z value of 0.89 for clinical cases.

4.3.2   ANN Biosignal Detection and Correction

Applications of signal detection techniques have been used in biological sciences to detect a single signal, or a group of signals, buried in various types of noise and nonrelevant biosignals for several decades. Applying pattern recognition techniques to spectroscopic data, for instance, has been used to help in structural elucidation of known molecules, and to significantly reduce the enormous duplicate work otherwise conducted in the area. Pattern recognition tools can therefore be employed to build search engines for spectral databases of various types of molecules.

An example of this can be seen in detecting the signature of one, or a group of complex carbohydrates in gas chromatography-electron impact mass spectroscopy (GC-EIMS), or nuclear magnetic resonance (NMR) spectra. [30]

Complex carbohydrates have been linked to biochemical functions of all cells, [35 – 37] such as cell recognition (e.g., initial steps in host pathogen and symbiotic relationships), intercellular adhesion (lectins and selectins), biosignal processes (oligosaccharins), developmental regulation, antibody binding, immune system modulation, and hormonal regulation. Consequently, complex carbohydrates, or the receptors that bind them, are also involved in many diseases, including autoimmune diseases, inflammatory diseases, and cancer. A tool to rapidly elucidate the chemical structures of complex carbohydrates can be instrumental in research to understand their biological functions. The presence of specific carbohydrates or their "uncommon" relatives, for instance, could be indicative of disease, the stage of a disease, or the presence of an antibody.

In this context, signal correction techniques could be used to correct the incoming biosignals and to compare them to a prerecorded clean library of signals. In this way, signal detection and correction techniques are used to discover and clean biosignals, and subsequently identify complex carbohydrates from which they originated. In this section, we discuss an artificial neural network solution to biochemical signal detection and identification, as well as biochemical signal correction for complex carbohydrates. [30]

Identification Of Complex Carbohydrate Structures from Their Spectral Signatures Using ANNs.

Structural and functional elucidation of complex carbohydrates is a key part of an increasing number of biomedical inquiries into these molecules. The structural determination of complex carbohydrates is the mandatory prerequisite to determining their functions. But the enormous chemical complexity and diversity of complex carbohydrates makes their structural elucidation a particularly challenging, lengthy task, and one that scientists would not wish to duplicate unnecessarily. Therefore, the primary need for the scientist faced with finding out the identity, chemical characteristics, and other attributes of a carbohydrate is to know whether that carbohydrate has already been analyzed by others and, if so, what is known about its chemistry, biology, and conformation. F. Valafar and H. Valafar [30] have developed a system for automated identification of complex carbohydrates using their chemical spectra that can provide this type of information.

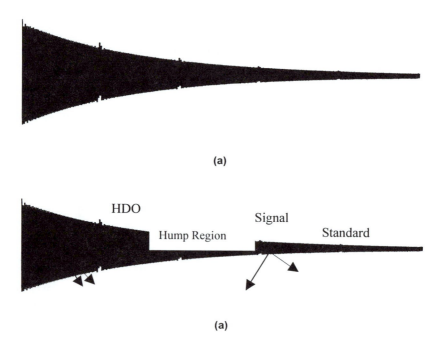

(a)

(a)

Figure 4.4   (a) A ^{1}H-NMR Time-Domain Signal of an *N*-linked Oligosaccharide. (b) The Fourier Transformed Frequency-Domain Spectrum of the Same Oligosaccharides.

In the following, we discuss Valafar's method in identifying complex carbohydrate structures from their ^{1}H-NMR spectra using artificial neural networks. In most classical signal processing methodology, the process of structural elucidation of a chemical compound from its ^{1}H-NMR spectrum first involves individual signal detection of elementary components (proton or ^{1}H signals). The second step in this process is the task of combining the detected individual signals in order to identify the structure of the carbohydrate in question. Valafar's use of ANNs in this process combines the two steps; the ANN performs both steps at the same time.

^{1}H-NMR spectra, in general, suffer from environmental, instrumental, and other types of variations that manifest themselves in a variety of aberrations. Low signal-to-noise ratio, [38 – 40] baseline drifts, [41 – 43] frequency shifts due to temperature variations, line broadening and negative peaks due to phasing problems, and malformed peaks (or peaks overlapped more than usual) due to inaccurate shimming are among the most common aberrations. Figure 4.4 demonstrates a clean ^{1}H-NMR spectrum of an *N*-linked complex carbohydrate.

As can be seen from Figure 4.4, large peaks not relevant to the structural elucidation of the complex carbohydrate usually dominate ^{1}H-NMR spectra of complex carbohydrates. These peaks include that of the solvent (heavy water in this case, HDO) and that of the standard. The proton signals (drifts) are typically in the order of 100 times weaker than the large peaks. Furthermore, most of these signals heavily overlap in the "hump" region of the spectrum, leaving the region unusable for structure elucidation.

For the purpose of automated identification of these spectra, elimination of the above mentioned aberrations becomes essential, as they can lead to erroneous identification. [41–45] A variety of signal processing techniques have been applied to "clean up" ^{1}H-NMR spectra. For instance, signal averaging[1] and apodization[2] have become standard ways of improving the signal-to-noise ratio. To correct baseline problems, a number of techniques have been used such as parametric modeling using *a priori* knowledge, [41,42] optimal associative memory (OAM), [42] spectral derivatives, [46] polynomial fitting, partial linear fitting, [47] and Bayesian analysis. [48] For peak detection (and solvent peak suppression), methods such as Bayesian analysis [48,49] and principal component analysis [50,51] can be mentioned. For signal-to-noise

---

[1] In signal averaging a spectrum is recorded several times. Each recorded signal is referred to as a "transient". The final spectrum is the arithmetic average of all the transients. The hope is that by using signal averaging the zero mean components of the noise present in the signal will be averaged out.[44]

[2] Apodization is a type of low (high) pass filtering performed in the time domain. Apodization is performed by speeding up or slowing down the rate of decay of time domain exponential functions. This is accomplished by multiplying the time domain signal by another function. This technique allows the improvement of the signal-to-noise ratio at the cost of the reduction in signal resolution (or vice versa).[44]

ratio problems, various types of filters (including adaptive filters such as matched filters [44,51]) in addition to standard apodization and signal averaging have also been used. A number of other mathematical techniques have also been introduced to address other specific types of aberrations encountered in ^{1}H-NMR spectra.

Although many of these signal processing techniques have enjoyed success, they remain solutions to specific types of aberrations. In order to produce sufficiently "clean" spectrum overall, one needs to use several of these methods to eliminate the aberrations present in a real spectrum. Furthermore, most of these techniques produce side effects that are magnified when improperly processed by a second signal processing algorithm, which can lead to false identification. Moreover, after the initial signal processing steps have been taken, the task of identifying the processed spectrum remains. This is not a trivial task as frequently the quality of the processed spectrum remains poor, requiring a sophisticated identification system.

Valafar and Valafar [30] have developed an artificial neural network system that addresses many of the above mentioned problems while identifying ^{1}H-NMR spectra of complex carbohydrates. Although the procedure still requires a minimal amount of preprocessing, it has significantly reduced the number of preprocessing steps while increasing the overall identification accuracy.

In this project, the authors developed an ANN system for a library of *N*-linked oligosaccharides, and one for xyloglucan oligosaccharides. While xyloglucans are plant cell wall oligosaccharides, the *N*-linked oligosaccharides are present in most animal biochemistry. Since the two systems used similar methods to develop an ANN identification system, we will only discuss here the development of the *N*-linked ANN identifier.

Preprocessing. Initial testing indicated that without preprocessing all selected methods for identification purposes would perform poorly. Therefore, it was decided to use some minimal preprocessing techniques to eliminate some aberrations before the identification stage. These preprocessing steps included baseline correction, high frequency noise reduction, and water and solvent peak elimination. These steps were respectively accomplished by a first derivative technique, a low-pass filter in the form of a specially designed averaging moving window, and a bin selection technique. The ANN eliminated the remaining aberrations in the process of identification. In other words, the ANN was able to learn during training to be insensitive to the remaining aberrations. Additionally, since each ^{1}H-NMR spectrum contained anywhere from 4K to 16K of data, an interpolation technique was used to normalize the length of all ^{1}H-NMR vectors to 5000. This would reduce (in most cases) the resolution of the spectrum to 2 points per Hertz, which is as low as Nyquist's theorem [51] would permit. The 5000-point vector covered the region between 1 and 5.5 ppm. The corrected spectra then were introduced to the ANN for training purposes.

Figure 4.5 shows the estimated *a posteriori* probability density functions [51] of the inter-[3] and intra-class[4] correlation coefficients between the raw (not processed) ^{1}H-NMR spectra of the *N*-linked data set[5] as defined by Bayes' theorem. [51] The required *a priori* density functions by Bayes' theorem were estimated using the nonparametric approach of Parzen density estimation. [52] Figure 4.6 shows the estimated *a posteriori* density functions of the preprocessed spectra from the same data set. As can be seen from the graphs, the overlap of the two density functions has been reduced from 56 to 43 %. This means that the "classical" signal preprocessing has simplified the identification task, and a Bayes' classifier, in combination with correlation coefficient analysis, now carries a 43 % uncertainty factor vs. the previous 56 %. Moreover, the probability density functions behave closer to expected (one large peak per density function, and smooth decay everywhere else in the function) after preprocessing.

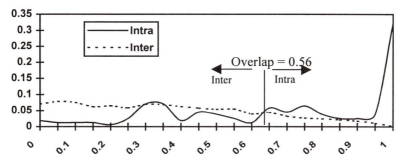

Figure 4.5 Estimated Distribution of Inter- and Intra-class Correlation Coefficients of Raw (Not Processed) ^{1}H-NMR Spectra of 109 ^{1}H-NMR Spectra of 23 *N*-linked Oligosaccharides.

ANN design
The authors used a two stage feedforward network with sigmoidal artificial neurons [33] in the hidden and output layers. The input layer of the network contained 5000 fan out neurons. The output layer contained 67 neurons

---

[3] By "class" we refer to the set of all spectra for a specific compound. In other words, each class in the xyloglucan experiment contained two spectra. In the *N*-linked database, 20 oligosaccharides were represented by five spectra, and the remaining three had three spectra, giving rise to five-member and three-member classes respectively. An "inter-class" correlation coefficient is the correlation coefficient between the spectra of two different oligosaccharides.
[4] "Intra-class" correlation coefficient is the correlation coefficient between two different spectra of the same oligosaccharide.
[5] The estimated Bayes' *a posteriori* distribution functions for the xyloglucan data set were similar to those shown here for the *N*-linked data set, and for space consideration are not shown here.

corresponding to the 67 oligosaccharides in the library. The number of the hidden neurons was empirically determined to be 27.

To develop the best performing ANN, several criteria were set forward: 1) the developed ANN was to have a very low FPF. In other words, if a spectrum of a complex carbohydrate was not present in the training library, the system should not try to find the closest match in the library. The outcome should be that the carbohydrate does not exist in the library; 2) the system needed to be tolerant of aberrations, and to be able to identify carbohydrates from its library even in the presence of relatively low signal-to-noise ratio. This translated into a high value for the area under the ROC curve, A_z. This also meant a high TPF value and 3) in the case of a mixture, the system was to indicate the carbohydrate of the highest ratio in the mixture.

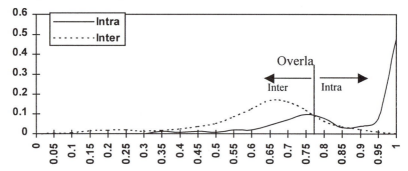

Figure 4.6 Histogram of the Correlation Coefficient Distribution of the Preprocessed ^{1}H-NMR Spectra of 109 ^{1}H-NMR Spectra of 23 *N*-linked Oligosaccharides.

With these goals in mind, a large number of training simulations were conducted. A large number of permutations were tried, namely, by varying the learning step size update policy, the number of hidden neurons, and the level of input noise. Valafar et al. dynamically manipulated the spectra during training by introducing some input noise in order to simulate the natural variability of these spectra. The noise simulated varying coupling constants due to temperature, line shape problems due to incorrect shimming, and minor baseline drifts. [53]

Table 4.1 shows the results of the best performing ANN in comparison with three other methods. The table shows the results of the experiments for both the *N*-linked and xyloglucan oligosaccharides.

Method A: Correlation coefficient analysis; Method B: Singular value decomposition; Method C: Correlation coefficient analysis and Bayesian classifiers; Method D: Backpropagation ANN.

The ANN system also showed less sensitivity to signal to noise degradation. Table 4.2 shows the degradation of identification accuracy of the four methods with increasing noise.

Table 4.1: Number of Correctly Identified *N*-linked and Xyloglucan Oligosaccharide Spectra (Total Number of Spectra is in Parentheses) by Four Different Identification Techniques after the Spectra Were Preprocessed as Described Above.

Method	*N*-linked Oligosaccharides		Xyloglucan Oligosaccharides	
	Training (67 spectra)	Testing (134 spectra)	Training (20 spectra)	Testing (20 spectra)
A	41	69	9	12
B	43	72	10	11
C	44	78	10	13
D	67	128	20	20

Table 4.2: Percentage Correct Identification of the Four Systems with Increasing Noise During Testing in the N-linked Oligosaccharide Database.

Method	Testing Noise Level				
	0%	5%	10%	15%	20%
A	51.49	41.86	37.21	34.88	27.91
B	53.73	46.51	39.53	34.88	25.58
C	58.21	53.49	46.51	37.21	32.55
D	95.52	95.35	81.40	62.79	39.53

4.3.3 Decision-making in Medical Treatment Strategies

Decision-making techniques can be used in medicine to solve various problems. Specifically, ANN pattern recognition engines have enjoyed significant success in medical decision-making. [1 – 20] Although, the ANN systems developed in this area demonstrate great potential benefit to the healthcare community, due to the numerous remaining challenges, the area remains one of the most active. To introduce the difficulties that researchers face in this area, we discuss here an ANN system designed to assist physicians in deciding on the best treatment strategy. Specifically, we will describe a research project conducted by H. Valafar et al. [32] to develop an ANN system to decide whether a beneficial, and yet at times harmful, medication (Hydroxyurea) should be prescribed in battling the symptoms of sickle cell anemia (SCA).

Predicting a sickle cell anemia patient's response to Hydroxyurea. Sickle cell anemia is a genetic disease mostly affecting African Americans in the US., although the disease is not limited to people from African origin worldwide. Treatment with Hydroxyurea (HU) partially alleviates disease symptoms in many patients with SCA.

Treatment with HU alleviates the clinical course in many patients with sickle cell anemia. [54] Most patients respond to HU with an increase in the fetal hemoglobin (HbF) concentration of blood by either increasing the amount of HbF in their F-cells and/or by increasing the proportion of F-cells. The response to HU varies from patient to patient. If the magnitude of the HU-elicited increase in the %HbF (with respect to the total Hb) of the patient's blood could be predicted, "non-responders" could be identified. Although Hydroxyurea is effective for many patients, it is ineffective, and at times harmful, for others. Therefore, it is desirable to devise a tool with which physicians can predict, with a high percentage of accuracy, the outcome of the treatment before the medication is administered. Hence, the ultimate goal of the project is to *predict the response level of a given patient to Hydroxyurea, using only the pretreatment data of a patient.*

To develop such a system, the first question that needs to be answered is: *What data should be used for the prediction/decision-making task?* In this particular project, the authors relied on the expertise and experience of the physicians who were involved in sickle cell anemia research. The final set of data to be used for prediction contained the results of a standard blood test, in addition to some genetic information. A detailed list of the parameters that were used can be seen in Table 4.3.

Selection of the parameters listed in Table 4.3 was based solely on educated guesses on the physicians' parts (such as the genetic information), and some earlier simple statistical analysis of various data. Therefore, it could be expected that some of the 23 parameters might not be relevant to the problem at hand. It is also quite possible that not all relevant parameters are included in the study.

Data preprocessing. Many medical databases, especially those that go years into patients' past history and treatment, are in printed or written form. The first step in this research was to create an electronic database usable by the modeling team. This process was accomplished at the Medical College of Georgia. All patients' data were entered into a widely available spreadsheet. These data then were sent to the modeling team for analysis.

Soon after the first round of analysis was completed, the following problems were observed:

1) *Missing data.* A quick look at the data revealed that much of the data was missing. For instance, if the patient was feeling well in that particular month, certain measurements (tests) were not conducted. Furthermore, there were instances where the patient simply did not show up for follow-up tests because he/she was feeling okay. In some instances, the paperwork containing the data for the early stages of the treatment was misplaced and lost. There were two types of missing data in our databases. In some instances, certain variables (pieces of data) were missing from a monthly record. In others, an entire monthly record was missing.

2) *Incorrect data.* Simple statistical correlation analysis revealed that there were some severe outliers. Most of these were traced back to human error. But there were also data that simply were off the chart, but not traceable to

any human error. All the human errors were corrected. However, the nontraceable extreme outliers were excluded from the study.

3) *Invalid or corrupt data.* In some cases, there were patients who became pregnant against the doctor's advice, or underwent a blood transfusion due to other complications in the middle of the treatment period. The data of such patients were excluded from the study as the effects of such events on a patient's blood chemistry and his or her ability to respond to Hydroxyurea was unclear.

Table 4.3 A Description of the 23 Parameters for Which Data was Obtained from the Patients. From H. Valafar, et al., [32].

Parameter	Description	Units
Age	Age of patient at the time of analysis	Days
Sex	Male/Female	F=1, M=2
NAGG	α Globin gene number	None
BAN	Number of BAN haplotypes	1,2, or 3* None
BEN	Number of BEN haplotypes	1,2, or 3* None
CAM	Number of CAM haplotypes	1,2, or 3* None
SEN	Number of SEN haplotypes	1,2, or 3* None
WGT	Weight of patient	Kg
%HbF	Fetal hemoglobin, as % total hemoglobin	None
HbF	Fetal hemoglobin, absolute value	g/L of blood
Hb	Total hemoglobin concentration	g/dL of blood
RBC	Red blood cell count	$\times 10^{12}$ / Liter
PCV	Packed cell volume (hematocrit)	Liter / Liter
RDW	% Variation in the size of red cells	None
Retic	Reticulocytes	$\times 10^{5}$
MCV	Mean cell (erythrocyte) volume	Femtoliters
MCH	Mean cell hemoglobin	Picograms
WBC	White cell count	$\times 10^{9}$ / Liter
Polys	Polymorphonuclear leukocytes	$\times 10^{9}$ / Liter
Plats	Platelet count	$\times 10^{9}$ / Liter
Bili	Bilirubin concentration in blood	mg / dL
NRBC	Nucleated red blood cells seen in peripheral blood	Number per WBC
Duration	Duration of treatment a patient received to arrive at the maximum %HbF level	Days

*The actual values were 0,1,or 2, but 0 could not be used (see last paragraph under ANN Analyses).

Problem definition. Further problems arose as the team prepared for the first round of modeling experiments. One of the more fundamental problems, and often one that is usually difficult to solve in medical decision-making problems, was with the definition of the problem (problem statement). After further close examination of the data, it was realized that the definition of the problem was inadequate and that the experiments were destined to either fail, or to produce results that were medically useless. The original statement of the problem was as follows: "Develop a system that can accurately distinguish positive responders from the nonresponders using pretreatment data." Furthermore, a "positive responder" was defined to be "a patient whose initial percentage HbF (%HbF) doubles at some point during the treatment."

After looking at the data, it was soon realized that while this definition may work for patients whose initial %HbF is, say 7%, or higher, it does not work so well for patients whose initial %HbF is 1% or 2%. In other words, while Hydroxyurea treatment might increase a patient's initial %HbF value from 1% to 2% at some point during treatment, it is not very likely that he/she would experience any benefits (reduced number of hospital visits, or reduced severity of symptoms) as a result of this minor increase. This meant that even in the bestcase scenario that a system with 100% accuracy (in separating the patients who can double their initial %HbF from those who cannot due to Hydroxyurea) was developed, its results would be clinically meaningless. This is because doubling the %Hbf value does not translate into reduced symptoms or hospital visits for many or all patients. A new definition for a "positive response" had to be devised.

After extensive study of published articles on Hydroxyurea and its alleviation of symptoms, two possible definitions were suggested:

1) *Dynamic patient threshold.* It was suggested that each patient has a different level of % HbF, beyond which his/her symptoms begin to taper off. A patient would be categorized as a positive responder if his/her %HbF level increased above this dynamic threshold as a result of the treatment. This dynamic level needs to be calculated or estimated for each patient via some type of computational means. Although this measure is probably the more accurate measure of positive response, it was soon realized that in order to estimate accurately each patient's threshold, one would need to have the response model in hand. Since the response model was the final goal of the project, this definition seemed impractical and was therefore abandoned.

2) *Static threshold.* The team agreed that the next best definition was that of a static threshold across all patients. This threshold was determined by consulting existing publications and the collaborators at MCG. All these sources seemed to agree that most patients experienced some type of relief of symptoms when their %Hbf rose about 15%. [55,57] Hence, if a patient's HbF concentration rose above 15% of total Hb during treatment, he/she was categorized as a positive responder, and all others as nonresponders. Three patients were excluded from this study, as their

initial % HbF was higher than 15. This threshold divides the final 83 patients included in the study into 58% responders and 42% non-responders.

Missing Data. The problem of missing data arises in medicine quite often. The most common causes of missing data are 1) patients who do not come into clinics for further tests when they start feeling better or, if they do come in, the nurses and the physicians who record the data are not as motivated to record all available information; and 2) data are commonly recorded on paper and, therefore, sometimes are misplaced and/or lost. While these are the two main causes of missing data, there are others that need not be mentioned here.

In general, regardless of the reason for missing data, the missing data can be categorized into two classes: 1) missing record: in some instances, the data for an entire record are missing. A common cause of this type of missing data in the case of SCA is due to patients who do not report to the clinic for their monthly tests when they experience some relief in their symptoms. In such cases, no data for that month are available for the patient; and 2) missing data points: In some instances, specific parameters in each record are missing. An example of this in the case of SCA would be when a patient who is feeling better reports to the clinic for a monthly test. In some such cases, not all the tests are conducted, or properly recorded. Human error is also a common source of this type of missing data.

The first type of missing data did not cause many problems in our experiments. This is because only the initial parameters of the patient (from before the beginning of the treatment) were used and the highest level of percentage HbF during treatment to train the artificial neural network. For this reason, missing intermittent data were not harmful to our experiments, except in cases when the highest percentage HbF was also missing. In the cases where the highest percentage HbF value was missing, all data of that patient were excluded from the study.

The second type of missing data could be potentially much more problematic, as it is much more likely for the value of some parameters to be missing at the initial recording before the beginning of the treatment. Since the initial values are vital information, all patients who were missing more than two initial parameters were dropped from the study. The patients whose data were missing one or two initial parameters were kept in the study as long as the level of initial percentage HbF was not missing. To fill in the missing parameters, some experiments were conducted with a few extrapolation algorithms. However, it was discovered that the best way to deal with the few missing parameters was to fill them in with zeroes. This is simply because Delta rule [33] and backpropagation algorithms were used to train our neural networks, and, as can easily be determined from weight update formulas, when the input parameter is zero, no learning is conducted in the first stage of the network. This was the best way to make use of the data without presenting the network with erroneous data.

Compliance. Compliance is one of the biggest problems in medical research. The simple cause of it is that some patients stop taking the medication, or at least reduce the dosage without instructions from the physician when they start feeling better. This can lead to corrupt data (for our purposes), as a patient could be falsely identified as a nonresponder. This was the case in our study. Our initial systems suffered from a relatively high FNF. From formula 4.6, it can easily be seen that this causes TPF to be reduced, and therefore A_z, the area under the ROC curve, to be lower than expected. As a result, it could lead to the false conclusion that the identification technique or system architecture is inadequate, while the source of the problem really lies in the data.

In the case of many medications, compliance can be measured by the variation in one or many biochemical parameters. This was the case with HU and SCA patients. One of HU's side effects is that it increases the volume of red blood cells. [58] Among the final 83 patients who were all categorized as compliant and were included in this study, the mean cell volume increased by an average of 22% as a result of HU treatment. This is in line with other studies. [55,56,58,59] The variable mean cell volume (MCV) is thus a good measure of compliance. This variable was analyzed for each patient. It was decided that six patients were not compliant and so their data were excluded from the study.

Figure 4.7 shows the bin distribution function of MCV before and after HU treatment. As can be observed, the distribution has clearly moved to higher values after the treatment and has a higher mean.

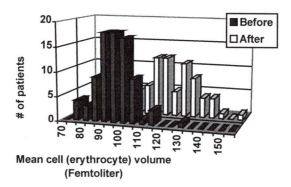

Figure 4.7: Distribution of Average Volume of the Red Blood Cells of 83 Sickle Cell Patients before and after Treatment with HU. From H. Valafar, et al., [32].

Neural network prediction model. An ANN using 23 input neurons, 4 hidden neurons, and 1 output neuron was used for the 15% threshold experiment. This neural network produced an output value higher than 0.5 if the patient was predicted to be a responder, and an output of less than 0.5 if the patient was predicted to be a nonresponder.

The threshold experiment was designed to eliminate the possibility that the ANN could simply "memorize" the values of the parameters of each patient. This was accomplished by training ANNs with the parameter values of 82 of the patients, and then using the values of the patient whose parameters had not been seen by the ANN, to test the ANN. This procedure was repeated 83 times and each time an ANN was trained. (A different patient was left out of the training each time) The result of this experiment is presented in Figure 4.8. Seventy patients were correctly classified as responders or nonresponders while 13 were misclassified. Thus, 84% of the responses were predicted correctly. This experiment was repeated five times with, on average, 86.6 correct predictions with a standard deviation of +/-2.0.

Variable selection. Researchers in the medical fields are also frequently faced with the problem of variable selection. In most cases, there is not enough information to select the relevant variables for a certain modeling/pattern recognition problem in medicine. Also, one of the reasons that researchers seek a mathematical model for a disease is to use it to determine the relevant variables. This information can be extremely helpful in understanding how the disease works, develops in the body, or is fought against by the body's immune system. In the latter case, if the immune system is failing to effectively fight the disease, information about relevant variables could lead to new medications that either help the body in eliminating the disease, or at least reduce its symptoms (e.g., the case of sickle cell anemia).

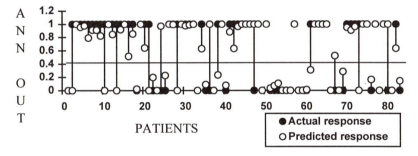

Figure 4.8: The Prediction by ANNs of Which Patients Would Respond to HU by an Increase in Their HbF Concentration to the Point Where it Accounts for 15% or More of Their Total Hb. ANNs Were Trained with the Values of the Parameters of 82 Patients and then Tested with the Values of the Parameters of the Patient that Had Not Been Used to Train the ANN. This Procedure Was repeated 83 times, each time Leaving Out a Different Patient and Training the ANN With the Data from the Other 82 Patients to Give the Values in the Figure. Patients Whose HbF Concentration Did Not Reach 15% of the Total Hb Should have generated an ANN "output" of less than 0.5, while patients whose HbF Concentration Exceeded 15% of the Total Hb Should Have Generated an ANN Output of More Than 0.5. From H. Valafar, et al., [32].

Valafar et al. designed their variable selection experiments in the SCA's case to identify which of the 23 parameters are most important or influential in assisting ANNs to predict those patients that will respond to HU treatment. Determining the importance of each of the 23 parameters was accomplished by employing two different methods. The first method consisted of a recursive elimination process in which a different set of parameters was taken out of the training set. The ANNs were trained with the values of the remaining parameters. The software measures the degradation of performance due to the missing parameters. This experiment is an exhaustive elimination process in which the removal of every combination of parameters (2^{23}-1=8,388,607 combinations) is evaluated. The degradation (or importance) of the parameters observed is the averages of ten experiments (two different seeds for the random number generator, and five runs per seed). The final effect of removing each set of parameters is calculated by averaging the performance degradation of the ten ANNs trained without that set of parameters.

The second method of parameter selection is an adaptive technique that takes into effect the synaptic connection strengths of each variable. This algorithm is initiated by setting equal values for each parameter. During the course of training, these values are updated to reflect the strength of the synaptic connection(s) associated with each parameter. This, in turn, is an indication of the contribution of each parameter towards the discovery of the correct answer. Thus, at the end of the training the contribution of each parameter reveals its importance in the solution of the problem. Each training session was repeated five times to eliminate any random behavior of the system.

Although the above two methods are distinctly different methods for parameter selection, both algorithms produced similar results in extracting the relevant parameters. For this reason, we will only discuss the results of the second method from this point forward.

The 23 parameters and their scores, which are proportional to their contributions in predicting the response to HU treatment, are listed in Table 4.4. This table contains the averaged data for over five different training sessions. The lack of any particularly influential contributors indicates that no one parameter contains the information needed to predict the response to HU. Therefore, based on the given contributions, it is reasonable to assume that the information needed for a successful classification is distributed among a number of parameters, perhaps even a fairly large number of parameters.

The ANNs whose testing results are shown in Figure 4.8 used the values of all 23 parameters. A separate experiment was carried out to determine if the values of just ten of the twenty-three parameters listed in the previous section could be used while maintaining the ANN's full ability to identify responders and non-responders. This experiment used the top ten parameters listed in Table 4.4. The ability to eliminate unnecessary parameters has the potential for reducing the problem size by more than 50%, and might assist in elucidating the mechanisms by which ANNs function.

Table 4.4 The Effectiveness of Each of the 23 Parameters to Assist ANNs in Predicting the Response of Patients to HU Treatment. From H. Valafar, et al., [32].

Parameter	Score
Duration	0.083
RDW	0.063
WBC	0.059
Plats	0.053
MCV	0.053
Polys	0.052
WGT	0.050
SSEN	0.045
Retic	0.043
Sex	0.042
SCAM	0.041
NAGG	0.041
Hb	0.040
SBAN	0.040
MCH	0.035
RBC	0.034
SBEN	0.034
Bili	0.034
Age	0.033
HbF	0.032
%HbF	0.031
PCV	0.031
SNBRC	0.030

The ANN trained only with the 10 selected variables had remarkably similar results to the one trained with all 23 variables. Except for 2 of the 83 patients, the results of the 2 networks were very similar. The network trained with ten variables produced outputs that were more clearly defined. The mean of the probability density function of the output z of the smaller network was higher for positive responders, and lower for nonresponders. By the same token, the standard deviation of both curves was smaller than those of the larger network. Furthermore, the two patients whose classification changed by using the smaller network were both marginally classified by the larger network. One was correctly classified as a responder, and one incorrectly as a nonresponder. With the smaller network, the first patient was incorrectly classified as a nonresponder; the second patient was correctly classified as a responder. Therefore, the TPF, FPF, and the ROC curves remained identical for both networks.

4.4   SUMMARY

Artificial neural networks have distinct features that can be advantageous in modeling natural phenomena in biology and medicine. Applications of ANNs in these fields are sure to help unravel some of the mysteries in various diseases and biological processes. In the SCA case, the ANN developed for the variable selection process helped pinpoint the parameters that possibly play an important role in understanding the works of SCA. This could lead to a significant increase in the life expectance of SCA sufferers.

Research in applications of ANNs in medicine and biological sciences currently remains strong. With more systematic data collection routines implemented in healthcare facilities, systems such as the ones described in this chapter are sure to find their way into doctors' offices and hospital laboratories.

REFERENCES

1.   Akay, M., Akay, Y.M., Welkowitz, W., Semmlow, J.L., and Kostis, J.B., Noninvasive Detection of Coronary Artery Disease Using Neural Networks, *Proc. of the Ann. Conf. on Eng. in Med. and Biol.,* 13(3), 1434 – 1435, Oct 31 – Nov 3, 1991.

2.   Akay, M., Noninvasive Diagnosis of Coronary Artery Disease Using a Neural Network Algorithm, *Biol. Cybern.,* 67: 361 – 367, 1992.

3.   Alpsan, D., Auditory Evoked Potential Classification by Unsupervised ART 2-A and Supervised Fuzzy ARTMAP Networks, *Int. Conf. on Neural Networks (ICNN '94)*, IEEE, Orlando, FL, 3512 – 3515, June 26 – July 2, 1994.

4.   Andrea, T.A. and Kalayeh, H., Applications of Neural Networks: Quantitative Structure-Activity Relationships of Dihydrofolate Reductase Inhibitors, *J. Med. Chem.,* 34:2824 – 2836, 1991.

5.   Andreassen, H., Bohr, H., Bohr, J., Brunak, S., Bugge, T., Cotterill, R.M.J., Jacobsen, C., Kusk, P., and Lautrap, B. Analysis of Secondary Structure of the Human Immunodeficiency Virus Proteins by Computer Modelling Based on Neural Network Methods, *J. Acquired Immune Deficiency Syndrome,* 3, 615, 1990.

6.   Apolloni, B., Avanzini, G., Cesa-Bianchi, N., and Ronchini, G., Diagnosis of Epilepsy via Backpropagation, *Proc. of the 1990 Int. Joint Conf. on Neural Networks*, Washington, DC, 2, 571 – 574, 1990.

7.   Armentrout, S.L., Reggia, J.A., and Weinrich, M., A Neural Model of Cortical Map Reorganization Following a Focal Lesion, *Artif. Intelligence in Med.,* 6(5), Oct 1994.

8.   Armstrong, W.W., Stein, B.A., Kostov, R., Thomas, M., Baudin, P., Gervais, P., and Popovic, D., Application of Adaptive Logic Networks and Dynamics to Study and Control of Human Movement, *Proc. of the Second Int. Symp. on 3D Anal. of Human Movement*, Poitiers, France, 81 – 84, June 30 – July 3, 1993.

9.   Armstrong, W.W., Kostov, A., Stein, R.B., and Thomas, M.M., Adaptive Logic Networks in Rehabilitation of Persons with Incomplete Spinal Cord Injury, *Workshop on Environmental and Energy Applications of Neural Networks*, Richland, WA, Pacific Northwest National Laboratory, March 30 – 31, 1995.

10.   Asada, N., Doi, K., MacMahon, H., Montner, S.M., Giger, M.L., Abe, C., and Wu, Y., Potential Usefulness of an Artificial Neural Network for Differential Diagnosis of Interstitial Lung Diseases: pilot study, *Radiology,* 177, Vol. 3, 857–60, December, 1990.

11.   Asada, N., Doi, K., MacMahon, H., Montner, S., Giger, M.L., Abe, C., and Wu, Y., Neural Network Approach for Differential Diagnosis of Interstitial Lung Diseases, *Proc. SPIE (Medical Imaging IV)*, 1233: 45 – 50, 1990.

12.   Ashenayi, K., Hu, Y., Veltri, R., Hurst, R., and Bonner, B., Neural Network Based Cancer Cell Classification, *Proc. of the World Congress on Neural Networks*, San Diego, CA, 1, 416 – 421 June 5 – 9, 1994.

13.   Astion, M.L. and Wilding, P., The Application of Backpropagation Neural Networks to Problems in Pathology and Laboratory Medicine, *Arch. Pathol. Lab. Med.*, 116:995 – 1001, 1992.

14.   Astion, M.L. and Wilding, P., Application of Neural Networks to the Interpretation of Laboratory Data in Cancer Diagnosis, *Clin. Chem.* (US) 38, 34 – 38, 1992.

15.   Avanzolini, G., Barbini, P., and Gnudi, G. Unsupervised Learning and Discriminant Analysis Applied to Identification of High Risk Postoperative Cardiac Patients, *Int. J. Bio-Med. Comput.,* 25, 207 – 221, 1990.

16.   Barski, L.L., Gaborski, R.S., and Anderson, P.G., A Neural Network Approach to the Histogram Segmentation of Digital Radiographic Images, *Intell. Eng. Sys. Through Artif. Neural Networks*, Dagli, Burke, Fernandez, and Ghosh, (eds.), 3, 375 – 380, ASME Press, NY, 1993.

17.   Bartels, P.H., Thompson, D., and Weber, J.E., Diagnostic Decision Support by Inference Networks, *In Vivo,* 7, 379 – 385, 1993.

18.   Baxt, W.G., Use of an Artificial Neural Network for Data Analysis in Clinical Decision-Making: the Diagnosis of Acute Coronary Occlusion, *Neural Computation,* 2, 480 – 489, 1990.

19.   Baxt, W.G., Use of an Artificial Neural Network for the Diagnosis of Myocardial Infarction, *Ann. of Intern. Med.,* 115, 843 – 848, 1991.

20.   Echauz, J. and Vachtsevanos, G., Neural Network Detection of Antiepileptic Drugs from a Single EEG Trace, *Proc. of the IEEE Electro/94 Int. Conf.,* 346 – 351, Boston, MA, May 10 – 12, 1994.

21.   Gibbons, R.J., Balady, G.J., Beasley, J.W., Bricker, J.T., Duvernoy, W.F., Froelicher, V.F., Mark, D.B., Marwick, T.H., McCallister, B.D., Thompson, P.D. Jr., Winters, W.L., Yanowitz, F.G., Ritchie, J.L.,

Gibbons, R.J., Cheitlin, M.D., Eagle, K.A., Gardner, T.J., Garson. A. Jr., Lewis, R.P., O'Rourke, R.A., and Ryan, T.J., ACC/AHA Guidelines for Exercise Testing, A Report of the American College of Cardiology/American Heart Association Task Force on Practice Guidelines (Committee on Exercise Testing), *J. of the Am. Coll. of Cardiol.*, 30(1), 260 – 311, July, 1997.

22.   Goodenough, D.J., Rossmann, K., and Lusted, L.B., Radiographic Applications of Receiver Operating Characteristic (ROC) Curves, *Radiology*, 110, 89 – 95, 1974.

23.   Hanely, J.A. and McNeil, B.J., The Meaning and Use of the Area Under a Receiver Operating Characteristic (ROC) Curve, *Radiology*, 143, 29 – 36, 1982.

24.   André, T.C.S.S. and Roque, A. C., A Neural Network System for the Diagnosis of Breast Cancer, *Proc. of the Int. Conf. on Math. and Eng. Techniques in Med. and Biol. Sci. 2000* (METMBS'00), Las Vegas, NV, 1, 1 – 6, June 26 – 29.

25.   Rodrigues, R.G.S., Pela, C.A., and Roque, A.C., Tomographic Image Reconstruction Using Neural Networks, *FFCLRP*, Brazil, V1 27 – 33.

26.   Chen, D., Chang, R.F., and Huang, Y.L., Breast Cancer Diagnosis Using Self-Organizing Map for Sonography, *Ultrasound. Med. Biol.*, 26(3), 405 – 11, March, 2000.

27.   Harbeck, N., Kates, R., Ulm, K., Graeff, H., Schmitt, M., Neural Network Analysis of Follow-Up Data in Primary Breast Cancer, *Int. J. Biol. Markers*, 15 Vol. 1, 116 – 22, January – March, 2000.

28.   Shi, L.M., Fan, Y., Lee, J.K., Waltham, M., Andrews, D.T., Scherf, U., Paull, K.D., Weinstein, J.N., Mining and Visualizing Large Anticancer Drug Discovery Databases, *J. Chem. Inf. Comput. Sci*, 40, Vol. 2, 367 – 79, March – April, 2000.

29.   Cherniak, R., Valafar, H., Morris, L.C., and Valafar, F., *Cryptococcus neoformans* Chemotyping by Quantitative Analysis of ^{1}H-NMR Spectra of Glucuronoxylomannans Using a Computer Based Artificial Neural Network, *J. of Clin. and Diag. Lab. Immunol.*, 5(2),146 – 159, March, 1998.

30.   Valafar, F. and Valafar, H., CCRC-Net: An Internet-Based Spectral Database for Complex Carbohydrates, Using Artificial Neural Networks Search Engines, *Trends in Anal. Chem.*, 18, 508 – 512, 1999.

31.   Guimaraes, G., The Discovery of Sleep Apnea with Unsupervised Neural Networks, *Int. Conf. on Math. and Eng. Techniques in Med. and Biol. Sci. (METMBS'2000)*, 1, 361 – 367, Las Vegas, NV, June 26 – 29, 2000.

32.   Valafar, H., Valafar, F., Darvill, A., Albersheim, P., Kutlar, A., Woods, C., and Hardin, J., Predicting the effectiveness of Hydroxyurea in Individual Sickle Cell Anemia Patients, *J. of Artif. Intell. in Med.*, 18 (2), 133 – 148, February, 2000.

33. Haykin, S., *Neural Networks: A Comprehensive Foundation*, Prentice Hall, NJ, 1999.

34. Wu, Y., Giger, M.L., Doi, K., Vyborny, C.J., Schmidt, R.A., and Metz, C.E., Artificial Neural Networks in Mammography: Application to Decision Making in the Diagnosis of Breast Cancer, *Radiology,* 187 Vol. 1, 81 – 7, April, 1993.

35. Varki, A., Biological Roles of Oligosaccharides: All the Theories are Correct, *Glycobiology,* 3, 97 – 130, 1993.

36. Goochee, C.F., Gramer, M.J., Andersen, D.C., Bahr, J.B., and Rasmussen, J.R., The Oligosaccharides of Glycoproteins: Factors Affecting Their Synthesis and Their Influence on Glycoprotein Properties, *Frontiers in Bioprocessing II.* (Todd, P., Sikdar, K., and Bier, M., eds.) 199 – 240, American Chemical Society, Washington, D.C., 1992.

37. Cook, G.M.W., Glycobiology of the Cell Surface: the Emergence of Sugars as an Important Feature of the Cell Periphery, *Glycobiology,* 5, 449 – 461, 1995.

38. Van Huffel, S., Enhanced Resolution Based on Minimum Variance Estimation and Exponential Data Modeling, *Signal Processing*, 33, 333 – 355, 1993.

39. Van den Boogaart, A., Howe, F.A., Rodrigues, L.M., Stubbs, M., Griffiths, J.R., *In Vivo* ^{31}P MRS: Absolute Concentrations, Signal-to-Noise and Prior Knowledge, *NMR in Biomed.*, 8, 87 – 93, 1995.

40. Angelidis, P.A., Spectrum Estimation and the Fourier Transform in Imaging and Spectroscopy, *Concepts Magn. Resonance*, 8 Vol. 5, 339 – 381, 1996.

41. Blumler, P., Greferath, M., Blumich, B., and Spiess, H.W., NMR Imaging of Objects Containing Similar Substructures, *Magn. Resonance*, Series A 103, 142 – 150, 1993.

42. Wabuyele, B.W. and Harrington, P., Optimal Associative Memory for Background Correction of Spectra, *Anal. Chem.*, 66, 2047 – 2051, 1994.

43. Wabuyele, B. W. and Harrington, P., Quantitative Comparison of Bidirectional Optimal Associative Memories for Background Prediction of Spectra, *Chemometrics and Intelligent Lab. Sys.*, 29, 51 – 61, 1995.

44. Angelidis, P. A., Spectrum Estimation and the Fourier Transform in Imaging and Spectroscopy, *Concepts Magn. Resonance*, 8(5), 339 – 381, 1996.

45. Goodacre, R., Timmins, E.M., Jones, A., Kell, D.B., Maddock, J., Heginbothom, M., Magee J. T., On Mass Spectrometer Instrument Standardization and Interlaboratory Calibration Transfer Using Neural Networks, *Analytica Chemica Acta*, 348, 511 – 532, 1997.

46.   Gerow, D.D. and Rutan, S.C., Background Subtraction for Fluorescence Detection in Thin–layer Chromatography with Derivative Spectrometry and the Adaptive Kalman Filter, *Analytica Chemica Acta*, 184, 53, 1986.

47.   Yu, K. M. and Jones, M.C., Local Linear Quantile Regression., *J. Am. Statistical Assoc.*, 93(441): 228 – 237, March, 1998.

48.   Whittenburg, S., Baseline Roll Removal in NMR Spectra Using Bayesian Analysis, *Spectroscopy Letters*, 28(8), 1275 – 1279, 1995.

49.   Whittenburg, S., Solvent Peak Removal in NMR Spectra Using Bayesian Analysis, *Spectroscopy Letters*, 29(3), 393 – 400, 1996.

50.   Harrington, P. B. and Isenhouer, T.L., Closure Effects in Infrared Spectral Library search Performance, *Appl. Spectrosc.*, 41, 1298, 1987.

51.   Papoulis, A., *Probability, Random Variables, and Stochastic Processes*, 3rd ed., McGraw-Hill, NY, 1991.

52.   Fukunaga, K., *Introduction to Statistical Pattern Recognition*, Second Edition. Academic Press, Boston, 255 – 268, 1990.

53.   Valafar, F., Valafar, H., and York, W.S., Identification of ^{1}H-NMR Spectra of Xyloglucan Oligosaccharide: A Comparative Study of Artificial Neural Networks and Bayesian Classification Using Nonparametric Density Estimation, *Int. Conf. Artif. Intelligence* 1999 (IC-AI'99), Las Vegas, NV, June 28 – July 1, 1999.

54.   Rodgers, G.P., Dover, G.J., Noguchi, C.T., Schechter, A.N., Nienhuis, A.W., and Nienhuis, M.D., Hematologic Responses of Patients with Sickle Cell Disease to Treatment with Hydroxyurea, *New England J. Med.*, 322 Vol. 15, 1037 – 1044, April, 1990.

55.   Charache, S., Terrin, M.L., Moore, R.D., Dover, G.J., Barton, F.B., Eckert, S.V., McMahon, R.P., and Bonds, D.R., Effect of Hydroxyurea on the Frequency of Painful Crises in Sickle Cell Anemia, *New England J. Med.*, 332, 1317 – 1322, May 18, 1995.

56.   Charache, S., Dover, G.J., Moore, R.D., Eckert, S., Ballas, S.K., Koshy, M., Milner, PF., Orringer, E.P., Phillips, G. Jr., and Platt, O.S., Hydroxyurea: Effects on Hemoglobin F Production in Patients With Sickle Cell Anemia, *Blood*, 79(10), 2555 – 2565, May 15, 1992.

57.   Powars, D.R., Weiss, J.N., Chan, L.S., and Schroeder, W.A., Is There a Threshold Level of Fetal Hemoglobin That Ameliorates Morbidity in Sickle Cell Anemia? *Blood*, 63(4), 921 – 926, April, 1984.

58.   Charache, S., Barton, F.B., Moore, R.D., Terrin, M.L., Steinberg, M.H., Dover, G.J., Ballas, S.K., McMahon, R.P., Castro, O., and Orringer, E.P., Hydroxyurea and Sickle Cell Anemia. Clinical Utility of a Myelosuppressive "Switching" Agent. The Multicenter Study of Hydroxyurea in Sickle Cell Anemia, *Med.*, 75, Vol. 6, 300–325, November, 1996.

59.   Steinberg, M.H., Lu, Z., Barton, F.B., Terrin, L.M., Charache, S., and Dover, G.J., Fetal Hemoglobin in Sickle Cell Anemia: Determinants of Response to Hydroxyurea, *Blood*, 89(3) 1078 – 1088, Feb, 1997.

5 APPLICATION OF NEURAL NETWORK IN DESIGN OF DIGITAL FILTERS

Dali Wang and Ali Zilouchian

5.1 INTRODUCTION

Any action on a signal that modifies the spectral content of the signal is called filtering. This includes the enhancement or suppression of certain features of the signal and is usually achieved by the use of linear time invariant systems. There are situations where the system may change with time in a particular manner; such systems are called adaptive filters. In this section, we describe fixed filters only.

There are two broad classes of digital filters. The first class is called finite impulse response (FIR) filters, since their response to an impulse dies away in a finite number of samples. FIR filters are developed as non-recursive structures and are inherently simpler to design.

The second class of digital filters is recursive filters. The impulse responses of recursive filters are composed of sinusoids that exponentially decay in amplitude. This makes their impulse responses infinitely long. Because of this characteristic, recursive filters are called infinite impulse response (IIR) filters.

An IIR filter can be represented by either difference equation or state space form. The state space form in general involves more numbers of coefficients than a transfer function unless it is represented as one of the canonical forms. However, there are many benefits from using a state space model in the analysis, design, and implementation of digital filters. First, the state space model, with the exception of canonical structures, is more robust than a transfer function representation. In other words, it exhibits less coefficient sensitivity. Second, various forms of state space models possess distinctive properties that are desirable in different applications. For instance, the balanced realization exhibits superior performance in the context of minimizing scaling and round-off noise. Third, the major part of modern control theory is based on the state space model. Furthermore, the difference function representation could be uniquely determined by the state space form representation. The reverse is not necessarily true. In this chapter, the state space model will be utilized for the IIR filter design.

In the above filter representations, all inputs, outputs or states are function of a single variable, which is time in most cases. We call these types of filters one-dimensional (1-D) filters. There are other types of filters in which the inputs, outputs and states are the function of more than one variable. One example is the filter used in image processing. Therein, the inputs and outputs values are the function of two variables, i.e., horizontal and vertical coordination. The digital filters used in this case are two-dimensional (2-D) filters. The same concept can be extended to M-D filters and signals. In this chapter, we will start with the design of 1-D IIR filter. The design process using neural networks (NN) is presented in detail for 1-D IIR filters. Then the concept is extended to 2-D filters. If the dimension of the filter is not explicitly specified, 1-D filters are implied in this work.

5.2   PROBLEM APPROACH

5.2.1   Neural Network for Identification

There are numerous techniques developed for digital filter design, both in frequency domain and in time domain. Most of these methods are analytical techniques. They work well with well-defined filter formats and the availability of accurate design data, such as the input and output of the filter. What if the data set used to design filter is noisy, or there is a need for customization in the filter's representation? This is where the NN based design technique comes into the picture.

The capability of neural networks as universal approximators has been extensively studied for system identification and modeling during the last two decades [1] – [20]. Most of the proposed methods are based on two types of NN architectures, back propagation and Hopfield recurrent neural network [1], [2], [10]. However, most of these identification techniques result in NN weight matrices which do not necessarily correspond to the parameters of the original system, such as in the works of Narendra and Parthasarathy [15], and Poggio and Girosi [16].

In this chapter, a novel NN architecture for design of recursive digital filters from input/output data in the state space form is presented. We use internal hidden neurons to encode the temporal properties of sequential inputs and outputs as the iterative states of the given process. The dynamic nature of the system is implicitly constructed within the internal neurons of the proposed model, which previous approaches have not addressed. Since the structure of the process is built into NN, we can obtain a particular state space structure as the result of the identification, such as controllability canonical, observability canonical   forms,   etc.   [8].   Such   flexibility   is   important   in   various implementations of linear discrete systems, such as computation complexity, memory requirement and overflow analysis. The significance of this work is in two fold. First, obtaining the state space model of a linear system is the basis for many engineering applications where a fast on-line, flexible, and robust solution is required [21]-[25]. Second, applying NN to this complex linear system modeling problems can be an aid to understanding and developing new architecture of NN for more general linear and nonlinear programming problems [12], [14]. In fact, the proposed identification scheme has been extended to general 2-D digital filters design problems in section 5.5 where an analytical solution is difficult to obtain.

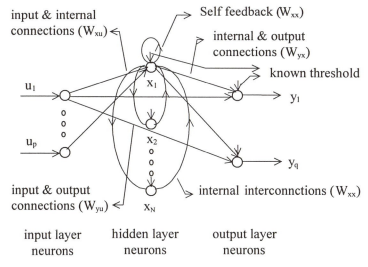

input & internal connections (W_{xu})

Self feedback (W_{xx})

internal & output connections (W_{yx})

known threshold

u_1

x_1

y_1

o
o
o

u_p

x_2

o
o
o

input & output connections (W_{yu})

x_N

internal interconnctions (W_{xx})

y_q

| input layer neurons | hidden layer neurons | output layer neurons |

Figure 5.1: A General Network Structure.

5.2.2 Neural Network Structure

The operation of an IIR filter can be specified by the equation

$$y(n) = \sum_{i=0}^{P} b_i x(n-i) - \sum_{i=1}^{Q} a_i y(n-i) \qquad (5.1)$$

The above difference equation provides us a procedure for determining the current output in terms of the present and past inputs as well as past output. An IIR filter can also be represented in state space form:

$$X(n+1) = AX(n) + Bu(n)$$
$$y(n) = CX(n) + Du(n) \qquad (5.2)$$

where $u \in \mathfrak{R}^p$, $y \in \mathfrak{R}^q$, $X \in \mathfrak{R}^N$ are input, output and state vectors, respectively. *A, B, C, D* are matrices of appropriate dimensions. The main objective of this work is to obtain *A, B, C, D* through NN by the training of an NN with available input/output data.

The proposed NN structure is a recurrent network from an error propagation viewpoint. The general network architecture is shown in Figure 5.1. For simplicity, the optional activation functions are not explicitly shown on the figure. The hidden neurons provide internal representations of the system via their self-feedback and connection with other neurons. These units memorize the status of the previous internal state, which are mapping information of previous states into present output. From such a viewpoint, the neural network is a recurrent network. However, the adjustment of the weights is based on the desired output values and actual outputs. Therefore, it can be considered as an error back propagation in the sense of training method.

In order to correlate the NN model with the state space model in Equation 5.2, an NN structure is proposed as shown in Figure 2. The association between various parameters of NN (weights, denoted as A_{NN}, B_{NN}, C_{NN}, D_{NN} for easy correlation) and the above state space model (A, B, C, D) can be observed from the proposed NN structure. A sub-section of A_{NN}, B_{NN}, C_{NN}, D_{NN} is shown in Figure 5.3 for a single neuron. The weights between hidden neurons (solid nodes) provide the representation of matrix A. The weights between input neurons (gray nodes) and hidden neurons and the weights between the hidden neurons and output neurons (empty nodes) represent the mapping of B and C, respectively. The weights between the input and output neurons map to D.

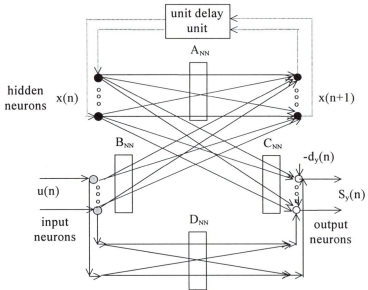

Figure 5.2: The Network Structure Designed for System Identification.

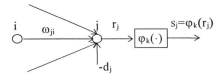

Figure 5.3: A Single Neuron.

5.3   A TRAINING ALGORITHM FOR FILTER DESIGN

The training objective is based on the instantaneous error value of a single input/output data pair. The algorithm can be implemented as a real time algorithm since the training could be accomplished as each input/output sample is fed to NN. The derivation of the algorithm is briefly presented in this section. It is different with conventional error back propagation since it possesses a recurrent process built into the network. The off line training algorithm can also be derived for system identification [8].

5.3.1   Representation

Consider an NN consisting of p external input connections, q external output connections and N hidden units. The various neurons can be classified into three categories: input neuron set $u \in R^p$ denoted as I, hidden neuron set $x \in R^N$ denoted as H, and output neuron set $y \in R^q$ denoted as O. At discrete time n, let $u(n)$ denote the p x 1 input vector, $x(n)$ denote the N x 1 vector as hidden neuron values, and $s_y(n)$ denote the q x 1 output vector of NN. As shown in Figure 3, a neuron j is either an output neuron or a hidden neuron prior to the delay. For such a neuron j which is connected to other neurons such as i, the corresponding activation value and output value are presented as follows:

$$r_i(n) = \sum_{\substack{j \in O}} \sum_{\substack{i \in H \cup I}} \omega_{ji}(n) s_i(n) - d_j(n) \qquad (5.3a)$$

$$r_i(n) = \sum_{\substack{j \in H}} \sum_{\substack{i \in H \cup I}} \omega_{ji}(n) s_i(n) \qquad (5.3b)$$

$$s_j(n) = \varphi_k\big(r_j(n)\big) \qquad (5.3c)$$
$$\scriptstyle j \in H \cup O$$

where $\omega_{ij}(n)$ is the weight between two neurons, $\varphi_k(\cdot)$ denotes the activation functions for hidden neurons ($\varphi_x(\cdot)$) and output neurons ($\varphi_y(\cdot)$), $d_j(n)$ is the desired outputs value at time n.

5.3.2   Training Objective

The on-line training objective is to minimize the mean-squared output of the NN at any instant discrete time n,

$$\varepsilon(n) = \frac{1}{2} \sum_{y \in O} S_y^{\ 2}(n) \qquad (5.3d)$$

where $s_y(n)$ is the output error at time n, which is the difference between the actual and desired outputs.

5.3.3   Weight Adjustment

A dynamic approach to minimize the cost function Equation 5.3d is to make the NN evolve its weight space along a trajectory that descends against the gradient of $\varepsilon(n)$. This condition implies that for all $i \in H \cup O$, $j \in H \cup I$:

$$\omega_{ii}(n+1) = \omega_{ii}(n) + \Delta\omega_{ii}(n) \tag{5.4a}$$

$$\Delta\omega_{ii}(n) = -\eta \frac{\partial\varepsilon(n)}{\partial\omega_{ij}(n)} \tag{5.4b}$$

where η is a learning rate which should be selected small enough to make weight change adiabatically and maintain the stability of the model.

The error gradient in Equation 5.4b could be obtained based on Equations 5.3c, and 5.3d:

$$\frac{\partial\varepsilon(n)}{\partial\omega_{ij}(n)} = \sum_{y \in O} s_y(n) \frac{\partial\varepsilon_y(n)}{\partial\omega_{ii}(n)} = \sum_{y \in O} s_y(n) \varphi_y'\big(r_y(n)\big) \frac{\partial r_y(n)}{\partial\omega_{ii}(n)} \tag{5.5}$$

The derivatives in the right hand side of Equation 5.5 are the gradients of output neuron value vs. NN weights. They are obtained using the following equations. From Equations 5.3

$$\frac{\partial r_y(n)}{\partial\omega_{ij}(n)} \Bigg|_{\substack{y \in O, \\ i \in H \cup O, j \in H \cup I}} = \sum_{k \in H \cup I} \cdot \frac{\partial(\omega_{yk}(n) \cdot s_k(n))}{\partial\omega_{ij}(n)}$$

$$= \sum_{k \in H \cup I} \left(\frac{\partial\omega_{yk}(n)}{\partial\omega_{ij}(n)} \cdot s_k(n) \right) + \sum_{k \in H \cup I} \left(\frac{\partial s_k(n)}{\partial\omega_{ij}(n)} \omega_{yk}(n) \cdot \right) \tag{5.6}$$

$$= \delta(i - y) \cdot s_j(n) + \sum_{k \in H} \frac{\partial s_k(n)}{\partial\omega_{ij}(n)} \omega_{yk}(n)$$

where $\delta(i-y)$ is the Kronecker delta function that equals to 1 when $i = y$ and 0 otherwise. The above derivation is based upon the following observation:

$$\frac{\partial\omega_{yk}(n)}{\partial\omega_{ij}(n)} = \begin{cases} 1 & \text{if } y = i \text{ and } k = i \\ 0 & \text{otherwise} \end{cases}$$

$$\frac{\partial s_k(n)}{\partial\omega_{ij}(n)} \Bigg|_{k \in I} = 0 \tag{5.7}$$

Thus, we can obtain the gradients of output neuron values v.s. NN weights as given in Equation 5.8, which are functions of neuron values, weights and the gradients of hidden neuron value v.s. weights at instant discrete time n.

$$\frac{\partial r_y(n)}{\partial \omega_{ij}(n)} = \delta(i-y) \cdot s_j(n) + \sum_{k \in H} \varphi_x'\big(r_k(n)\big) \frac{\partial r_k(n)}{\partial \omega_{ij}(n)} \omega_{yk}(n) \quad (5.8)$$

$$\begin{array}{c} y \in O, \\ i \in H \cup O, j \in H \cup I \end{array}$$

The gradients of hidden neuron value v.s. weights in Equation 5.8 are obtained as follow

$$\frac{\partial r_x(n)}{\partial \omega_{ij}(n)} = \sum_{k \in H \cup I} \frac{\partial(\omega_{xk}(n-1)s_k(n-1))}{\partial \omega_{ij}(n)} =$$

$$\begin{array}{c} x \in H, \\ i \in H \cup O, j \in H \cup I \end{array} \quad (5.9)$$

$$\delta(i-x) \cdot s_j(n-1) + \sum_{k \in H} \varphi_x'\big(r_k(n-1)\big) \frac{\partial r_k(n-1)}{\partial \omega_{ij}(n-1)} \omega_{xk}(n)$$

The observation similar to Equation 5.7 is also applied here. The η is assumed to be sufficiently small such that $\omega_{ij}(n) \approx \omega_{ij}(n-1)$.

The iterative process defined by Equation 5.9 provides the values needed in Equation 5.8. In sequel, the derivative value required in Equation 5.5 can be obtained. The weight update process in Equation 5.4 is accomplished with all the neuron value and derivative values at discrete instant time n.

5.3.4   The Training Algorithm

Based on the previous discussion, the proposed algorithm can be summarized as follows:
1. Initialize NN by random assignment of initial weights, zero value for all the weight gradients and hidden neuron values.
2. Present an input, desired output vectors pair to the NN.
3. Calculate the activation level of all neurons, including hidden neurons and output neurons.
4. Calculate the output error using Equation 3.3d.
5. Calculate weight gradients using Equation 5.9, 5.8, and 5.5.
6. Update the NN weights by equations 5.4a and 5.4b.
7. Repeat steps 2 to 6 for a new input/desired output pair. Multiple epochs may be required until the error criterion is bounded to a pre specified value.

5.4   IMPLEMENTATION ISSUES

5.4.1   Identifying a System in Canonical Form

There are infinite state-space structures with the same transfer function for a linear system or digital filter. The representation of Equation 5.2 can be transformed into different forms, such as controllability canonical form, observability canonical form, normal structure or balanced structure. These special forms can be built into NN by utilizing special network structures. By selection or elimination of certain weight connections in advance, we can obtain the system representation in such a particular form. This, on the other hand, simplifies the network design and reduces the number of free parameters compared to a fully connected network.

5.4.2   Stability, Convergence, Learning Rate and Scaling

The stability of recurrent networks has been extensively studied [4]. In general, for the asymptotically convergence of the network, the learning rate η should be assigned a small value. However, for fast convergence and local minimum avoidance, a large learning rate η is preferred. To resolve such two conflicting requirements, an adaptive learning rate scheme may be adapted similar to NN MATLAB Toolbox[27]. There are advantages by starting with a low learning rate and adaptively changing it. In order to improve the stability and convergence of the network, the input and desired output data are scaled to a proper range of value before being fed into the network.

5.5   2-D FILTER DESIGN USING NEURAL NETWORK

5.5.1   Two-dimensional Signal and Digital Filters

There are many signals that are inherently two-dimensional (2-D) in nature and for which 2-D signal processing techniques are required. Included in this group of signals are photographic data, medical X-rays, seismic data, gravity and magnetic data, etc. Many of fundamental ideas of 1-D signal processing may readily be extended to 2-D case. However, there are some very important concepts of 1-D systems that are not directly extendible to 2-D systems.

One major difference between 1-D and 2-D systems is that we can introduce global and local state in the 2-D cases. The global state (which is of infinite dimension in general) preserves all the past information, while the local state gives us a size of recursion to be performed at each step by a 2-D system. This leads to the definitions of global as well as local controllability, observability and as a result, the minimality of 2-D systems.

Similar to their 1-D counterparts, the 2-D recursive digital filters have the advantage of computation efficiency and memory reduction capabilities in comparison with non recursive digital filters. The 2-D state space models have been mainly used for the spatial domain representation of the 2-D causal recursive digital filters (CRDF). Kung et al. [38] have shown that the Roesser's model [37] is the most general form and the other representations can be imbedded in the Roesser's model.

Roesser's local state space (LSS) model divides the local state into a horizontal and a vertical state which are propagated in horizontal and vertical directions respectively. It is defined by the equations

$$\begin{bmatrix} x^k(i+1,i) \\ x^k(i,i+1) \end{bmatrix} = \begin{bmatrix} A_1 & A_2 \\ A_3 & A_4 \end{bmatrix} \begin{bmatrix} x^h(i.j) \\ x^v(i.j) \end{bmatrix} + \begin{bmatrix} B_1 \\ B_2 \end{bmatrix} u(i,j) \equiv AX + BU$$

$$y(i,j) = \begin{bmatrix} C_1 & C_2 \end{bmatrix} \begin{bmatrix} x^h(i.j) \\ x^v(i.j) \end{bmatrix} + Du(i,j) \equiv CX + DU$$

(5.10)

where;

i is an integer-valued vertical coordinate,

j is an integer-valued horizontal coordinate,

$x^h(i, j) \in R^{n1}$ is the horizontal state vector,

$x^v(i, j) \, R^{n2}$ is the vertical state vector,

$u(i, j) \in R^p$ is the input vector,

$y(i, j) \in R^q$ is the output vector,

and A_1, A_2, A_3, A_4, B_1, B_2, C_1, C_2, D are real matrices of appropriate dimensions.

5.5.2 Design Techniques

During the last two decades various design techniques have been proposed for 2-D recursive digital filters, either in frequency domain or in spatial domain [28], [30]-[36], [40]. However, most of those techniques are for a special class of 2-D filters called as separable-in-denominator digital filters (SDDF) [31-33], [36]. This is due to the fact that a SDDF filter shares some important properties of 1-D counterpart such as stability, minimality conditions and absence of singularity of the second kind. Therefore, many 2-D spatial design techniques have been developed using SDDF as the extensions of corresponding 1-D techniques [31-33]. There are relatively few techniques developed on identification and design of general 2-D recursive digital filters. One of the earliest methods was proposed by Shanks [39] et al., and Aly and Fahmy [30]. However, the problem of general 2-D identifications using an analytical solution has not been addressed due to its mathematically complex nature.

The NN approach designed for a 1-D recursive filter could be extended for general 2-D recursive digital filters. By a similar measure, an NN model has been developed to approximate an arbitrary 2-D system response and obtain the LSS model parameters from NN structure. The distinction of the proposed identification technique in comparison with existing methods lies in its two fold flexibility. First, the filter's input and resulting output could be selected arbitrarily by the designer in spatial domain. In other words, the proposed technique can be uniformly applied for identification of a 2-D filter with an impulse response, a step response or a response to a random 2-D input signal. Second, the method is applicable to a general Roesser's LSS model as well as specific classes of 2-D filters, such as separable in denominator filters.

5.5.3   Neural Network Approach

By using a similar NN structure proposed for a 1-D recursive filter as shown in Figure 5.1, we could develop a technique for 2-D recursive filter design. Consider the general Roesser's LSS model (5.10), an NN structure, which combines recurrent and feedforward processes similar to an LSS 2-D model. In order to correlate the proposed NN model with LSS model (5.10), an NN structure is shown in Figure 5.4. Hidden neurons are classified into two different types related to the vertical and horizontal states with their self feedback loops and connections. The correlation between various coefficients in model (5.10) and weight connections as shown in Figure 5.4 can be easily observed. The weights between input neurons and hidden neurons (ω_{uh}, ω_{uv}) are represented by matrices B_1, B_2 respectively. The weights between similar hidden neurons (ω_{hh}, ω_{vv}) are established by matrices A_1 and A_4 respectively. The weights between two different classes of hidden neurons (ω_{hv}, ω_{vh}) are provided by the interconnection matrices A_2 and A_3 respectively. The weights between hidden neurons and output neurons (ω_{yh}, ω_{yv}) are represented by matrices C_1 and C_2. Finally, the weights between input and output neurons (ω_{yu}) are related to each other by the elements of matrix D. Therefore, by proper generation of various weights in the proposed NN model, the identification of LSS model (5.10) can be achieved.

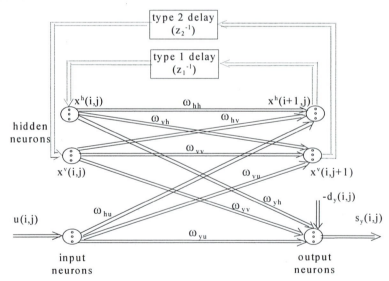

Figure 5.4: The Neural Network Structure for a 2-D System Identification.

A general 2-D system identification algorithm is developed based on the NN structure. It is a pattern mode learning since the weights are updated after the presentation of each training sample data. The technique distinguishes itself from an ordinary NN training algorithm in two aspects. First, there are two classes of hidden neurons in the proposed NN structure. They are related to each other via weight connection but develop their values in distinct ways. Second, the neuron outputs are the function of two independent variables, instead of one variable, as is related to a 1-D case. Due to the feedback of hidden neurons, the NN architecture is a recurrent one. In addition, there are feedforward information processes such as the direct path from input to hidden neurons and hidden to output neurons. The adjustments of the weights are based on the desired output values and NN actual outputs. Therefore, it can also be considered as a supervised learning network in the sense of training method. The details of the algorithm are presented here. Interested readers can refer to Wang [8].

5.6   SIMULATION RESULTS

5.6.1   1-D Filters

Three numerical examples are provided herein; each emphasizes different aspects of the proposed algorithm. The following L2 and L∞ norm [24], [26] error criteria are defined for error analysis.

$$\varepsilon_\infty = \max_{n \in T} |H - H_{NN}| \Big/ \max_{n \in T} |H|$$

$$\varepsilon_2 = \left[\sum_{n \in T} (H - H_{NN})^2 \right]^{1/2} \Big/ \left[\sum_{n \in T} (H)^2 \right]^{1/2}$$

where H and H_{NN} are the impulse responses of the original system and identified system, respectively, and T is the given trajectory (from discrete time n_0 to n_1) along which the error norms are calculated.

Example 5.1:
    The system to be identified is a 5th order Chebyshev type I filter with 0.8 decibel of ripple in the passband and 0.5 as cutoff frequency [26]. The transfer function of the filter is given as:

$$T(z) = \frac{0.0247 z^5 + 0.1237 z^4 + 0.2473 z^3 + 0.2473 z^2 + 0.1237 z + 0.0247}{1.0000 z^5 - 1.0925 z^4 + 1.6014 z^3 - 1.1520 z^2 + 0.6420 z - 0.2074}$$

First, we generate 200 random input data whose amplitudes are uniformly distributed in the range of [-1, 1] and obtain the corresponding output. An NN with single input, single output and five neurons in hidden layer is trained using input/desired output pairs for 200 epochs. The final training mean squared error in Equation 5.3d is 1.0169E-03. The identified filter tusing NN is:

$$A_{NN} = \begin{bmatrix} 0.0579 & 0.3237 & 0.0910 & -0.8210 & 0.0541 \\ -0.1457 & 0.4156 & 0.5757 & 0.0925 & -0.0298 \\ 0.1962 & -0.4314 & 0.1408 & 0.0872 & -0.5384 \\ 0.7494 & 0.3956 & -0.1125 & 0.2536 & 0.0094 \\ 0.0246 & -0.2153 & 0.5508 & -0.0211 & 0.2364 \end{bmatrix}$$

$$B_{NN} = \begin{bmatrix} 0.3272 & 0.6079 & 0.4382 & -0.1568 & 0.0296 \end{bmatrix}^T$$

$$C_{NN} = \begin{bmatrix} -0.2413 & 0.4809 & 0.0930 & 0.6673 & 0.0279 \end{bmatrix}$$

$$D_{NN} = 0.0247$$

The transfer function of the identified filter is:

$$T_{NN}(z) = \frac{0.02469\,z^5 + 0.1231\,z^4 + 0.246\,z^3 + 0.2431\,z^2 + 0.1196\,z + 0.02023}{z^5 - 1.104\,z^4 + 1.607\,z^3 - 1.163\,z^2 + 0.6461\,z - 0.209}$$

The impulse responses of both the original system (H) and NN identified system (H_{NN}) are obtained for 40 samples. The two error values are $\varepsilon_2 = 0.39$ percent and $\varepsilon_\infty = 0.277$ percent respectively.

For comparison, the system is identified with the same set of data by two other well known methods, least square [24], and subspace [27]. The comparison is shown in Table 5.1. To verify the robustness of the proposed method, the same system is identified in two noisy conditions. In the first case, the measurement contains zero-mean white noise whose variance is 5 percent of the maximum amplitude of the response. In the second case, in addition to white noise, the measurement also contains 5 percent density of wild (spike) noise whose amplitude is equal to 10 percent of the maximum amplitude of the response. The error norms of the proposed identification technique in comparison to available techniques are shown in Tables 5.2 and 5.3. The results show that the proposed technique provides more robust solutions under noise, especially wild noise condition.

Table 5.1: Error Norms of Example 5.1 under Noise Free Conditions.

	N. N.	Lease Square	Subspace
ε_2	3.90e-03	3.67e-15	1.93e-15
ε_∞	2.77e-3	3.77e-15	2.52e-15

Table 5.2: Error Norms of Example 5.1 under White Noise Conditions.

	N. N.	Lease Square	Subspace
ε_2	3.52e-02	8.30e-2	4.95e-2
ε_∞	3.56e-2	7.15e-2	3.71e-2

Table 5.3: Error Norms of Example 5.1 under White Noise +
Spike Noise Conditions.

	N. N.	Lease square	Subspace
ε_2	5.34r-2	1.21e-01	9.07e-2
ε_∞	5.29e-2	8.98e-2	6.48e-2

Example 5.2:

This example is presented in order to demonstrate the use of an observability canonical state space form as the result of identification. By selection and elimination of some weights in advance, the observability canonical form is obtained. The filter to be identified is governed by the following state space model.

$$A = \begin{bmatrix} -0.0051 & 0.2043 & -0.7014 \\ 0.5641 & 0.0923 & 0.3789 \\ 0.4642 & -0.6482 & -0.3021 \end{bmatrix} \qquad B = \begin{bmatrix} 0.4121 & 0.8415 & 0.2693 \end{bmatrix}^T$$

$$C = \begin{bmatrix} 0 & 0.5373 & 0.4676 \end{bmatrix}$$

The corresponding transfer function is as follows:

$$T(z) = \frac{0.5781\ z^2 + 0.1419\ z\ + 0.1103}{z^3 + 0.2149\ z^2 + 0.4291\ z - 0.3562}$$

In order to obtain the observability canonical form, some weight connections between hidden neurons were eliminated in advance. In addition, some of the connections were taken out of update process by assigning a unity weight in the beginning of the training phase. The NN is trained with 300 random generated inputs. The identified system is given as:

$$A_{NN} = \begin{bmatrix} 0 & 0 & 0.3568 \\ 1 & 0 & -0.4248 \\ 0 & 1 & -0.2139 \end{bmatrix} \qquad B_{NN} = \begin{bmatrix} 0.1099 & 0.1413 & 0.5781 \end{bmatrix}^T$$

$$C_{NN} = \begin{bmatrix} 0 & 0 & 1 \end{bmatrix}$$

The corresponding transfer function matrix is presented as:

$$T_{NN}(z) = \frac{0.5781\ z^2 + 0.1413\ z\ + 0.1099}{z^3 + 0.2139\ z^2 + 0.4284\ z - 0.3536}$$

The two error values for 50 samples impulse response are $\varepsilon_2 = 0.03686$ percent and $\varepsilon_\infty = 0.03827$ percent respectively. For comparison, the same system is identified with least square and subspace methods. In addition, two noise conditions are considered similar to the above example. The error norm comparisons are shown in the Tables 5.4, 5.5 and 5.6.

Table 5.4: Error Norms of Example 5.2 under Noise Free Conditions.

	N. N.	Lease square	Subspace
ε_2	3.69e-4	1.47e-15	1.26e-15
ε_∞	3.83e-4	1.20e-15	9.60e-16

Table 5.5: Error Norms of Example 5.2 under White Noise Conditions.

	N. N.	Lease Square	Subspace
ε_2	1.39e-02	3.26e-2	1.45e-2
ε_∞	1.40e-2	2.46e-2	1.40e-2

Table 5.6: Error Norms of Example 5.2 under White Noise +
Spike Noise Conditions.

	N. N.	Lease Square	Subspace
ε_2	2.86e-2	1.01e-01	4.18e-2
ε_∞	2.69e-2	6.88e-2	4.67e-2

Example 5.3:

This example is provided in order to emphasize the effectiveness of the proposed model for multi-input and multi-output systems. The filter to be identified is a two inputs, two outputs system governed the following state space form as provided by Taylor [26]:

$$A = \begin{bmatrix} -0.5484 & 0.4138 & 0.2432 \\ -0.4776 & -0.5864 & 0.0900 \\ 0.0472 & -0.2550 & -0.2294 \end{bmatrix} \quad B = \begin{bmatrix} 0.6010 & 0.1577 \\ 0.1769 & 0.9879 \\ 0.8284 & 0.2572 \end{bmatrix}$$

$$C = \begin{bmatrix} 0 & 0.2194 & 0.6960 \\ 0.1016 & 0.6347 & 0.7948 \end{bmatrix} \quad D = \begin{bmatrix} 0.6962 & 0.6695 \\ 0.7529 & 0.2500 \end{bmatrix}$$

The corresponding transfer function matrix can be derived as:

$$T(z) = \begin{bmatrix} \dfrac{0.6962z^3 + 1.5652z^2 + 1.1769z + 0.3906}{z^3 + 1.3642z^2 + 0.7910z + 0.09359} & \dfrac{0.7529z^3 + 1.8589z^2 + 1.3596z + 0.4297}{z^3 + 1.3642z^2 + 0.7910z + 0.09359} \\ \dfrac{0.6695z^3 + 1.3091z^2 + 0.7197z + 0.1067}{z^3 + 1.3642z^2 + 0.7910z + 0.09359} & \dfrac{0.2500z^3 + 1.1885z^2 + 0.7510z + 0.1146}{z^3 + 1.3642z^2 + 0.7910z + 0.09359} \end{bmatrix}$$

The NN is trained with 200 random generated inputs for 200 epochs. The system identified by NN is:

$$A_{NN} = \begin{bmatrix} -0.8939 & 0.3002 & 0\ 3030 \\ -0.5009 & -0.3405 & 0.2023 \\ -0.0768 & -0.7551 & -0.1298 \end{bmatrix} \quad B_{NN} = \begin{bmatrix} 0.8840 & 0.4722 & 1.0046 \\ 1.0047 & 1.1123 & 1.2534 \end{bmatrix}$$

$$C_{NN} = \begin{bmatrix} 0.3337 & -0.5623 & 0.5832 \\ 0.3683 & -0.2553 & 0.6237 \end{bmatrix} \quad D_{NN} = \begin{bmatrix} 0.6962 & 0.6695 \\ 0.7529 & 0.2500 \end{bmatrix}$$

The corresponding transfer function matrix is:

$$T_{NN}(z) = \begin{bmatrix} \dfrac{0.6962z^3 + 1.565z^2 + 1.177z + 0.3906}{z^3 + 1.364z^2 + 0.791z + 0.09358} & \dfrac{0.7529z^3 + 1.859z^2 + 1.36z + 0.4297}{z^3 + 1.364z^2 + 0.791z + 0.09358} \\ \dfrac{0.6695z^3 + 1.309z^2 + 0.7197z + 0.1067}{z^3 + 1.364z^2 + 0.791z + 0.09358} & \dfrac{0.25z^3 + 1.189z^2 + 0.7509z + 0.1146}{z^3 + 1.364z^2 + 0.791z + 0.09358} \end{bmatrix}$$

The two error values measured for the first 50 samples of impulse response are calculated in vector form:

$$e_2 = \begin{bmatrix} 7.082 & 9.105 \\ 4.320 & 9.211 \end{bmatrix} \cdot 10^{-4} \qquad e_\infty = \begin{bmatrix} 5.817 & 7.546 \\ 2.995 & 8.033 \end{bmatrix} \cdot 10^{-4}$$

5.6.2 Two-dimensional Filters

Two numerical examples are provided for 2-D recursive filter, each emphasis different aspects of the proposed algorithm. The following L_2 and L_∞ norm [29]-[32] error criteria are defined for error analysis:

$$\varepsilon_\infty = \max_{(i,j)\in\Lambda'} |H(i,j) - H_{NN}(i,j)| \Big/ \max_{(i,j)\in\Lambda'} |H(i,j)|$$

$$\varepsilon_2 = \left[\sum_{(i,j)\in\Lambda'} (H(i,j) - H_{NN}(i,j))^2 \right]^{1/2} \Big/ \left[\sum_{(i,j)\in\Lambda'} (H(i,j))^2 \right]^{1/2}$$

MNR = Maximum Negative Ripple

where $H(i,j)$ and $H_{NN}(i,j)$ are the impulse responses of the original system and identified system, respectively, $\Lambda'=\{(i, j) \mid 0 \leq i \leq M', 0 \leq j \leq N'\}$ is the given region where the error norms are calculated.

Example 5.4: First Quarter Gaussian Filter

The prototype model used by Aly and Fahmy in [30] for designing a 2-D causal recursive filter is presented here. It is a first quadrant Gaussian 2-D scalar filter described by the following impulse response:

$$H(i, j) = 0.256322 \exp\{-0.103203 \cdot [(i-4)^2 + (j-4)^2]\}$$

with most of its energy in the first-quadrant. The selected region for identification consists of $\Lambda=\{(i, j) \mid 0 \leq i \leq 10, 0 \leq j \leq 10\}$. The same region was used for error norm calculation: $\Lambda'=\Lambda$.

The proposed NN consists of one input neuron, one output neuron, and two groups of hidden neurons, each with three neurons. After 80 epochs of training, the identified 2-D filter in Roesser's LSS model of order (3,3) is given as,

$$A_{NN} = \left[\begin{array}{ccc|ccc} 3.0059e0 & -1.8840e0 & 2.2325e0 & 8.8268e\text{-}1 & -3.1194e\text{-}1 & -6.4903e\text{-}1 \\ 2.0812e0 & -9.6880e\text{-}1 & 2.0168e0 & 1.4857e0 & -5.2382e\text{-}1 & -1.0898e0 \\ -9.1445e\text{-}1 & 6.9213e\text{-}1 & -1.3040e\text{-}1 & 3.5259e\text{-}1 & -1.2248e\text{-}1 & -2.5830e\text{-}1 \\ \hline -1.7607e\text{-}1 & 1.2241e\text{-}1 & -1.5569e\text{-}1 & 4.8882e\text{-}1 & -2.7540e\text{-}1 & -1.6637e\text{-}1 \\ 4.4500e\text{-}1 & -2.8219e\text{-}1 & 3.2107e\text{-}1 & 8.4165e\text{-}1 & 1.2692e0 & 4.7849e\text{-}1 \\ -4.5596e\text{-}1 & 3.1093e\text{-}1 & -3.8792e\text{-}1 & -8.6954e\text{-}1 & -8.7703e\text{-}1 & 1.4460e\text{-}1 \end{array} \right]$$

$$B_{NN} = \begin{bmatrix} -6.6533e\text{-}3 & -1.1209e\text{-}2 & -2.6681e\text{-}3 \mid -1.0466e\text{-}1 & 6.0182e\text{-}2 & -1.4947e\text{-}1 \end{bmatrix}^T$$

$$C_{NN} = \begin{bmatrix} 3.8730e+1 & -2.9244e+1 & 1.8910e+1 \mid -1.2433e0 & 4.3619e\text{-}1 & 9.1152e\text{-}1 \end{bmatrix}$$

D_{NN} = 9.4009e - 03

Table 5.7 is presented to compare the error measurements of our design to that of Aly and Fahmy [30]. Notice that the total order realization of our design (3 + 3 = 6) is the same theirs (4 + 2 = 6).

Table 5.7: Simulation Experiments for Example 5.4.

	$\epsilon 2$ %	ϵ_∞ %	MNR
Our Design	3.71	5.16	Always positive
Design [Aly and Fahmy]	10.78	9.19	0.04479

Example 5.5: A (2, 2) 2-D Digital Filter

This example is presented to illustrate the identification of a 2-D system using various responses. The random input response as well as the impulse response are utilized to identify the given 2-D filter. The 2-D filter to be identified is governed by the following state space model (D=0):

$$A = \left[\begin{array}{cc|cc} 1.0000\text{e-}1 & 2.0000\text{e-}1 & 0 & -1.0000\text{e-}1 \\ -1.0000\text{e-}1 & 0 & 1.0000\text{e-}1 & 0 \\ \hline 1.0000\text{e-}1 & 0 & 2.0000\text{e-}1 & 0 \\ 0 & 1.0000\text{e-}1 & 1.0000\text{e-}1 & 1.0000\text{e-}1 \end{array} \right]$$

$$B = \begin{bmatrix} 1.0000\text{e}0 & 1.0000\text{e}0 & | & 5.0000\text{e-}1 & 1.0000\text{e}0 \end{bmatrix}^T$$

$$C = \begin{bmatrix} 1.0000\text{e}0 & 5.0000\text{e-}1 & | & 5.0000\text{e-}1 & 1.0000\text{e}0 \end{bmatrix}$$

First, we generated 50 x 50 random input data within region $\Lambda = \{(i, j) \mid 0 \le i \le 49, 0 \le j \le 49\}$ whose amplitude was uniformly distributed in the range of [-1, 1] and then obtained the corresponding output. An NN with single input, single output and a total of four (two for each type) hidden neurons is trained using generated input/desired output pairs for 40 epochs. The identified filter is:

$$A_{NN} = \left[\begin{array}{cc|cc} 4.9556\text{e-}2 & -7.4836\text{e-}2 & -5.5614\text{e-}2 & 7.9612\text{e-}2 \\ 2.2048\text{e-}1 & 3.7549\text{e-}2 & 5.5330\text{e-}2 & 6.1638\text{e-}2 \\ 4.5853\text{e-}3 & -6.2183\text{e-}2 & 5.4698\text{e-}2 & 9.2342\text{e-}2 \\ -1.7303\text{e-}2 & 8.1130\text{e-}2 & 7.5142\text{e-}2 & 1.1919\text{e-}1 \end{array} \right]$$

$$B_{NN} = \begin{bmatrix} 1.0399\text{e}0 & 7.2060\text{e-}1 & | & 7.7643\text{e-}1 & 5.8653\text{e-}1 \end{bmatrix}^T$$

$$C_{NN} = \begin{bmatrix} 7.6376\text{e-}1 & 9.7999\text{e-}1 & | & 9.9556\text{e-}1 & 8.1350\text{e-}1 \end{bmatrix}$$

In the second phase, the same NN was trained with the impulse response defined in the region $\Lambda = \{(i, j) \mid 0 \le i \le 9, 0 \le j \le 9\}$ for 40 epochs. The state-space form of the identified filter is as follow:

$$A_{NN} = \left[\begin{array}{cc|cc} 1.0641\mathrm{e}\text{-}1 & 7.7004\mathrm{e}\text{-}2 & -1.2062\mathrm{e}\text{-}2 & 5.3092\mathrm{e}\text{-}2 \\ 9.1873\mathrm{e}\text{-}2 & 8.5081\mathrm{e}\text{-}2 & 2.9856\mathrm{e}\text{-}2 & 7.1189\mathrm{e}\text{-}2 \\ \hline -3.8449\mathrm{e}\text{-}3 & 6.9885\mathrm{e}\text{-}2 & 8.1948\mathrm{e}\text{-}2 & 8.9764\mathrm{e}\text{-}2 \\ -6.8676\mathrm{e}\text{-}4 & 7.1595\mathrm{e}\text{-}2 & 8.3222\mathrm{e}\text{-}2 & 9.0390\mathrm{e}\text{-}2 \end{array}\right]$$

$$B_{NN} = \left[1.2555\mathrm{e}0 \quad 3.1189\mathrm{e}\text{-}1 \mid 1.1109\mathrm{e}0 \quad 4.4088\mathrm{e}\text{-}1\right]^{T}$$

$$C_{NN} = \left[1.0784\mathrm{e}0 \quad 4.6248\mathrm{e}\text{-}1 \mid 8.4716\mathrm{e}\text{-}1 \quad 6.9503\mathrm{e}\text{-}1\right]$$

The region for error norm calculation is $\Lambda' = \{(i, j) \mid 0 \le i \le 19, 0 \le j \le 19\}$ for both of designed filters. In Table 5.8, a comparison of the error analysis of two different training results is shown. It is observed that a random input response provides a more accurate model in comparison to an impulse response. A similar conclusion is obtained based on other simulation results, due to the fact that the responses generated by a large amount of random inputs contains more information compared to the impulse responses.

Table 5.8: Results of Example 5.5

	$\varepsilon2$ %	ε_∞ %	MNR (Original -5.5e-3)
Design with a random response	0.166	0.115	-5.8e-03
Design with an impulse response	2.82	3.14	always > 0

5.7   CONCLUSIONS

In this chapter, a novel NN technique is introduced for the design of recursive digital filters in the state space form. Instead of using spatial representation of time by delayed input/output feedback, we use hidden neurons to encode the temporal properties of the system. Through the self feedback of hidden neurons as well as the interconnection between the neurons in the input, hidden, and output layers, the proposed NN structure mimics the dynamics of a linear discrete system or digital filter. The proposed method also provides flexibility in selection of various state-space forms such as controllability and observability canonical forms as an identification model.

The NN approach is also extended for the design of general 2-D recursive digital filters where an analytical solution is not necessarily available. An attractive feature of the proposed algorithm is that the LSS model structure to be identified could be predefined in the design stage. This feature not only provides us with flexibility in selection of the structure of a 2-D filter, but also facilitates analyses on several implementation issues of 2-D filter, such as computation efforts and memory requirement. Furthermore, the proposed method herein places no limitation on the type of response to be approximated. Namely, any type of responses with sufficient data points could be used as a training sample for filter identification.

The effectiveness as well as robustness of this method have been demonstrated by simulations experiments for both single input/single output and multi-input/multi-output digital filters.

REFERENCES

1.   Hopfield, J.J., Neural Networks and Physical Systems with Emergent Collective Computational Abilities, *Proc. Nat. Acad. Sci.*, Vol. 79, 2554 – 2558, April 1982.

2.   Hopfield, J.J., Neurons with Graded Response have Collective Computational Properties Like Those of Two State Neurons, *Proc. Nat. Acad. Sci.*, Vol. 81, 3088 – 3092, May 1984.

3.   Elman, J.L., Finding Structure in Time, *Cognitive Science,* Vol. 14, 179 – 211, 1990.

4.   Pineda, F.J., Dynamics and Architecture for Neural Computation, *J. Complexity,* Vol. 4, 216 – 245, 1988.

5.   Pineda, F.J., Recurrent Back Propagation and the Dynamical Approach to Adaptive Neural Computation, *Neural Computation,* Vol. 1, 161 – 172, 1989.

6.   Robinson, A.J. and Fallside, F. A Recurrent Error Propagation Network Speech Recognition System, *Computer Speech and Language 5,* 259 – 274, 1991.

7.   Irwin, K., Warwick G.W. and Hunt, K.J., *Neural Networks for Control and Systems*, IEE Publication, 1992.

8.   Wang, D. Identification and Approximation of 1-D and 2-D Digital Filters, Ph.D Dissertation, Florida Atlantic University, Boca Raton, FL, May, 1998.

9.   Wang, D. and Zilouchian, A., Identification of Discrete Linear Systems in State Space Form Using Neural Network, *Proc. of Second IEEE Int. Caracas Conf. on Devices, Circuits and Syst.,* Venezuela, 338 – 342, March, 1998.

10.   Rumelhart, D.E. and McClelland, J.L.(eds.), *Parallel Distributed Processing: Explorations in the Microstructure of Cognition*, Vol. 1, MIT Press, Boston, MA, 1986.

11.   Galvan, J.B. and Perez-Ilzarbe, M.J., Two Neural Networks for Solving the Linear System Identification Problem, *Proc. of IEEE Conf. on Neural Networks,* 3226 – 3231, 1993.

12.   Cichocki, A. and Unbehauen, R., Neural Networks for Solving Systems of Linear Equations and Related Problems, *IEEE Trans. on Circuits and Syst.,* Vol. 39, No.2, 124 – 137, Feb., 1992.

13.   Mammone, R.J. and Zeevi, Y., *Neural Networks, Theory and Application*, Academic Press, NY, 1990.

14. Lippman, M.P. and Chua, L.O., Neural Networks for Nonlinear Programming, *IEEE Trans. on Circuits and Syst.*, Vol. 35, No.5, 554 – 562, May 1988.

15. Narendra, K.S. and Parthasarathy, K., Identification and Control of Dynamic Systems Using Neural Networks, *IEEE Trans. on Neural Networks*, Vol. 1, No. 1, 4 – 27, March, 1990.

16. Poggio, T. and Girosi,F., Network for Approximation and Learning, *Proc. of IEEE*, 1481 – 1495, Sept., 1990.

17. Jamshidi, J. (ed.), *Circuits, Systems & Information*, TSI Press, Albuquerque, NM, 1991.

18. Horton, M.P., Real-time Identification of Missile Aerodynamics Using a Linearised Kalman Filter Aided by an Artificial Neural Network, *IEE Proc. Control Theory Appl.*, Vol. 144, No. 4, 299 – 308, July, 1997.

19. Hampel, F.R., Ronchetti, E.M., Roussew, P., and Stahel, W.A., *Robust Statistics - the Approach Based on Influence Functions*, John Wiley & Sons, NY, 1987.

20. Wang, D. and Zilouchian, A., Identification of 2-D Recursive Digital Filters in State-Space Form Using Neural Network, *Int. J. of Intelligent Automation and Soft Computing* .

21. Silverman, L.M., Realization of Linear Dynamic Systems, *IEEE Transaction on Automatic Control*, AC-16, 554 – 567, 1971.

22. Wang, D. and Zilouchian, A., Model Reduction of Discrete Linear Systems via Frequency Domain Balanced Structure, *IEEE Trans. on Circuits and Syst.* Vol. 47, No. 6 , 830–838, July 2000.

23. Moonen, M., Moor, B. D., Vandenberghe, L., and Vandewalle, J., On- and Off-line Identification of Linear State-Space Models, *Int. J. Control*, Vol. 49, No. 1, 219 – 232, 1989.

24. Ljung, L., *System Identification, Theory for the User*, Prentice-Hall, Inc., Englewood Cliffs, NJ, 1987.

25. Wang, D. and Zilouchian, A., Model Reduction of 2-D Separable-in-Denominator Systems via Frequency Domain Balanced Realization, *Proc. of 37th IEEE Conf. on Decision and Control*, Tampa, FL, 2179 – 2184, 1998.

26. Taylor, F.J., *Digital Filter Design Handbook*, Marcel Dekker Inc., NY, 1983.

27. *MATLAB Toolbox*, The Mathwork Inc., Boston, MA, 1998.

28. Ramos, J., A Subspace Algorithm for Identifying 2-D Separable in Denominator Filters, *IEEE Trans. on Circuits and Syst.*, Vol. 41, No. 1, 63 – 67, January, 1994.

29. Hinamoto, T. and Maekawa, S., Spatial-Domain Design of a Class of Two-Dimensional Recursive Digital Filter, *IEEE Trans. on ASSP*, Vol. 32, No. 1, 153 – 162, February, 1984.

30.   Aly, S.H. and Fahmy, M.M., Spatial-Domain Design of Two-Dimensional Recursive Digital Filters, *IEEE Trans. on Circuits and Syst.*, Vol. 27, No. 10, 892 – 901, October, 1980.

31.   Lashgari, B., Siverman, L.M., and Abramatic, J., Approximation of 2-D Separable in Denominator Filters, *IEEE Trans. on Circuits and Syst.*, Vol. 30, No. 2, 107 – 121, February 1983.

32.   Hinamoto, T. and Maekawa, S., Design of 2-D Separable in Denominator Filters Using Canonic Local State-Space Models, *IEEE Trans. on Circuits and Syst.*, Vol. CAS-33, No. 9, 922 – 929, September, 1986.

33.   Lin,L., Kawamata, M., and Higuchi,T., Design of 2-D Separable-Denominator Digital Filter Based on the Reduced-Dimensional Decomposition, *IEEE Trans. on Circuits and Systems*, Vol. CAS-34, No. 8, 934 – 941, August, 1987.

34.   Raymond, D.M., and Fahmy, M.M., Spatial-Domain Design of Two-Dimensional Recursive Digital Filters, *IEEE Trans. on Circuits and Syst.*, Vol. 36, No. 6, 901 – 905, June, 1989.

35.   Bose, T. and Chen, M., Design of Two-Dimensional Digital Filters in the Spatial Domain, *IEEE Trans. on Signal Processing*, Vol. 41 No. 3, 1464 – 1469, March, 1993.

36.   Attasi, S., Modeling and Recursive Estimation for Double Indexed Sequences, in *System Identification: Advances and Case Studies*, Mehra, R.K., and Lainiotis, D.G., (eds.), Academic Press, NY, 1976.

37.   Roesser, R.P., A Discrete State-Space Model for Linear Image Processing, *IEEE Trans. on Automatic Control*, Vol. AC-20, 1 – 10, February 1975.

38.   Kung, S., Levy, B.C., Morf, M., and Kailath, T., New Results in 2-D Systems Theory, Part II: 2-D State-Space Models - Realization and the Notions of Controllability, Observability, and Minimality, *Proc. of the IEEE*, Vol. 65, No. 6, 945 – 959, June, 1977.

39.   Shanks, J.L., Treitel, S., and Justice, J.H., Stability and Synthesis of Two-Dimensional Recursive Filters, *IEEE Trans. on Audio Electro-Acoust.*, Vol. AU-20, 115 – 128, June, 1972.

40.   Hinamoto, T., Realizations of a State-Space Model from Two-Dimensional Input-Output Map, *IEEE Trans. on Circuits and Syst.*, Vol. CAS-27, No. 1, 36–44, Jan., 1980.

6 APPLICATION OF COMPUTER NETWORKING USING NEURAL NETWORK

Homayoun Yousefi'zadeh

6.1   INTRODUCTION

This chapter investigates the application of perceptron neural networks in modeling traffic sources in packet based computer communication networks. It is motivated by recent measurement studies that indicate the presence of significant statistical features in packet traffic belong to the fractal nature of the processes rather than their stochastic nature. The chapter first provides an illustration of the statistical features of the measured traffic over the Internet. It then outlines a learning scheme based on back propagation algorithm for a class of perceptron neural networks that can be used to capture several of the fractal properties observed in actual data. The most important conclusion of this chapter is that, despite the existence of numerical difficulties, neural networks may allow building of accurate models to predict the behavior of packet traffic sources.

6.2   SELF SIMILAR PACKET TRAFFIC

Teletraffic analysis of the computer communication networks is one of the most important applications of mathematical modeling and queuing theory. Recently, the widespread deployment of packet switching has generated a set of challenging problems in queuing theory. The problem of bursty traffic packet arrival modeling is considered one of the most important problems in this category. Given that performance models are only reliable when their underlying assumptions are satisfied, the problem of obtaining an accurate model of packet traffic is particularly important in all packet based networks. Although numerous models of packet arrival processes have been proposed during the past few years, there is still a lack of complete understanding of the features in packet traffic. This is partly due to uncertainties in the traffic characteristics of the emerging networks and services, and partly due to the difficulties in characterizing the traffic arrival models and resource usage patterns in the emerging networks.

Analyses of traffic data from networks and services such as ISDN traffic, Ethernet LANs, common channel signaling network (CCSN) and variable bit rate (VBR) video have convincingly demonstrated the presence of features such as self-similarity, long range dependence, slowly decaying variances, heavy-tailed distributions and fractal dimensions. These features, indeed, are more characteristic of fractal processes than those of conventional stochastic processes. Conventional traffic processes from regular telephone traffic or the Poisson and Poisson-based models seem to be Markovian in nature, characterized by exponential decays. The types of packet traffic with the above mentioned characteristics are interpreted to be bursty in nature. To be more specific, Leland and Wilson from Bellcore research center have presented a preliminary statistical analysis of Ethernet traffic, on the presence of "burstiness" across a wide range of time scales [2]: traffic spikes ride on the longer term ripples that, in turn, ride on longer term swells, etc. This is also explained in terms of self-similarity, i.e., self-similar phenomena show structural similarities across all or at least a very wide range of time scales [3-5]. The degree of self-similarity measured via the Hurst parameter typically depends on the utilization level of the transmission medium and can be used to measure burstiness of the traffic.

As another important difference between the aggregated bursty traffic and the so called Poisson-like conventional models, it could be mentioned that the aggregated traffic is expected to become less bursty or smoother as the number of traffic sources increases based on the conventional models, but it has very little to do with the reality. In fact, contrary to commonly held views, it has been observed that the burstiness of LAN traffic intensifies as the number of traffic sources increases. Conventional characterizations suppose that packet traffic consists of alternating active and silent periods with well-defined statistics. On the contrary, measurement studies have noted that there is no actual burst length, and bursts occur over many time scales. At every step, examination of the data shows that the bursts resolve into bursts over smaller time scales. This burst-within-burst structure captures the fractal properties observed in actual traffic data.

6.2.1   Fractal Properties of Packet Traffic

The main objective of the current section is to establish a foundation for a statistically well-defined property of time series called self-similarity. Intuitively, self-similar phenomena display structural similarities across too many time scales. Measuring a single parameter called the Hurst parameter usually specifies the degree of self-similarity. The following discusses mathematical and statistical properties of the self-similar processes.

Second-Order Self-similarity
Let

$$X = (X_t : t = 0,1,2,...) \tag{6.1}$$

be a covariance stationary stochastic process with mean μ, variance σ^2, and autocorrelation function $\tau(k)$, $k \geq 0$. In particular suppose X has an autocorrelation function of the form

$$\tau(k) \sim a_1 k^{-\beta}, \qquad as \qquad k \to \infty \tag{6.2}$$

where $0 < \beta < 1$ and constants $a_1, a_2, ...$ denote finite positive integers. For each m = 1, 2, 3, ... let

$$X^{(m)} = (X_k^{(m)} : k = 1,2,3,\cdots) \tag{6.3}$$

denote the new covariance stationary time series with corresponding autocorrelation function $\tau^{(m)}$ obtained by averaging the original series X over nonoverlapping blocks of size m, i.e., for each $m = 1,2,3,..., X^{(m)}$ is given by

$$X_k^{(m)} = 1/m(X_{km-m+1} + \cdots + X_{km}), \qquad k \geq 1 \tag{6.4}$$

The process X is called exactly second-order self-similar with self-similarity parameter H=1-β/2 if the corresponding $X^{(m)}$ has the same correlation structure as X, i.e., $\tau^{(m)}(k) = \tau(k)$ for all m = 1, 2, 3,... and k = 1, 2, 3, ... X is called asymptotically second-order self-similar with self-similarity parameter H=1-β/2 if $\tau^{(m)}(k)$ agrees asymptotically with $\tau(k)$ given by (6.2), for large m and k. In other words, X is exactly or asymptotically second-order self-similar if the aggregated processes $X^{(m)}$ are the same as X or become indistinguishable from X with respect to their correlation functions. Fractal Gaussian noise (FGN) is a good example of an exactly self-similar process with self-similarity parameter H, 1/2 < H < 1. Fractional Arima processes with the parameters (p, d, q) such that 0 < d < 1/2 are examples of asymptotically second-order self-similar processes with self-similarity parameter d + 1/2. Mathematically, self-similarity manifests itself in a number of equivalent ways as follow.

(1) The variance of sample mean decreases more slowly than the reciprocal of the sample size. This is called slowly decaying variance property meaning.

$$\text{var}(X^{(m)}) \sim a_2 m^{(-\beta)}, \qquad m \to \infty, 0 < \beta < 1 \tag{6.5}$$

(2) The auto-correlation decay hyperbolically rather than exponentially fast, implying a nonsummable autocorrelation function $\sum_k \tau(k) = \infty$. This is called long range dependence property which means $\tau(k)$ satisfies relation (6.2).

(3) The spectral density f(.) obeys a power law near the origin. This is the concept of 1/f noise with the meaning

$$f(\lambda) = k\lambda^{-\gamma} \tag{6.6}$$

as $\lambda \to \infty$ with $0 < \gamma < 1$ and $\gamma = 1 - \beta$.

It looks like the most striking feature of self-similar processes is that their aggregated process $X^{(m)}$ possesses a nondegenerate correlation function as $m \to \infty$. This is in stark contrast to typical packet traffic models considered in literature, all of which have the property that their aggregated processes $X^{(m)}$ tend to second order pure noise, i.e., $\tau^{(m)} \to 0$ as $m \to \infty$. As an equivalent method of description, they may be characterized by the following properties:

- The sample mean variance decreases like the reciprocal of the sample mean.
- The autocorrelation function decreases exponentially fast, implying a summable autocorrelation function. This, in fact, is equivalent to the short range dependence property.

- The spectral density is bounded at the origin.

The concept of self-similar processes provides a very elegant explanation of an empirical law commonly referred to as the Hurst effect. In order to describe the Hurst effect, it should be mentioned that for a given set of observations $X_k = (X : k = 0,1,2,...,n)$ with sample mean $\overline{X}(n)$ and sample variance $S^2(n)$, the rescaled adjusted range or the R/S statistic is given by

$$\frac{R(n)}{S(n)} = \frac{1}{S(n)}[\max(0, W_1, W_2, \cdots, W_n) - \min(0, W_1, W_2, \cdots, W_n)]$$

$$W_k = (X_1 + \cdots + X_k) - k\overline{X}(n), \qquad k \geq 1$$

(6.7)

While many naturally occurring time series appear to be well represented by the relation $E[R(n)/S(n)] \sim k_1 n^H$, as $n \to \infty$, with Hurst parameter H typically about 0.73, observations X_k from a short range dependent model are known to satisfy $E[R(n)/S(n)] \sim k_1 n^H$, as $n \to \infty$. This discrepancy is usually referred to as the Hurst effect.

Degree of Self-similarity

In this part, methods of estimating self-similarity degree are introduced based on the properties of covariance stationary second-order self-similar processes, namely slowly decaying variances, long-range dependence, and a spectral density obeying a power-law. Hence the problem may be approached in three ways:

- Time-domain analysis based on the R/S statistic;
- Analysis of variances of the aggregated processes;
- Periodogram-based analysis in the frequency domain.

The objective of the first method is to estimate the Hurst parameter H via the Hurst effect. Briefly, the approach consists of plotting $\log(R(n)/S(n))$ vs. $\log(n)$ in the logarithmic scale that results in a plot called "rescaled adjusted range plot" or the "pox diagram of R/S." For a well-defined parameter H, a typical rescaled adjust range plot starts with a transient zone showing the nature of short range dependence and continues with a steady state part which is a straight line with a certain slope. There are also some fluctuations around that line. In fact, if such asymptotic behavior appears, then graphical R/S analysis may be used to estimate the self-similarity degree. An estimate $\hat{H}$ of self-similarity parameter H is given by the line's asymptotic slope, which can take any value between ½ and 1. The most useful feature of the R/S analysis is its relative robustness against changes of marginal distribution.

In the second method, the variances of the aggregated second-order self-similar processes $X^{(m)}, m \geq 1$, decrease linearly in log-log plots against m, with slopes arbitrarily flatter than m. This behavior is, in fact, seen for the large values of m as the representative of time. The so called variance time plots are obtained by plotting $\log(\mathrm{var}(X^{(m)}))$ against $\log(m)$ and by fitting a simple least squares line through the resulting points in the plane. Values of the estimate $\hat{\beta}$ of the asymptotic slope between -1 and 0 suggest self-similarity with a degree of $\hat{H} = 1 - \hat{\beta}/2$.

In contrast to the previous two methods, the third method takes advantage of the presence of limit law for a more refined data analysis like the existence of confidence levels for H. This is simply done by using maximum likelihood types estimates (MLE) based on the periodogram-based analysis in the frequency domain. As an example, Whittle's approximate MLE may be mentioned to be used for the approximate Gaussian processes. A combination of an MLE-type approach and the one above of the mentioned aggregation methods lead to an operational procedure for obtaining confidence intervals for the self-similarity parameter H. Plots of the point estimates $\hat{H}^{(m)}$ of $H^{(m)}$ vs. m with their specified confidence level will typically vary a lot for small aggregation levels but will stabilize after a while and fluctuate around a constant value, the final estimate of self-similarity parameter H. For a complete discussion, see Leland and Wilson [2].

Mathematical Explanation of Self-similarity
Mathematically, self-similarity in measurements from aggregated traffic of Ethernet, ISDN, CCSN, and VBR traffic can be explained by a simple aggregation argument: aggregating many elementary renewal reward processes representing individual user traffic produces self-similarity in limit as the number of users increases. First, let us define the concept of infinite variance syndrome. A random variable is said to exhibit an infinite variance syndrome or is called heavy tailed if

$$P[U \geq u] \sim u^{-\alpha} L(u) \tag{6.8}$$

where L(u) is a slowly varying function at infinity and $0 < \alpha < 2$. The crucial property that distinguishes the renewal reward process source model from the commonly assumed source model is that the interrenewal arrivals, i.e., the lengths of the active/inactive periods, are heavy tailed or, in terms of Mandelbort terminology, exhibit the infinite variance syndrome. A number of evidence supports the existence of infinite variance syndrome in packet traffic measurements. Hellstern and Wirth[9] have observed that the extreme variability of ISDN data cannot be adequately captured using traditional packet traffic models but instead is best described by the concept of heavy-tailed distributions. Duffy and Willinger[14] have observed the same evidence in the CCSN traffic studies. They have noticed that the call holding time distribution for calls originating during high traffic periods is heavy tailed with an estimated value of about 2.0, and for calls originating during light traffic periods, the estimated value drops down to about 1.0. Erramilli et al., [10] first proposed the idea of using fractal dimensions to characterize the fractal-like nature of the traffic measurements. Intuitively, a dimension is an indication of the extent to which a set, e.g., arrival times, fills the space in which it is embedded [11–13]. As an example, the so-called correlation dimension associated with a measure, known as correlation integral, is an appropriate tool to characterize the behavior of self-similar sets.

6.2.2   Impacts of Fractal Nature of Packet Traffic

Fractal characterization is, in fact, applicable to many aspects of teletraffic systems such as arrival, service time, buffering, quality of service, and queuing. Although, theoretically, classical Markovian models can always be used to describe any finite set of traffic measurements, the resulting systems are very complex and highly parameterized in case of fractal processes. Hence, it is better to use simpler and more effective models. In this section, the major findings from the most recent real network environment measurements are summarized.

Heavy-Tailed Service Densities

Heavy-tailed densities as a characteristic of fractal processes are suitable for modeling a number of applications such as call holding times [15], and individual call records [16]. In general, they are expected to be seen in switched data services as well as packet based services when there are resources that need to be held for duration of a call or a session. As an example, constant bit rate (CBR) services in ATM networks may be mentioned. From the practical point of view, there are numerous difficulties in accurately engineering these services even when the well known insensitivity of the Erlang-B results is used to characterize the service time. The major problem here is the very slow rate of convergence that allows considerable deviations from the theory over time scales of engineering interest. For a more detailed discussion see Smith[16–22].

Assuming there is a convergence, the rate of convergence problem may be resolved by extending the length of period; however, for long interval observations, the assumptions about the stationarity of arrival processes do not hold and hence the Erlang-B results are not applicable. Intuitively, it looks like the service rate over smaller time intervals can be much greater than the long term and rate conditioned on a departure; hence fractal scaling of the service processes should be applicable here.

Packet Loss

Packet loss processes are very well known to be highly bursty although usually characterized by their long term rates. The limitations of using long term rates in order to describe bursty processes and the problem of serial correlation in losses have been identified to be due to the periodicities in the arrival process. The work was done by Ramaswami et al [6], Erramilli et al [10], and Mandelbort [17]. Briefly, any packet loss rate measurement is likely to be arbitrary over a wide range of time scales, and the long term rate is probably too low to be meaningful. On the contrary, with the cases of transmission errors and packet arrivals, fractal characterizations are applicable in describing packet loss processes. In order to illustrate the above-mentioned point, Erramilli et al., have analyzed the loss processes in simulations driven by Ethernet traffic traces. The study has relied on correlation analysis for different data sets. It has measured the burstiness of the loss process using the fractional correlation dimension. The study has shown that when the packet loss occurs, it occurs at much higher rates than the long term rates, and hence there will be a considerably more impact on the applications than that indicated by the long term rate. In addition, other fractal parameters such as the Hurst parameter are also applicable to the loss process. Please see Erramili et al., [3] for further details.

Fractal Queuing

The presence of fractal properties in actual arrival, service time, and QoS processes may serve as a motivation for the development of the fractal queuing to analyze the performance implications of the processes with long range dependence. One possibility is that if fractal properties impact performance indirectly by biasing the long term traffic measurements, then they can be counted on to transform inputs to conventional queuing models. The direct analysis of models that use fractal characterizations as the input is another possibility, although the lack of a Markovian structure makes such models extremely difficult to analyze. There are, however, three promising approaches: the first one is based on a self-similar stochastic model, specifically, fractal Brownian motion[17], the next one is based on dynamical system approach using chaos theory, and the last one based on the neural networks theory. While the first two approaches are only mentioned briefly here, the last one is the main focus of this chapter and will be discussed in detail.

6.3   NEURAL NETWORK MODELING OF PACKET TRAFFIC

Neural networks as a class of nonlinear systems are able to learn and to perform tasks done by other systems. They are suitable for speech and signal processing, pattern recognition, system modeling, and servomechanism control. They acquire requisite information based on the examples supplied to them.   The various kinds of neural networks generally have energy functions. The learning procedure of neural networks is, indeed, nothing more than decreasing these energy functions until reaching local minimum levels. Neural networks are robust in the sense that if there is a relatively small error in the system, the network will continue its desired action.   This characteristic of the neural networks makes them quite suitable for the traffic modeling task discussed below. In this chapter, perceptron neural networks, along with their learning algorithm back propagation, are utilized as the traffic modeling tool.

6.3.1   Perceptron Neural Networks and Back Propagation Algorithm

The perceptron network is arguably the most popular neural network architecture, and certainly the trigger of the current widespread explosion of activity in the field. The function of the perceptron network is to reproduce certain target output patterns at the last layer of nodes. The task is achieved by adjusting the weighting functions of each interconnecting link according to a rule which compares the activity patterns at output nodes with the desired target patterns and propagates the difference back through the network leading to a small adjustment to each link's weighting function. A simple feedforward perceptron network does not have any feedback connection between two different layers or a layer with itself. In this situation, the input data from the input layer appears in the output layer via the interface of hidden layers. Feedforward networks with no feedback connection between two different layers are generally considered because of their nonlinear properties. Figure 6.1 shows a typical perceptron network.

Perceptron neural networks can be used to model teletraffic patterns. The modeling procedure relies on attempting to predict the dynamical behavior of the describing system after learning corresponding dynamics. The network usually obtains the information required for the learning procedure from a number of available samples.

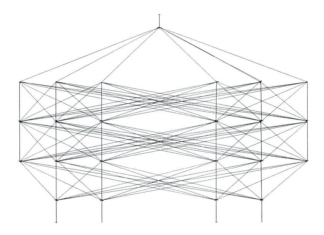

Figure 6.1: A Typical Perceptron Neural Network.

In the following section, an approach capable of dealing with the fractal properties of the aggregated traffic is introduced. This approach takes advantage of perceptron neural networks with back propagation learning algorithm. It provides an elegant solution for self-similar traffic modeling and has the advantage of simplicity compared to the previously mentioned approaches namely stochastic and deterministic nonlinear chaotic map models. It is, motivated by the desire of having a relatively simple model of the complex packet traffic generation process. As opposed to stochastic and chaotic modeling approaches, it does not introduce a parameter that describes the fractal nature of traffic and hence need not cope with the complexity of estimating Hurst parameters or fractal dimensions. The approach simply takes advantage of using a fixed structure nonlinear system that is able to predict either the number of packets generated by a traffic source or the number of arrived packets in a buffer after getting trained by accessing to a number of samples of the generation or arrival pattern.

The back propagation algorithm (BPA) performs simple gradient descent to reduce the mismatch between the desired and actual outputs. The BPA uses all of the Processing Elements (PEs) and adjusts their total interconnections by propagating the output layer error to the preceding layer via the existing connections. The operation is then repeated until reaching the input layer. In other words, output error moves from each layer to the preceding layer - just opposite the direction of the movement of the original information - until reaching the input layer.

The back propagation network used for the task of modeling consists of an input layer with up to eight neurons, four hidden layers with twenty neurons in each layer, and an output layer with one neuron. The inputs of the network are eight consecutive samples of the traffic pattern and the output of the network is the ninth sample, which is supposed to be predicted. In the learning phase information may come back through the network in order to update the weighting functions. The network may also be heteroassociative or auto associative. The following notation briefly describes the traffic modeling task using back propagation algorithm for the choice of sigmoid output transfer function. Please see references [23,24] and Chapter 3 for a complete discussion about perceptron networks and back propagation algorithm.

lc : The learning coefficient

$x_j[s]$: The present output state of the j-th PE from the layer s

$w_{ij}[s]$: Weighting function of the connection between layer $s-1$ i-th PE and layer s j-th PE

$I_j[s]$: The combined input of the j-th PE from layer s

$f(I_j[s])$: The output transfer function of the j-th PE from layer s

$e_j[s]$: The derivative of the absolute error function with respect to the combined input of the j-th PE from layer s

- Propagate the input I in the forward direction through the network until reaching to the output o. During propagation of this information through the network, all of the combined inputs I_j and output states x_j for each PE are set.
- For each PE in the output layer calculate the scaled local error $(d_k - o_k)$ and obtain the variations of weighting functions from relations (6.9) and (6.10), respectively.

$$e_k(o) = (d_k - o_k) \bullet f(I_k) \bullet (1 - f(I_k)) \tag{6.9}$$

$$\Delta w_{ij}[s] = lc \cdot e_j[s] x_i[s-1] \tag{6.10}$$

- For each PE in layer s, which is located below the output layer and above the input layer, obtain the scaled relative error and the variation in weighting functions from relations (6.11) and (6.10), respectively.

$$e_j[s] = x_j[s](1 - x_j[s])\sum_k \{e_k[s+1]w_{kji}[s+1]\} \qquad (6.11)$$

- Update all of the weighting functions by adding the variations to the old values.

Inserting momentum terms, derivative corrections, and fast back propagation techniques are also deployed to enhance the convergence speed of the algorithm. The number of samples required for the training procedure in general depends on the complexity of the source and network dynamics. The use of neural networks provides a simpler approach for the task of modeling because it works based on indirect learning of the source or network dynamics. The learning schema relies on the information available in a number of samples.

6.3.2 Modeling Individual Traffic Patterns

In the proposed approach, a fixed structure perceptron neural network is used for the task of modeling. The network consists of an input layer with up to eight neurons, four hidden layers with twenty neurons in each layer, and an output layer with one neuron. The inputs of the network are eight consecutive samples of the traffic pattern and the output of the network is the ninth sample, which is supposed to be predicted. Based on the richness of the dynamic, it might be possible to reduce either the number of the neurons in the input layer or the number of hidden layers, but as the standard structure, the above mentioned structure is used unless otherwise stated. In the following, three different approaches based on the type of input samples used for training of the neural network are introduced.

The first method makes direct use of the available traffic samples. In order not to deal with very large numbers, the sample with value one is inserted to the neural network when the source is active and the sample with value zero is inserted in the neural network when the source is passive. This, indeed, is the normalized version of the peak packet generation rate divided by the peak rate.

The method suffers from a major drawback though. Since the samples provided for the network are discrete values equal to either zero or one, the network learning speed is very low. In fact, having a continuously distributed sample spectrum over the interval $[0,1]$ leads to having a much faster learning procedure.

The second method can be used in cases of generating artificial traffic patterns by chaotic maps. The approach simply accomplishes the task of modeling by inserting a level of indirection, i.e., the neural network concentrates on predicting consecutive samples of the chaotic map as the packet traffic generator. It is motivated by the fact that the evolution of a state variable over a discrete time period for simple classes of chaotic systems can be used to model packet traffic sources. The modeling task relies on establishing a relationship between the state variable and the source activity. One elegant approach is to consider the source to be active and generating traffic at a peak rate if the state variable exceeds a threshold, and to be idle otherwise. This is specially attractive when one tries to predict the behavior of ON-OFF source models. Please see Yousefizadeh[1] for further details. By modeling the chaotic map and generating its samples, it would be very easy to generate the same artificial traffic pattern using the same threshold value as far as the neural network is able to follow the corresponding chaotic map. This, indeed, is a combination of the approaches introduced in prior research work as neural network modeling of chaotic maps and as chaotic modeling of bursty traffic [7,8].

The third method provides a sophisticated and elegant learning approach for well-behaved sources. A well-behaved source is defined as a source that does not generate more than a specified number of packets in a time frame, i.e., there is an upper limit on the number of packets generated by the source. The most significant point about this approach is that it uses the real traffic samples where the samples have been arranged to create a continuous range of numbers distributed in $[0,1]$ interval. Suppose that the source generates no more than a specified number of traffic packets, say P in a period of time T. Then, considering an origin for the time, the cumulative distribution function of the traffic pattern for the period T is defined as the number of packets generated since the beginning of the time divided by the maximum number of packets p/P. Obviously, this is a monolithic increasing function starting at zero and ending at one. Note that if the source generates a number of packets less than the maximum number, a monolithic function ending at a value less than one will be observed. The samples of these functions can be used to provide the desired sample set. At the end of the period, the desired output is compared with the network output and if the value of error has not entered the acceptable bound, the training procedure is repeated. Relying on these three training algorithms, the fixed structure neural network is used to model a number of artificial traffic patterns generated by single and double intermittency chaotic maps described as:

Single:

$$x_{n+1} = \begin{cases} \varepsilon + x_n + cx_n^m & : \quad (0 \le x_n \le d) \\ \dfrac{x_n - d}{1-d} & : \quad (d \le x_n \le 1) \end{cases} \quad \text{Where } c = \frac{1-\varepsilon-d}{d^m} \quad (6.12)$$

Double:

$$x_{n+1} = \begin{cases} \varepsilon + x_n + cx_n^m & : \quad (0 \le x_n \le d) \\ -\varepsilon_2 + x_n - c_2(1-x_n)^m & : (d \le x_n \le 1) \end{cases} \quad \text{Where } c_1 = \frac{1-\varepsilon_1-d}{d^m} \quad (6.13)$$

Single and double intermittency maps represent a class of piecewise linear/nonlinear maps that can be used to generate an artificial self-similar traffic pattern. In either case, the source is generating packets at a maximum rate for as long as the map is in the active period $d \le x_n \le 1$. The main objective of the packet generation model is the steady state behavior of the idle periods that are related to many fractal properties observed in actual data and corresponded to self-similar patterns. Interestingly enough, both single and double intermittency maps show fractal properties namely, slowly decaying variances, long range dependence, and 1/f noise. Yousefizadeh[1] includes further details. The generated traffic patterns can, hence, be considered self-similar patterns. Figures 6.2 and 6.3 show the single and double intermittency maps.

The use of artificial traffic patterns provides the possibility of comparing the results obtained from all three approaches. Figures 6.4 and 6.5 show the modeling results in the case of single and double intermittency maps for initial conditions $x_0 = 0.1$ and $x_0 = 0.3$, respectively. The number of samples required for training of the neural network before reaching the sync stage is 698,500 in the case of single intermittency map and 889,710 in the case of double intermittency map. Comparing all three approaches, it seems that the second approach provides the best results in terms of tracking. Comparing the first and third approaches, it is easy to observe that the third approach provides more reliable results as it is able to follow the traffic in a longer period of time.

As can be seen from the figures, the familiar ON-OFF follow-up learning pattern is observed, i.e., the neural net learns to follow the traffic pattern after approximately 700,000 and 890,000 iterations in cases of single and double intermittency maps, respectively, and is able to stay within the acceptable error bound for the next 60 samples in the case of the third learning algorithm. The network then goes out of sync and needs to be trained again in order to be able to follow the pattern properly.

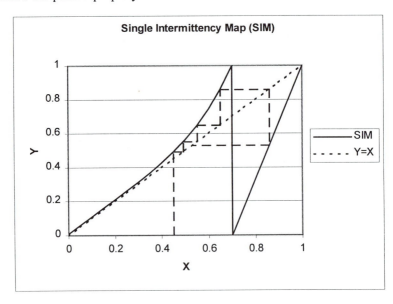

Figure 6.2: Single Intermittency Map Shown for $m = 5$ and $d = 0.7$.

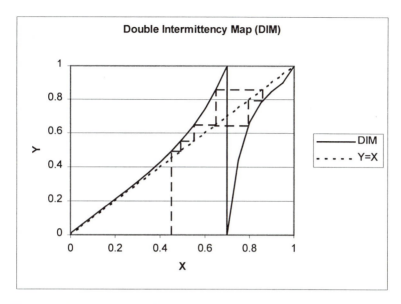

Figure 6.3: Double Intermittency Map Shown for $m = 5$ and $d = 0.7$.

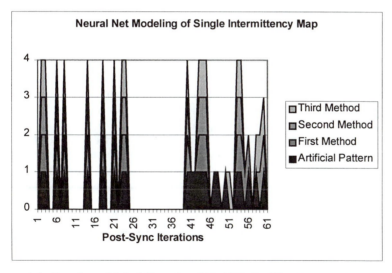

Figure 6.4: Results of Modeling the ON-OFF Traffic Pattern Generated by Single Intermittency Map for the Initial Condition $x_0 = 0.1$. The Horizontal Axis Displays the Time while the Vertical Axis Displays the Normalized Packet Generation at a Peak Rate.

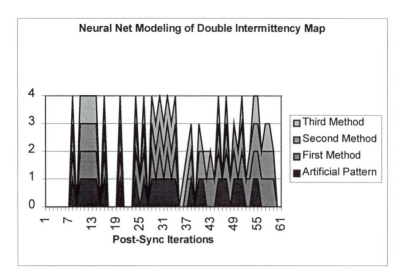

Figure 6.5: Results of Modeling the ON-OFF Traffic Pattern Generated by Double Intermittency Map for the Initial Condition $x_0 = 0.3$. The Horizontal Axis displays the Time while the Vertical Axis Displays the Normalized Packet Generation at a Peak Rate.

6.3.3 Modeling Aggregated Traffic Patterns

In the following discussion, each source is assumed to generate packets with a pattern following the double intermittency map packet generation model described in the previous section. By using different initial conditions and/or different threshold values, different traffic patterns are obtained for different sources.

For example, the model may be considered as an ATM queuing system with a number of Virtual Channels (VCs) with each VC belonging to a traffic source. In these models, the queuing behavior is separated into burst and cell scale components as the result of relying on cell rates rather than interarrival time.

There is a finite capacity buffer corresponding to each source, which keeps the generated packets before they get transmitted. The occupancy of each buffer is determined by the flow of cells from the corresponding source and the rate at which the cells are serviced. In this model a queue is identified by its buffer capacity $C_{\max}$, and its server capacity $O_{\max}$. In each queue, the generation rate is compared with the service rate to determine whether the size of the queue is increasing or decreasing as well as whether the queue is losing cells.

Here, the objective is to model the traffic pattern of the queue. Using the following notation

$I(i,k)$: The input rate of the i-th channel at time k.

$O(i,k)$: The output rate of the i-th channel at time k.

$Q(i,k)$: The queuing rate of the i-th channel at time k.

$L(i,k)$: The loss rate of the i-th channel at time k.

$C(i,k)$: The queue size of the i-th channel at time k.

the state of the queue for each channel is specified by

$$I(i,k) = O(i,k) + Q(i,k) + L(i,k) \tag{6.14}$$

at any instant of time as shown in Figure 6.6. Note that besides Q values that could be positive or negative, all the other values are always positive.

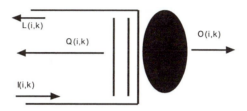

Figure 6.6: The Queuing Diagram of the I-th Source at Time k.

Originally, all of the queues are empty. A queue begins to form when the source input rate exceeds the service rate. Hence, the queue rate $Q(i,k)$ and the loss rate $L(i,k)$ remain zero as far as the input rate is less than or equal to the service rate, i.e.,

$$O(i,k) = \begin{cases} I(i,k): & I(i,k) < O_{max} \\ O_{max}: & I(i,k) = O_{max} \end{cases} \tag{6.15}$$

The queue size $C(i,k)$ begins to increase as soon as the input rate exceeds the service rate S_{max}. While the queue is not empty, the output rate is always equal to the queue server capacity and the total queuing rate is the difference between the input rate and queue server capacity. The loss rate is zero at this stage.

The queue size keeps increasing and finally becomes full if the input rate remains higher than the queue server capacity. In that situation, the queuing rate is zero and the excess input rate is the cell loss rate as

$$L(i,k) = I(i,k) + O(i,k) \tag{6.16}$$

with $O(i,k) = O_{max}$. The effect of a change in the input rate is not immediately apparent if there are packets in the queue waiting to be transmitted. It is the queuing rate that changes according to

$$Q(i,k+1) = Q(i,k) + I(i,k+1) - I(i,k) \tag{6.17}$$

The queue size begins to decrease in size when the input rate becomes less than the server capacity, i.e., $I(i,k) < O_{max}$, and the queuing rate goes below zero as the result, i.e., $Q_{max} < 0$. The queue becomes empty if this situation lasts. The output rate is obtained from the following equation,

$$O(i,k) = \begin{cases} O_{max} & C(i,k) \geq 0 \\ I(i,k) & C(i,k) = 0 \end{cases} \tag{6.18}$$

After providing a brief queuing analysis for individual queues, now it is time to take look at the system from a high level point of view. For the rest of this section and the following two sections, it is assumed that a number of sources are sharing the total bandwidth available from the main channel. Each source has an ON-OFF model and is generating traffic at a peak rate when it is active. The source becomes active as soon as the state variable of the describing chaotic map goes beyond the threshold value and becomes passive as soon as the state variable goes below the threshold. The double intermittency map is chosen to be the chaotic map used for the packet generation as it generates a self-similar traffic pattern. The traffic pattern of each source is separated from the other one by choosing a different initial condition.

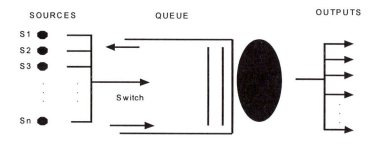

Figure 6.7: A sample network used to demonstrate the modeling power of neural networks for modeling aggregated level bursty traffic.

The following scenario illustrates neural network modeling of bursty traffic at the aggregated level. In order to be able to access an aggregated traffic pattern, a system consisting of 100 individual sources and a queue as indicated in Figure 6.7 is considered. This might be realized as an example of a real network with a number of nodes sending their packets to the network gateway. The traffic pattern might include a variety of different packets such as telnet, ftp, rlogin, mail, etc. The arrived packets are stored in a relatively large size buffer before being forwarded to corresponding destinations. In order to be able to simulate the real network, each individual source is replaced by an artificial generator following an ON-OFF pattern. The generated traffic, hence, can be considered self-similar. It is important to note that the objective here is merely to predict the traffic pattern arrived at the gateway. A fixed structure perceptron neural network is used for the task of modeling. The network consists of an input layer with eight neurons, three hidden layers with twenty neurons in each layer, and an output layer with one neuron. This is the same typical structure as indicated in Figure 6.1. The inputs of the network are eight consecutive samples of the traffic pattern and the output of the network is the ninth sample that is supposed to be predicted.

The traffic pattern of each source is obtained from double intermittency map and is distinguished from the other sources by assigning a different threshold value to the corresponding map. Figure 6.8 shows the result of a neural network modeling task. Again, the familiar tracking period followed by a divergent behavior is observed. The only difference is that self-similarity increases the speed of convergence at the aggregated level. It is important to note that the burstiness of the aggregated level traffic increases as the trained network can follow the pattern for a smaller number of samples before going out of sync. In this example, the neural network learning algorithm converges approximately after 280,000 iterations and is able to follow the main pattern for the next 55 arrivals.

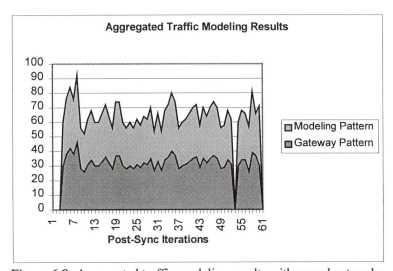

Figure 6.8: Aggregated traffic modeling results with neural networks.

Knowing that a statistically self-similar traffic pattern exhibits a fractal-like behavior in the sense that aggregated streams of such traffic pattern typically intensify burstiness instead of smoothing it, the observed result is very interesting. The result shows self-similarity provides an extra source of information that can be interpreted as some kind of correlation among the generated traffic patterns. The conclusion is that the simple nonlinear dynamic of neural networks is able implicitly to capture self-similarity and hence neural networks may be viewed as suitable generators of self-similar traffic.

6.4   APPLICATIONS OF TRAFFIC MODELING

In this section, the applications of the modeling scheme introduced earlier are investigated. Consider a system that consists of a number of sources sharing the space available in a central buffer and generating packets following an ON-OFF source model. Figure 6.9 shows the structure of a multiple source queuing system. The challenge is the dynamic assignment of the buffer space such that the probability of loss is minimized. In this study, two different scheduling algorithms are considered. These are fixed time division multiplexing (FTDM) and statistical time division multiplexing (STDM). In FTDM each source takes advantage of a fair portion of the buffer space and there is no sharing, while in STDM the unused portion of buffer space assigned to each source might be used to service packets generated by other sources.

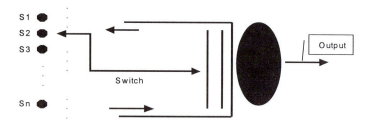

Figure 6.9: The structure of a multiple source queuing system.

6.4.1   Packet Loss Prevention

The first application of this section introduces a dynamic buffer management algorithm relying on the modeling power of neural networks. There are a different number of buffer management algorithms studied in literature. The simplest method is complete sharing in which the buffer space is shared among all the existing sources without enforcing any capacity allocation mechanism. This method introduces an unfair consumption of buffer by greedy sources while providing the lowest loss rates. The second method is called complete partitioning in which the available capacity of the buffer is equally shared among the existing sources. This method has the best fairness characteristic while it greatly suffers from efficiency degradation by introducing the highest loss rate. In the presence of a specified scheduling algorithm, a threshold buffer management algorithm is introduced as the third alternative solution. In a threshold method, each source has its own fixed portion of the buffer space that can only be used for buffering packets generated by that specific source. There is also an additional portion of the buffer completely shared among the existing sources. This method is called partial sharing. A dynamic buffer management algorithm is classified under the threshold methods with the ability to adjust the buffer size of each source dynamically. In order to show the performance of the modeling approach, four different buffer management scenarios are compared in the presence of FTDM and STDM scheduling algorithms.

- The first scenario happens when complete sharing (CS) mechanism is enforced. This is a simple queuing mechanism in which all of the generated packets are directly sent to the central buffer and wait there until getting transmitted. This method introduces maximum efficiency for the available buffer space. The drawback is that the space may not be used fairly. Hence, a source with a high output rate is able to consume a big portion of the buffer space and cause the queue to overflow.
- The second method is a simple implementation of complete partitioning (CP) scheme in the presence of FTDM and STDM in which the capacity of the central buffer is distributed equally among the sources. The most important characteristic of the method is that the buffer space is distributed fairly. FTDM suffers from a possible low efficiency rate compared to STDM, i.e., sources with lower generation rates may not use the whole portion of the bandwidth assigned to them while sources with higher generation rates have packets ready to be transmitted. This does not happen when STDM is employed as the unused portion of the server bandwidth is used if there is any packet ready to be transmitted.
- The third method is a simple implementation of static partial sharing (SPS) scheme that has equal portions for the sources with an additional shared portion that can be shared among all the sources.
- The fourth method, known as dynamic neural sharing (DNS), is the dynamic assignment of the buffer space relying on the results obtained from the perceptron network prediction algorithm, i.e., adjusting the buffer space according to the packet generation pattern of each source. This is a generalization of the third method, keeping the shared portion size fixed and adjusting the buffer space size of each source dynamically.

It is important to mention that for the last three methods, there is a separate queue for each source, which holds the packets generated by that source. The difference between the third and the fourth scenario is that, in the third scenario, the buffer space assigned to each source is fixed and each source is able to send its generated packets to either its own buffer or the shared buffer if space is available, while in the fourth scenario, the portion of the buffer space assigned to the source with a higher packet generation rate is increased in case other sources are not generating enough packets to use their assigned share of the buffer space.

In order to investigate the performance of the method, a triple source system is used. The traffic patterns of the first, second, and third source consist of an artificial pattern generated by 30, 40, and 50 individual double intermittency map packet generators, respectively. The traffic generated by each source is collected and sent to the corresponding buffer in a round robin manner. It is especially important to note that there is a slight difference among the number of packets generated by each source as the result of having a different number of ON-OFF packet generators per source. In order to evaluate the performance of different methods, the overall and per-source loss probability of the system for different choices of buffer size with a fixed service rate are compared. The buffer space can be shared among all of the sources or may be divided into equal portions for individual source usage. The server bandwidth may also be used according to FTDM or STDM scheduling mechanisms.

Figures 6.10 through 6.13 show the total and single source packet loss probability vs. packet size diagram for the triple source queuing system in the presence of FTDM and STDM scheduling algorithms. The single source is the source with the lowest generation rate to compare the fairness of different schemes. The simulation results have been obtained from an iterative algorithm with a total number of ten million iterations per choice of buffer size. Applying a continuous learning algorithm, the fixed structure neural network has been able to follow the traffic pattern within the specified error range between 20 and 30 times covering an average of 50 samples per time. Worth mentioning is that the performance of various methods is very different as the result of applying different methods for traffic management of a heavily utilized system. It is clearly observed from the figures that, for both FTDM and STDM using neural sharing scheme, the total loss rate compared to complete partitioning scheme as well as per-source loss rate compared to complete sharing scheme are reduced. The results may be interpreted as a sign that the neural sharing scheme has come up with a solution between the two extreme cases. Comparing SPS and DNS results shows the higher efficiency of the latter method. This is a significant improvement compared to the other three schemes.

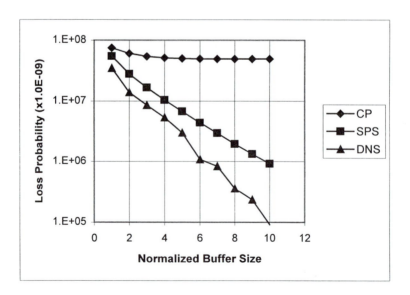

Figure 6.10: The Total Packet Loss Probability vs. Buffer Size Diagram for the Triple Source Queuing System Using CP, SPS, and DNS in Presence of FTDM.

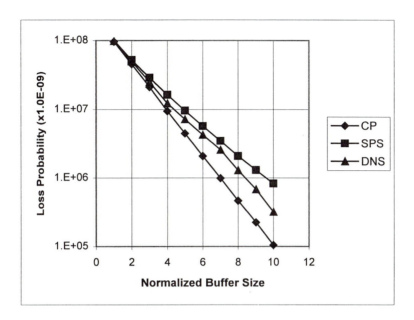

Figure 6.11: The Single Source Packet Loss Probability vs. Buffer Size Diagram for the Triple Source Queuing System Using CP, SPS, and DNS in Presence of FTDM.

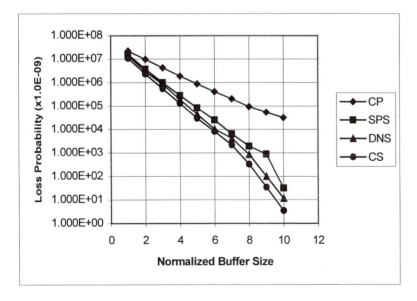

Figure 6.12: The Total Packet Loss Probability vs. Buffer Size Diagram for the Triple Source Queuing System Using CP, SPS, DNS, and CS in Presence of STDM.

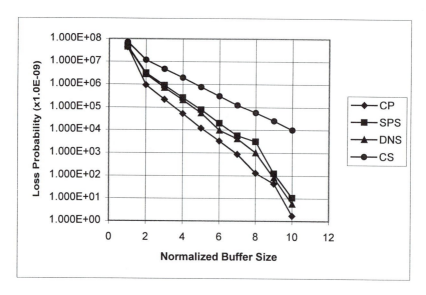

Figure 6.13: The Single Source Packet Loss Probability vs. Buffer Size Diagram for the Triple Source Queuing System Using CP, SPS, DNS, and CS in Presence of STDM.

6.4.2   Packet Latency Prediction

The second application introduces a neural based approach for predicting the queuing delay as the dominant delay factor observed in a finite buffer used as an interface for transmitting a number of packets generated by a number of sources in a multiple source system. Hence, it might be considered as a part of packet scheduling algorithms that is able to estimate packet latency. Packet latency prediction is addressed by counting on the predictive power of neural networks directly. Packet latency is defined as the time each packet spends in the queue before getting transmitted.

Again, consider the above triple source system sharing the buffer space of the central buffer following complete sharing scheme in the presence of STDM scheduling. For the case of packet latency estimation, the system load varies based on the value of service rate. The case chosen as the objective here is to determine the queuing delay of the generated packets. Supposing each packet carries a sequence number indicating the order in which it was sent to the central buffer, the objective is then to predict the average number of time units a packet spends in the queue before leaving the buffer. The task is approached by applying the neural network modeling scheme to predict the total number of generated packets. Knowing the buffer service rate, average latency can be calculated directly from the arrival rate of the buffer. The dominant average latency factor in most of the cases is the queuing latency. This, indeed, is a problem of predicting the state of the queue. Again the prediction tool is the fixed structure neural network that is supposed to learn the dynamic of the arrival pattern of the buffer.

The real and estimated average latency vs. service time diagram for the triple source queuing system is shown in Figure 6.14. The typical system consists of 120 sources generating traffic according to an ON-OFF pattern and sending the generated packets to a central buffer. The buffer size is assumed to be fixed and large enough to prevent loss. The average latency has been calculated over all of the time periods in which the neural network is able to follow the arrival pattern of the central buffer.

As can be observed from Figure 6.14 the estimation results are quite acceptable within the three percent error range as long as the averaging period is long enough in order for the neural network to be able to follow the traffic pattern a number of times within the specified error bounds and as long as the buffer service rate does not exceed an existing threshold value. As a matter of fact, it is observed that the average packet latency drops very sharply with an order of ten or more choosing a value beyond the threshold value. As the result of having very small average latencies, the neural network latency estimation findings are not acceptable for service rate values beyond the threshold value. The value of the threshold generally depends on the dynamics of the system and for the triple source system is the normalized value 13. Rememeber that for this complete sharing case, the modeling scheme relies on the combined dynamics of all of the sources to achieve the latency estimation as there is only one queue in the system. It is important also to note that the same qualitative approach may be used in cases of having separate queues.

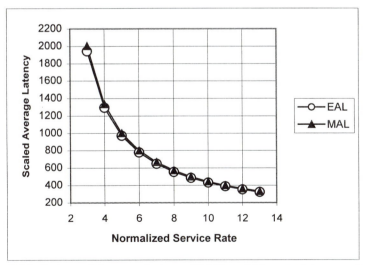

Figure 6.14: Estimated Average Latency (EAL) and Measured Average Latency (MAL) of the Packets vs. Service Rate Diagram for the Triple Source Queuing System. CS Buffer Management and STDM Scheduling Have Been Employed.

6.4.3   Experimental Observations

The following section briefly discusses some of the practical findings in the implementation of the algorithm, which are the direct conclusion of dealing with complicated nonlinear dynamics.

The specific problem can be explained by the chaotic nature of traffic, i.e., that traffic patterns with self-similar characteristics have been shown to exhibit chaotic behavior. Please see Erramilli et al.[3] for a detailed discussion. To explain the problem, it can be simply said that, although it is possible to reach to a very small network error at some steps during the learning phase, if the network error is studied for further samples, it is observed that this error begins to grow as time proceeds. The reason is summarized in the chaoticness of the system, i.e., since the nonlinear network wants to model a chaotic system, it becomes chaotic itself. In this situation, a small error may be considered as a small difference between two close initial conditions of the desired output and the network output and as a characteristic of a chaotic system. The error begins to grow very soon, which is nothing more than high sensitivity to the variations of initial conditions. As a matter of fact, this can be interpreted as a good sign for the network that has been trained to model a chaotic map and has become chaotic itself. One way to relieve the effect of having an error component that grows over time is to repeat periodically the learning phase followed by the recalling phase; otherwise the results exceed the acceptable error range. In practice, after the first learning phase with a several million examples, the neural network can predict less than 100 samples before requiring repetition of the learning phase as in the previous recalling phase and so on to bring the results within the acceptable error range.

Based on the same line of reasoning, all of the convergence results are affected strongly by the choice of initial conditions. It can be said that the initial values of the parameters play a crucial role in the convergence of the algorithm. It is even possible to have a divergent algorithm, if the initial values of the parameters are chosen unsuitably. As a practical result, it is better to set the initial values of the parameters on small numbers, e.g., $w_{ij} = 0.01 \;\; \forall i, j$.

Finally, it has to be mentioned that the choice of initial conditions has another important impact on the performance of the application. The issue can be addressed by saying that in case of loss prevention application, it is very important for the algorithm to be able to predict the traffic pattern generated by all the sources. The very wild behavior of a self-similar traffic pattern generally leads to an out-of-sync prediction power for different initial conditions, i.e., the acceptable error bounds are not reached after having the same number of iterations for all the sources. The only way of dealing with this problem is to choose the initial conditions suitably and find the time intervals by which all of the traffic patterns can be predicted by the neural network.

6.5   SUMMARY

This chapter was dedicated to the application of neural networks in modeling self-similar traffic patterns of computer communication networks. Neural networks rely on the information available in a number of samples in order to capture complex dynamics of packet traffic phenomena. This feature makes them practical for the task of modeling, as they do not need to deal with analytical complexities involved with stochastic and chaotic systems' approach.

Neural networks are used to predict the behavior of the ON-OFF source models based on some threshold levels. This, in particular, can be related to the prior research work in which neural networks are used to predict the complex behavior of a class of discrete-time chaotic maps. Artificial load generation is, hence, an elegant application of neural networks in this area. The approach allows the generation of complex traffic patterns using relatively simple models that can be incorporated into traffic streams. There is a scope to investigate efficient software implementation. The statistical features of generated traffic can be compared to those of real traffic in order to show that there are some match points for the generated traffic.

Neural networks are additionally used for modeling of the single and aggregated traffic patterns. A traffic pattern may either correspond to a single source or a number of sources. Multiplexing and splitting of traffic streams where each source is modeled by a neural network is the method used for dynamic buffer management and packet loss reduction. This approach can, in fact, introduce an interesting application in ATM networks supporting a number of virtual channels. Neural networks are also used to predict the latency time for the packets generated by a traffic source. The latter includes the time each packet spends in the queue before transmission.

REFERENCES

1. Yousefizadeh, H., Performance modeling of a class of queuing systems with self-similar characteristics, Ph.D. Dissertation, University of Southern California, 1997.

2. Leland, W.E. and Wilson J., Statistical analysis and stochastic modeling of self-similar data traffic, *ITC,* Vol. 14, 319, 1994.

3. Erramilli, A., Gordon, J., and Willinger, W., Applications of fractals in engineering for realistic traffic processes, *ITC,* Vol. 14, 35, 1994.

4. Leland W.E. and Wilson J., On the self-similar nature of Ethernet traffic, *IEEE/ACM Trans. on Networking*, Vol. 2, No. 1, 1, Feb. 1994.

5. Erramilli, A. and Singh, R.P., An application of deterministic chaotic maps to model packet traffic, Bellcore Technical Memorandum, 1992.

6. Ramaswami, V., Traffic performance modeling for packet communication: whence, where, and whither, *Proc. of Australian Teletraffic Seminar*, Nov. 1988.

7. Yousefizadeh, H. and Jonckheere, E.A., Neural network modeling of discrete-time chaotic maps, Submitted to: *IEEE Trans. On Circuit and Syst., Part II.*

8.   Yousefizadeh, H., Shafiee, M., and Zilouchian, A., Chaotic arrays modeling with neural networks, *Proc. of the Iranian Conf. of Electrical Eng.*, Vol. 4, 667, May 1993.

9.   Hellstern, K.M. and Wirth, P., Traffic models for ISDN data users: office automation application, *Proc. ITC-13*, Denmark, 1991.

10.   Erramilli, A., Gosby, D., and Willinger, W., Engineering for realistic traffic: a fractal analysis of burstiness, *Proc. of ITC Special Congress*, India, 1993.

11.   Fowler, H.J. and Leland, W.E., Local area network traffic characteristics, with implications for broadband network congestion management, *IEEE JSAC*, Vol. 9, No. 7, Sep. 1991.

12.   Sriram, K. and Whitt, W., Characterizing superposition arrival processes in packet multiplexers for voice and data, *IEEE JSAC*, Vol. SAC-4, NO. 6, Sep. 1986.

13.   Heffes, H. and Lucantoni, D.M., A Markov modulated characterization of packetized voice and data traffic and related statistical multiplexer performance, *IEEE JSAC*, Vol. 9, No. 7, Sep. 1991.

14.   Duffy, D.E. and Willinger, W., Statistical analysis of CCSN/SS7 traffic data from working CCS subnetworks, *IEEE JSAC*, 1994.

15.   Bolotin, V.A., Modeling call holding time distributions for CCS network design and performance qnalysis, Preprint 1993.

16.   Smith, D.E., On the holding times of data calls, Bellcore Technical Memorandum, 1986.

17.   Mandelbort, B., The fractal geometry of nature, Freeman, NY, 1983.

18.   Erramilli, A., Singh, R.P., and Pruthi, P., Chaotic maps as models of packet traffic, *ITC* Vol. 14, 329, 1994.

19.   Beran, J., et al., Variable bit rate video traffic and long range dependence, *IEEE/ACM Trans. on Networking*, Vol. 2, No. 3, Apr. 1994.

20.   Pitts, J.M., et al., An accelerated simulation technique for modeling burst scale queuing behavior in ATM, *ITC* Vol. 14, 777, 1994.

21.   Self-similarity in high speed packet traffic: analysis and modeling of Ethernet measurements, *Statistical Science*, 1994.

22.   Mandelbort, B. and Taqqu, M.S., Robust R/S analysis of long run serial correlation, *Proc. 42nd Edition ISI*, 69, 1979.

23.   Van Ooyen, A. and Neihuis, B., Improving the convergence of back propagation algorithm, *Neural Networks*, Vol.5, No.3, 1992.

24.   Guo, H. and Gelfland, S., Analysis of gradient descent learning algorithms for multilayer feedforward neural networks, *IEEE Trans. on Circuit & Syst.*, Vol. 38, No.8, Aug. 1991.

7 APPLICATION OF NEURAL NETWORKS IN OIL REFINERIES

Ali Zilouchian and Khalid Bawazir

7.1   INTRODUCTION

In response to demands for increasing oil production levels and more stringent product quality specifications, the intensity and complexity of process operations at oil refineries have been exponentially increasing during the last three decades. To alleviate the operating requirements associated with these rising demands, plant designers and engineers are increasingly relying upon automatic control systems. It is well known that model based control systems are relatively effective for making local process changes within a specific range of operation [1]. However, the existence of highly nonlinear relationships between the process variables (inputs) and the product stream properties (outputs) have bogged down all efforts to come up with reliable mathematical models for large scale crude fractionation sections of an oil refinery. In addition, the old inferred property predictors are neither sufficiently accurate nor reliable for utilization of advanced control applications [2]. On the other hand, the implementation of intelligent control technology based on soft computing methodologies such as neural network (NN), fuzzy logic (FL), and genetic algorithms (GA) can remarkably enhance the regulatory and advanced control capabilities of various industrial processes such as oil refineries [3]-[11].

Presently, in the majority of oil refineries (such as Ras Tanura located in Saudi Arabia), product samples are collected once or twice a day according to the type of analysis to be performed and supplied to the laboratory for analysis. If the laboratory results do not satisfy the specifications within an acceptable tolerance, the product has to be reprocessed to meet the required specification [2]. This procedure is costly in terms of time and dollars. In the first phase, an off-line specification product should be first routed to a holding facility. In the second phase, the process should be tuned before any further processing is carried out. In order to resolve this problem in a timely fashion, a continuous on-line method for predicting product stream properties and consistency with and pertinence to column operation of the oil refinery are needed.

In general, on-line analyzers can be strategically placed along the process vessels to supply the required product quality information to multivariable controllers for fine tuning of the process. However, on-line analyzers are very costly and maintenance intensive. To minimize the cost and free maintenance resources, alternative methods should be considered.

In this chapter, the utilization of artificial neural network (ANN) technology for the inferential analysis of a crude fractionation section of the Ras Tanura Oil Refinery at Dhahran is presented. The implementation of several neural network models using back propagation algorithm based on collection of real-time data for a three-months operation, of the plant is presented. The proposed neural network architectures can accurately predict various properties associated with crude oil production. The simulation results for modeling of several products such as naphtha 95% cut point and naphtha Reid vapor pressure are analyzed. The results of the proposed work can ultimately enhance the on-line prediction of crude oil product quality parameters for the crude fractionation processes of various oil refineries.

The chapter is organized as follows. Section 7.2 covers various steps pertaining to collection of plant data that are used during the training and verification phases of the neural network program. A systematic procedure to construct a NN model is also presented in this section. In section 7.3, selection of appropriate data sets as well as data analysis procedures are discussed. Section 7.4 is devoted to various steps in the implementation phase of neural network models in the crude oil fractionation process. In sections 7.5 and 7.6, the training procedures as well as the results of modeling for naphtha 95% Cut-point and naphtha Reid vapor pressure products are analyzed. It is shown that the proposed NN models predict products qualities well within the specified error goals in both training and verification phases. Various implementation issues such as model building, model data analysis, effects of neuron distribution on training, and model robustness are also discussed in this section. Finally, section 7.6 summarizes the contributions of this chapter.

7.2   BUILDING THE ARTIFICIAL NEURAL NETWORK

The mathematical algorithms developed to model neurons can be adapted for many useful predictions in processing plants. The complexity of the pattern to be recognized dictates the complexity of the required algorithm. Some very useful predictions can be constructed in processing plants using algorithms whose coefficients are discovered through training. Figure 7.1 is a graphical representation of the artificial neural network structure.

A neural network predictor is built by discovering the weights as shown in Figure 7.1. N_1K_1 through N_1K_n are the corresponding weights of the first neuron. The output Q_p is the predicted inferred process stream property (%H_2S, 95% cut-point, etc.)The coefficients of the model are discovered by training a neural network program using back propagation algorithms [12, 13, 17]. The inputs of NN consist of plant data such as temperature, flow rate and pressure where, the respective product quality is considered as desired output of the program model. The neural network program will be trained by adjusting the weight coefficients until the difference between the predicted product quality and the measured product quality is within acceptable limits. When the coefficients have been determined, they should be tested by comparing the predicted quality to the measured quality for data sets which were not used in finding the coefficients. The process of finding the ANN coefficients is called training the network [13], [14].

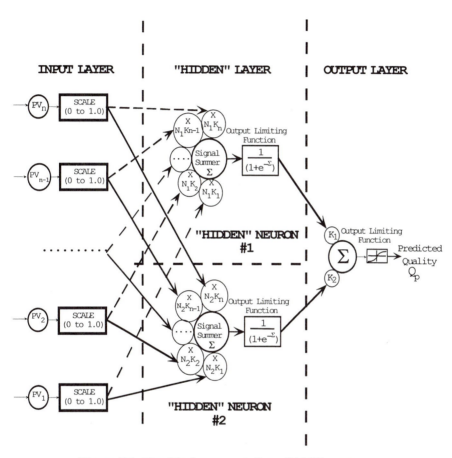

Figure 7.1: Graphical representation of ANN structure.

7.2.1 Range of Input Data

Neural networks will not be an accurate predictor if the operating inputs/output data are outside their training data range. Therefore, the training data set should possess sufficient operational range including the maximum and minimum values for both input/output variables.

7.2.2   Size of the Training Data Set

A minimum of two valid data sets is required for each coefficient in the training algorithm. A large number of valid data sets provide much better accuracy in the prediction phase. However, some training data sets are not valid either due to the dynamic nature of a process or as the result of inaccuracies in data acquisition techniques. A large data set will average out various inaccuracies within a system.

7.2.3   Acquiring the Training Data Set

The least intrusive technique for obtaining the training data set is to take data during the course of normal operations. This procedure probably will not satisfy the required variations in some process variables. However, plant tests can be accomplished by varying the process variables within the region of interest to complete the gaps within the required data. In general, it is not necessary to have field analyzers to develop a neural network predictor for a stream quality. Samples can be taken and sent to the laboratory for analysis at the same time that data (flow, temperature, and pressure) are taken from field transmitters.

7.2.4   Validity of the Training Data Set

In many industrial applications accuracy is not as important as repeatability. For example, a network trained for a pressure transmitter with a 15 pound per square inch (PSI), zero shift will predict accurately - unless the transmitter is re-calibrated; however, the lack of repeatability exhibited in data taken from hand-written shift logs has proven too unrepeatable to use as a training set. The training data set can be taken from the Distributed Control System (DCS) or the supervisory control and data acquisition computer.

Many of the processes have significant time constants and dead times [8]. Unless it is desired to include these time constants and dead times in the prediction, the process should have been operating at steady-state for a period equal to at least two time constants before including the operating data in the data set. Flow and pressure inputs should be averaged to eliminate the problem of signal noise.

7.2.5   Selecting Process Variables

Initial process variable selection is not critical; almost anything upstream of the measurement point could be useful. As many process variables should be included as can be handled. The training process will automatically determine which are important and which can be deleted from the calculation.

For example, the process variables shown in Figure 7.2 are selected to predict Reid vapor pressure (RVP) in the bottom of a stripper column. Their relative importance, determined by neural network training, is shown in Table 1. If those process variables chosen initially do not give the required accuracy of prediction, less important variables should be dropped and other parameters added.

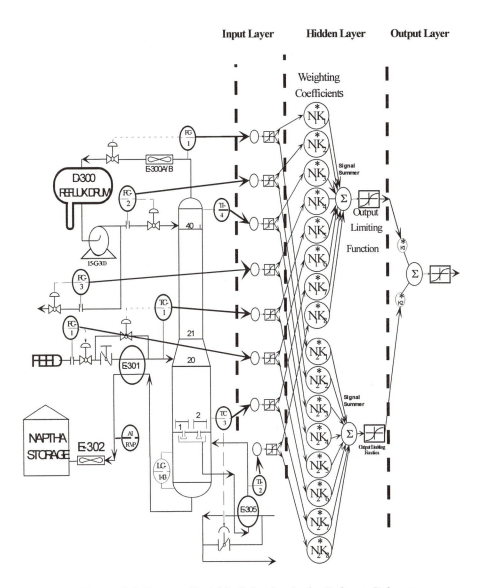

Figure 7.2: Process Variable Selection in the Stripper Column.

Table 7.1: Relative Importance of Process Variables Determined by NN.

FEED FLOW RATE	F-1	4 %
REFLUX FLOW RATE	F-2	8 %
OVERHEAD PRESSURE	P-1	6 %
FEED TEMPERATURE	T-1	13 %
REBOILER RETURN TEMPERATURE	T-2	27 %
TRAY 2 TEMPERATURE	T-3	36 %
COLUMN TOP TEMPERATURE	T-4	5 %
DISTILLATE FLOW RATE	F-3	2 %

7.3   DATA ANALYSIS

The first step in data analysis is to ensure that all column parameters are collected properly. Data unavailable due to transmitter downtime or calibration at the time of data collection should be identified. Since artificial neural networks require that all model parameters be available all the time, unavailable data for any of the parameters force the elimination of the complete data set that is collected at that time. This includes lab data, which is not collected at the scheduled sampling time. The definition of a complete data set includes all process parameters plus one lab value. Lab analyzed properties (95% cut-point, sulfur content, freeze point, etc.) are analyzed individually to generate neural network models.

Once a complete set of parameters is collected, the neural network model can then be used to do a complete data analysis. the neural network model allows the user to specify which data set will be used for model building (teaching phase), and which for model verification (testing phase). A statistical method can be used to eliminate a suspected bad lab data set. The main assumption of the statistical method is that there has to be a correlation between model inputs (process parameters) and model output (lab value).

7.3.1   Elimination of Bad Lab Values

Bad lab values can be identified as follows: The neural network model is given three data sets for model verification (out of 180 data sets), and the rest of the data sets are used for model building. All data sets are switched between model verification mode to model building mode until all data are tested. At any point during the above process, if any of the three model verification points fall outside the lab repeatability, the degree of deviation from repeatability is recorded. At the end of this analysis, all deviant points are completely removed from model building. Thus, it can be established that all remaining data sets conform to the general trend of the correlation.

As a final step, each of the deviant points is again individually added to the model and tested as a verification point by itself. If this point still falls outside lab repeatability, then it is permanently eliminated. Otherwise, the point is returned to the model.

The elimination of data sets during this step does not necessarily reflect only bad lab value. It is possible that the lab analysis is done correctly; however, either the snapshot of the process values taken do not coincide with the time of sampling by operators, or the plant is not operating at steady state conditions at the time of sampling.

7.3.2   Process Parameters and Their Effect on NN Prediction

All identified process parameters do not necessarily have an effect on each of the lab values (properties). The final step of data analysis is to identify the most important process parameters that have a significant effect on the inferred analysis and eliminate those parameters which have little or no effect. Two methods can be used to perform the elimination process. The first is using engineering judgment to realize which process parameters can have little or no effect on the model. An example of this is to remove all naphtha stabilizer parameters when the network is being used to model riesel sulfur.

The second method is to utilize the neural network model itself. The neural network program can generate an analysis of the final weights given to each of the process parameters to fit the data. This method of elimination, however, is not as straightforward as one might expect. The neural network model relies more on process parameters with a large degree of variance. It is possible that the most important parameter that affects a particular lab data set keeps the same value in all the generated data sets. The neural network program will ignore such a parameter. Thus elimination should not include variables which, from an engineering point of view, should have a contribution on the inferred analysis.

For example, a fuel gas density analyzer in the plant under investigation gives density measurements (used for heating value) about five minutes after they would be useful to improve furnace (and the affected crude column) stability. Data for the fuel gas supply pressure, burner gas pressure (P-236), burner gas pressure (P-120), heater coil outlet temperature (T-178), and gas pressure controller output to valve (PC-120 VO) are used as the inputs for a neural network training set – with the output of the density analyzer (A-156) as the stream quality to be predicted. The block diagram of the plant is shown in Figure 7.3. The data sets are used to train a neural network predictor with five hidden sigmoidal neurons. The training result can be found in reference [3]. However, the output of NN model indicates that the algorithm does not predict the stream quality (gas density) with enough accuracy to be useful. It was concluded that an input which would prevent the predicted quality from varying when the measured quality is constant is not in the training set.

If the missing input (fuel gas flow rate) is added to the data set, the algorithm predicts gas density by using the difference in deferential pressure resulting from flow through a fixed and a variable orifice – and becomes useful for eliminating upsets introduced by rapid fuel gas density (and heat of combustion) fluctuations. The simulation results indeed have shown an accurate model prediction after adding the fuel gas flow rate to the data set [3]. Note that it is necessary to align data from different inputs to get a data set whose elements occur simultaneously.

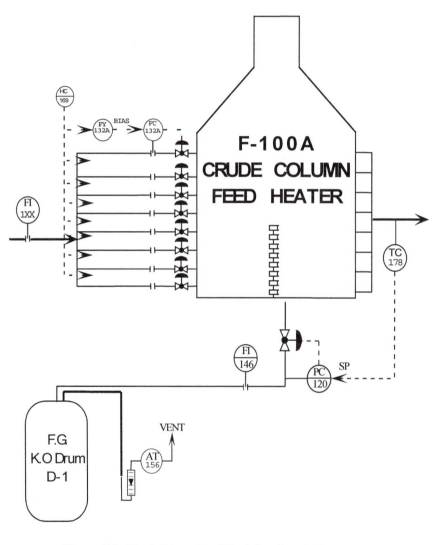

Figure 7.3: Block Diagram of Fuel Gas Supply System.
(F.G. K.O.= Fuel Gas Knock Out)

7.4   IMPLEMENTATION PROCEDURE

The major steps that are involved in implementing the ANN predictor are shown in Figure 7.4.

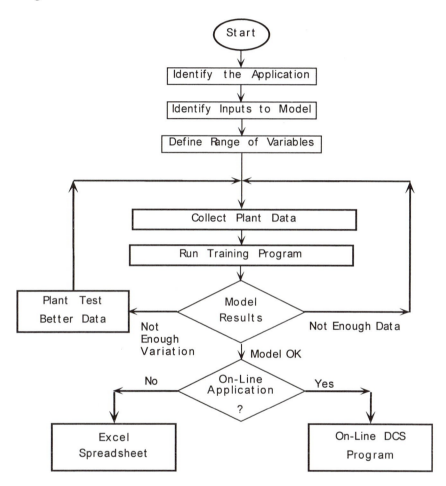

Figure 7.4: Major Steps for Implementing of ANN in the Crude Oil Fractionation Process.

7.4.1   Identifying the Application

The first step for construction of an NN model is the appropriate identification of a potential application. For example, suppose Ethane is burned as fuel gas and propane with 2% ethane can be sold for $18/barrel in the market. The operating objectives of a de-ethanizer unit will be to minimize the propane in the overhead product and maintain slightly less than 2% Ethane in the column bottoms.

On the other hand, if ethane, with 1.5% propane sells for $16/barrel and propane is used as fuel gas, the operating objectives of the de-ethanizer will be different from the above case: to minimize ethane in the column bottoms and keep the propane in the overhead as near 1.5% as possible, without exceeding the sales limit. To achieve either set of the above objectives, a continuous measurement of propane in the column overhead and ethane in the bottoms is necessary.

7.4.2   Model Inputs Identification

The neural network algorithm will not match a random number set. For prediction model to work, there must be some relationships between input/output variables [9], [11]. Training will quantify such a relationship. If a neural network will not train with a good data set, a significant variable may not have been included in the data set. If a rigorous mathematical equation can be written between the inputs and the output, a neural network is unnecessary.

As an example, a double product (overhead and bottoms) distillation column as shown in Figure 7.5 has two variables: the heat balance and the material balance, which determine its separation capability, and two product quality variables: heavy key in the overhead (propane in the de-ethanizer overhead product) and the light key in the bottoms (ethane in the de-ethanizer bottoms). Those process measurements, flows, pressures, and temperatures, which could be used to calculate the heat and material balances, should be chosen as inputs to the neural network. To predict propane in the de-ethanizer overhead we start with the input variables as shown in Table 7.2.

Table 7.2: Process Variables for Propane Prediction.
(MBD = Million barrels per day; MPPH = Thousand pounds per hour)

Process Variable	Tag	Range
Tray 27 temperature	TI-6	100-300 °F
Overhead Temperature	TI-5	100-300 °F
Reflux Temperature	TI-7	100-300 °F
Feed Temperature	TI-4	0-200 °F
Reflux Rate	FI-2	0-20MBD
Distillate Rate	FI-3	0-10MBD
Feed Rate	FI-6	0-80MBD
Reboiler Steam	FI-4	0-30MPPH
Bottoms Product	FI-5	0-80MBD

A good starting point for ethane prediction in the bottoms would use the process variables as shown in Table 7.3.

Table 7.3: Process Variables for Ethane Prediction.
(MBD= Million barrels per day, MPPH= Thousands pound per hour)

Process Variable	Tag	Range
Tray 1 temperature	TI-1	100-400 °F
Bottoms Temperature	TI-3	100-400 °F
Tray 18 Temperature	TI-8	100-300 °F
Feed Temperature	TI-4	0-200 °F
Re-boiler Return Temperature	TI-2	100-400 °F
Distillate Rate	FI-3	0-10MBD
Feed Rate	FI-6	0-80MBD
Re-boiler Steam	FI-4	0-30MPPH
Bottoms Product	FI-5	0-80MBD

If the plant data include significant variation in each of these process variables and the neural network coefficients for a process variable are very small, that process variable can be dropped from the model. If the network will not train, and other conditions are met, other process variables based on engineering experience should be included in the model.

7.4.3   Range of Process Variables

The range of the process variables in the training data set should include the entire operating range. The data set should include data for each process variable, evenly distributed throughout the range for which prediction is desired.

7.5   PREDICTOR MODEL TRAINING

For the naphtha 95% cut point and naphtha Reid vapor pressure stream properties, the plant data, including the stream quality desired to predict, are collected in a Microsoft Excel™ spreadsheet to facilitate data manipulation. The data are then scaled to a fraction of the transmitter range so that they are confined to a sub-interval of $[0\ldots1]$. A practical region for the data is chosen to be $[0.1\ldots0.9]$. In this case each input or output parameter p is normalized as p_n before being applied to the neural network according to:

$$p_n = [(\,0.9 - 0.1\,)\,/\,(\,p_{max} - p_{min}\,)] * (p - p_{min}) + 0.1$$

where p_{max} and p_{min} are the maximum and minimum values, respectively, of data parameter p. The spreadsheet file is then converted to text file and loaded into the MATLAB™ neural network toolbox [14]. The MATLAB software program uses a back propagation training algorithm to adjust the weights of the network in order to minimize the sum-squared error of the network. This is done by continually changing the values of the network weights in the direction of steepest descent with respect to the error [11]-[17]. The change in each weight is proportional to that element's effect on the sum-squared error of the network.

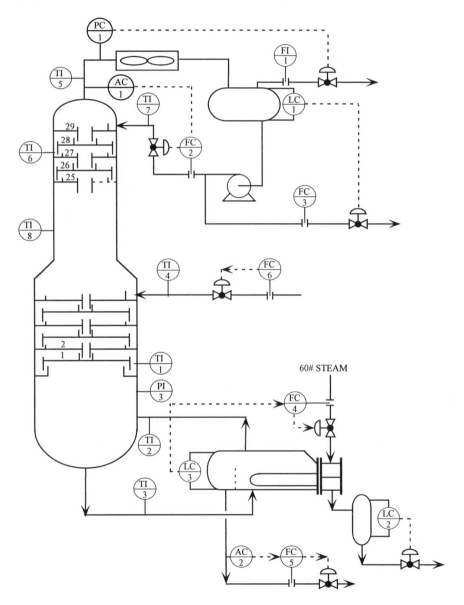

Figure 7.5: De-ethanizer Column Process Flow Diagram.

Initially, one hidden layer with five neurons is built (additional neurons and/or layers can be added if necessary) and all weights are randomly initialized to small numbers. Next, training parameters are defined. These parameters include the following: maximum number of training iterations and acceptable error between desired and predicted values.

The neural network program using back propagation training algorithm starts training and through this process it will look for the specified error on a multidimensional surface. By selecting the minimum error to be a very small number (10^{-3} for example) the program will end up in one of the following states:

1. Minimum error goal is matched before exceeding the limit on maximum allowed iterations. In this case, the objective of the training is successfully met.
2. Program cannot achieve this minimum error but, in the process, it locates the global minimum (optimum solution). In this case, the number of hidden neurons and/or the number of hidden layers can be increased to achieve the desired minimum error.
3. Training diverges. The error increases as the training process continues. (Training data sets are not valid.) In this case it is necessary to construct valid data sets.

7.6   SIMULATION RESULTS AND DISCUSSIONS

As discussed earlier, the objective of the proposed work is to eliminate the dependency on laboratory and/or on-line sample analyzers for sampling of product qualities. The goal can be achieved by the construction of neural networks to predict those particular product qualities to meet the more stringent market specifications. In doing so, the neural network model, from a practical viewpoint, should adhere to two constraints: The optimization of process control and the   reduction on the cost of maintenance and operations, which would ultimately result in an increase in profit [3].

First, the neural network model accuracy of prediction should be consistent and within the defined acceptable tolerance of the desired product quality it is set to predict. It is highly crucial to have a neural network that provides accurate predictions. It is a plant requirement to have the neural network predicted output fed as one of the inputs to a multivariable controller. This will provide the controller with the knowledge of the final product quality, and how close to or far from the desired set point it is. With the aid of this knowledge, the controller will act promptly to keep the process in its targeted path, thus eliminating any off-specs product from taking place.

Secondly, it is a requirement to have the neural network running on-line with fast execution time during both training and prediction phases. The multivariable controller is gathering information about the process and at the same time it is looking at the neural network to provide its prediction. The controller will perform its tight control actions as long as the neural network prediction is made available to the controller at the right moment, not a couple of minutes late. Also, operational objectives often change to meet market needs and in doing so the desired process set points have to change as well to provide the desired product specifications. Retraining the neural network on the new sets of process variables and desired product properties is inevitable. The faster the neural network program predicts after retraining, the faster it provides its output to the controller [3], [9].

7.6.1   Naphtha 95% Cut Point

Modeling of the naphtha 95% cut point property was carried out using a back propagation neural network algorithm. Various configurations, in terms of the number of hidden layers and the number of hidden neurons, have been tested. For the application presented here, two-layer networks consisting of a single hidden layer and an output layer have proved to be adequate. Although a three-layer network is theoretically capable of modeling more general and arbitrary functions than a two-layer network [17], the naphtha 95% cut point data used in training and verification modes were sufficiently well behaved that three-layer networks did not perform better than the ones consisting of two layers.

To demonstrate the modeling capability of a back propagation network, 85 data sets were analyzed. Each data set consisted of 33 process variables as inputs to the model and one product quality (naphtha 95% cut point) as an output. A total of 70 data sets were used in the training phase and 15 data sets were used in the verification phase. Table 7.4 summarizes the simulation results.

For the first case, a single hidden layer consisting of five neurons was utilized. The model could not achieve the desired error goal of 0.01 after performing 10,000 iterations, which was the maximum allowed number of iterations. A maximum error of 7.84°F at training phase was obtained. . In the verification phase, a maximum error of 11.59°F. was detected. Figures 7.6 and 7.7 shows the results of the training and verification phases, respectively.

For further investigation, a first-momentum term was added to the back propagation algorithm. However, the model still could not achieve the desired error goal after 10,000 iterations as shown on the table. Finally, with the same model as above, an adaptive learning rate was added, and the neural network model achieved the desired sum squared error goal of 0.01 in 3180 iterations.

Next, the number of hidden neurons was increased. Table 7.5 also shows the training and verification results using eight neurons in the hidden layer. The model was able to achieve an acceptable error in the training phase of 1.35°F but failed to achieve comparable results in the verification phase where the maximum absolute error was 6.82°F. Further increase in the number of hidden neurons only improved the results in the training phase. The verification phase continued to show error values too significant to be accepted for good prediction of the Naphtha 95% cut point property.

The next step in the simulation was to increase the number of hidden layers. Two hidden layers were selected and the number of neurons in each layer was varied. Figures 7.8 and 7.9 show the training and verification results using eight neurons in the first hidden layer and four neurons in the second hidden layer. The result shows slight improvement in the verification phase but more accurate prediction is still required. It can be noticed that in the training phase the models performed well, however, in the verification phase all the tested models could not predict with enough accuracy. It was suspected that the neural network models were memorizing the relationship between the inputs and the output since they were trying to adhere to a very small error goal in the training phase.

Table 7.4: Initial Simulation Results for Naphtha 95% Cut Point.

Hidden Neurons	Training Phase				Verification Phase	
	Error goal	Iterations	Final SSE	Max. Error $^\circ$F	Final SSE	Max. Error $^\circ$F
5 BP	0.01	10000	0.045	7.84	0.45	11.59
5M	0.01	10000	0.031	1.57	0.22	7.92
5	0.01	3180	0.01	2.17	0.28	5.49
8	0.01	4563	0.01	1.35	0.25	3.29
10	0.01	2088	0.01	1.83	0.27	7.95
8-4	0.01	4302	0.01	1.33	0.14	4.81

It is important to prevent the neural network model from memorizing the input/output relationship. A neural network with enough hidden neurons given enough iterations and a very small error goal will actually memorize a given relationship between model inputs and outputs. In other words, a network memorizes relationships between outputs and inputs when the model building points are allowed to conform to a degree much less than lab repeatability. It means that an acceptable sum squared error goal in the training phase should generate a degree of accuracy very close to lab repeatability. A typical value used for lab repeatability for the naphtha 95% cut point is 3.6°F. If one insists on achieving a degree of accuracy greater than lab repeatability, the network memorizes the relationship during the training process; this is known as overfitting. When overfitting occurs, each data point during the training is fit perfectly but the network is not able to predict with the same accuracy during the verification phase. A two-layer network with 12 neurons in the hidden layer was trained with an error goal of 0.05 (Table 7.5) to yield a maximum error of 1.7 $^\circ$F in the training phase. The maximum error in the verification phase was 5.79°F. The network could not generalize. It memorized the relation between inputs and outputs in the training phase and did not follow the general trend of the relation between inputs and targets.

Table 7.5: Simulation Results for Naphtha 95% Cut Point.

Hidden Neurons	Training Phase				Verification Phase	
	Error goal	Iterations	Final SSE	Max. Error $^\circ$F	Final SSE	Max. Error $^\circ$F
5	0.1	374	0.1	3.21	0.091	4.17
	0.3	196	0.3	6.61	0.432	13.76
8	0.1	238	0.1	2.87	0.097	3.29
	0.05	681	0.05	1.8	0.111	3.99
	0.01	5686	0.01	1.37	0.161	4.7
12	0.1	385	0.1	3.18	0.117	4.87
	0.05	1237	0.05	1.7	0.145	5.79
5 - 2	0.1	269	0.1	3.03	0.098	5
	0.2	211	0.2	3.81	0.127	5.45

Table 7.5 shows a summary of the simulation results. The best model architecture (in terms of better prediction in both training and verification modes) consists of eight neurons in one hidden layer. Both hidden and output layers use sigmoidal activation functions as the nonlinear element for their neurons. The model is trained to achieve a sum squared error goal of 0.1. The sum squared error goal in the verification mode is 0.097.

In this application it is important that the neural network output is equal to or less than the acceptable error. As mentioned earlier, the acceptable error value is based on lab repeatability. For the naphtha 95% cut point, this value is 3.6°F. Further data analysis is performed to look at the absolute error in each data set in both the training and verification modes. The maximum absolute error in the training data sets is 2.87°F, whereas in the verification mode the maximum absolute error is 3.29°F.

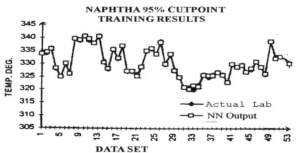

Figure 7.6: L_{inf} Error Norm in the Training Phase.
(Hidden Layer Neurons: 5, Error Goal=0.01, Max. Error=7.84°F)

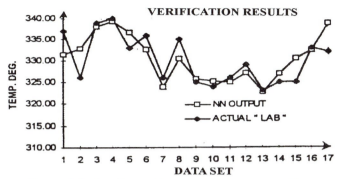

Figure 7.7: L*inf* Error Norm in the Verification Phase.
(Hidden Layer Neurons = 5, Error Goal = 0. 01, Max. Error = 7.84°F)

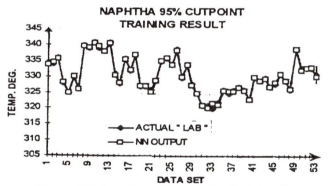

Figure 7.8: L*inf* Error Norm in the Training Phase.
(Hidden Layers Neurons = 8,4; Error Goal = 0.01, Max. Error =1.33°F)

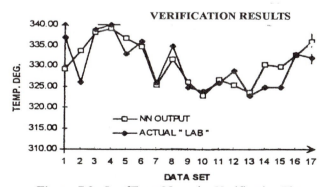

Figure 7.9: L*inf* Error Norm in Verification Phase.
(Hidden Layers Neurons = 8,4; Error Goal = 0.01, Max. Error = 4.81°F)

7.6.2   Naphtha Reid Vapor Pressure

To demonstrate the modeling capability of a back propagation algorithm for RVP prediction parameters, 83 data sets were analyzed. Each data set consisted of seven process variables (Table 7.6) as inputs to the model and one product quality (Reid vapor pressure) as an output. A total of fifty five data sets were used in the training phase and 28 data sets are used in the verification phase.

Table 7.6: Inputs to the RVP Neural Network Model.

Process Variables
FEED FLOW RATE
REFLUX FLOW RATE
OVERHEAD PRESSURE
FEED TEMPERATURE
REBOILER RETURN TEMP.
TRAY 2 TEMPERATURE
COLUMN TOP TEMPERATURE

The best model architecture (in terms of better prediction in both training and verification modes) consisted of five neurons in one hidden layer. Both hidden and output layers use sigmoidal activation functions as the nonlinear element for their neurons. The model is trained to achieve a sum squared error goal of 0.1. The sum squared error goal in the verification mode is 0.097. The maximum absolute error in the training data sets is 0.21 psi, whereas in the verification mode the maximum absolute error is 0.48 psi. The detail work can be found in reference [3].

7.7   CONCLUSIONS

In this chapter, various neural network architectures are proposed for the prediction of product quality of an oil refinery. The important parameters involved in acquiring valid data sets are considered. Close attention is paid to the proper selection of the input data. Finally, two product quality properties, namely, naphtha 95% cut point and naphtha Reid vapor pressure, were successfully modeled using neural network.

After the generation of the neural network models, the central processing computer system of an oil refinery may use them on-line. Using the NN model on-line is straightforward except for one point of caution. The network was trained within a specific range for the different process variables and the lab data. It is important to realize that while neural network models are excellent interpolators, they can be bad extrapolators due to the non-linearity of the correlation generated. It is, therefore, important to check process parameters used in the prediction and to make sure that these parameters fall within the range that was used to create the model. If parameters fall out of range, then the predicted lab value is questionable. Lab data collected while process parameters are outside the range can be used to further expand the window of operation of the neural network model. As the variability in plant operation increases, and the network window expands, the generation models can become more and more reliable.

REFERENCES

1. Antsaklis, P. J. and Passino, K.M. (eds.), *An Introduction to Intelligent and Autonomous Control*, Kluwer Academic Publishers, Norwell, MA, 1993.
2. *Ras Tanura Refinary Facilities Manual*, Ras Tanura Refinary: Refining Division, Section 3, 2nd ed., 1995.
3. Bawazeer, K. H., Prediction of Crude Oil Product Quality Parameters Using Neural Networks, MS Thesis, Florida Atlantic University, Boca Raton, FL, August, 1996.
4. Bawazeer, K.H. and Zilouchian, A., Prediction of Crude Oil Production Quality Parameters Using Neural Networks, *Proc. of 1997 IEEE Int. Conf. on Neural Networks.*, New Orleans, 1997.
5. Borman, S., Neural Network Applications in Chemistry Begin to Appear, *Chemical Eng. News*, Vol. 67, No. 17, 24-29, 1989.
6. Parlos, A. G., Chong, K.T., and Atiya, A.F., Application of Recurrent Neural Multilayer Perceptron in Modeling Complex Process Dynamic, *IEEE Trans. on Neural Networks*, Vol. 5, No. 2, March, 1994.
7. Nekovie, R., and Sun,Y., Back Propagation Network and its Configuration for Blood Vessel Detection in Angiograms, *IEEE Trans. on Neural Networks*, Vol. 6, No. 1, 1995.
8. Berkan, R.C., Upadhyaya, B., Tsoukalas, L., Kisner, R., and Bywater, R. Advanced Automation Concepts for Large-Scaled Systems, *IEEE Control Syst. Mag.*, Vol. 11, No. 6, 4–13, Oct., 1991.
9. Draeger, A., Engell, S., and Ranke, H., Model Predictive Control Using Neural Networks, *IEEE Control Mag.*. Vol. 15, No.5, 61-67, 1995.
10. Ray, W., Polymerization Reactor Control, *IEEE Control Syst. Mag.*, Vol. 6, No. 4, 3–9, August, 1986.
11. Bhat, N., Minderman, P., McAvoy, T., and Wang, N., Modeling Chemical Process Systems via Neural Network Computation, *IEEE Control Syst. Mag.*, Vol. 10, No.3, 24–31, April, 1990.
12. Rosenblatt, A., *Principles of Neurodynamics*, Spartan Press, Washington, DC, 1961.
13. Fausett, L., *Fundamentals of Neural Networks*, Prentice-Hall, Englewood Cliffs, NJ, 1994.
14. Demuth, H. and Beale, M., *Neural Network Toolbox for Use with MATLAB* , the Math Works Inc. , Natick, MA, 1998.
15. Miller, W.T., Suton, R., and Werbos, P. (eds.), *Neural Networks for Control*, MIT Press, MA, 1990.

16. Lippmann, R.P., An Introduction to Computing with Neural Network; *IEEE Acoustic, Speech, and Signal Proc. Mag,* 4–22, April, 1987.

17. Kosko, B., *Neural Networks and Fuzzy Systems,* Prentice-Hall, Englewood Cliffs, NJ, 1991.

8 INTRODUCTION TO FUZZY SETS: BASIC DEFINITIONS AND RELATIONS

Mo Jamshidi and Aly El-Osery

8.1   INTRODUCTION

One of the more popular new technologies is "intelligent control," which is defined as a combination of control theory, operations research, and artificial intelligence (AI). Judging by the billions of dollars worth of sales and close to 2000 patents issued in Japan alone since the announcement of the first fuzzy chips in 1987, fuzzy logic still is perhaps the most popular area in AI. Thanks to tremendous technological and commercial advances in fuzzy logic in Japan and other nations, today fuzzy logic continues to enjoy an unprecedented popularity in the technological and engineering fields including manufacturing. Fuzzy logic technology is being used in numerous consumer and electronic products and systems, even in the stock market and medical diagnostics. The most important issue facing many industrialized nations in the next several decades will be global competition to an extent that has never before been posed. The arms race is diminishing and the economic race is in full swing. Fuzzy logic is but one such front for global technological, economical, and manufacturing competition.

In order to understand fuzzy logic it is important to discuss fuzzy sets. In 1965, Zadeh [1] wrote a seminal paper in which he introduced fuzzy sets, i.e., sets with unsharp boundaries. These sets are generally in better agreement with the human mind that works with shades of gray, rather than with just black or white. Fuzzy sets are typically able to represent linguistic terms, e.g., warm, hot, high, low. Nearly ten years later Mamdani [2] succeeded in applying fuzzy logic for control in practice. Today, in Japan, U.S.A, Europe, Asia and many other parts of the world fuzzy control is widely accepted and applied. In many consumer products like washing machines and cameras, fuzzy controllers are used in order to obtain intelligent machines (Intelligent Machine Quotient-- MIQ®) and user friendly products. A few interesting applications can be mentioned: control of subway systems, image stabilization of video cameras, image enhancement and autonomous control of helicopters. Although the U.S and Europe hesitated in accepting fuzzy logic, they have become more enthusiastic about applying this technology.

Fuzzy set theory is developed comparing the precepts and operations of fuzzy sets with those of classical set theory. Fuzzy sets will be seen to contain the vast majority of the definitions, precepts, and axioms that define classical sets. In fact, very few differences exist between the two set theories. Fuzzy set theory is actually a fundamentally broader theory than current classical set theory, in that it considers an infinite number of "degrees of membership" in a set other than the canonical values of 0 and 1 apparent in classical set theory. In this sense, one could argue that classical sets are a limited form of fuzzy sets. Hence, it will be shown that fuzzy set theory is a comprehensive set theory.

Conceptually, a fuzzy set can be defined as a collection of elements in a universe of information where the boundary of the set contained in the universe is ambiguous, vague, and otherwise fuzzy. It is instructive to introduce fuzzy sets by first reviewing the elements of classical (crisp) set theory.

This chapter is organized as follows. Section 8.2 briefly describes classical sets, followed by introduction to classical set operations in section 8.3. Properties of classical sets are given in section 8.4. Section 8.5 is a quick introduction to fuzzy sets. Fuzzy set operations and properties are given in sections 8.6 and 8.7, respectively. Section 8.8 presents fuzzy vs. classical relations. Finally, a conclusion is given in section 8.9.

8.2   CLASSICAL SETS

In classical set theory, a set is denoted as a so-called *crisp set* and can be described by its characteristic function as follows:

$$\mu_C : U \to \{0,1\} \tag{8.1}$$

In Equation 8.1, U is called the universe of discourse, i.e., a collection of elements that can be continuous or discrete. In a crisp set each element of the universe of discourse either belongs to the crisp set ($\mu_C = 1$) or does not belong to the crisp set ($\mu_C = 0$).

Consider a characteristic function μ_{Chot} representing the crisp set hot, a set with all "hot" temperatures. Figure 8.1 graphically describes this crisp set, considering temperatures higher than 40°C as hot. (Note that for all temperatures T, we have $T \in U$).

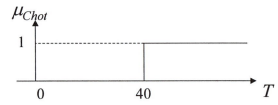

Figure 8.1: The Characteristic Function μ_{Chot}.

8.3   CLASSICAL SET OPERATIONS

Let A and B be two sets in the universe U, and $\mu_A(x)$ and $\mu_B(x)$ be the characteristic functions of A and B in the universe of discourse in sets A and B, respectively. The characteristic function $\mu_A(x)$ is defined as follows:

$$\mu_A(x) = \begin{cases} 1, & x \in A \\ 0, & x \notin A \end{cases} \tag{8.2}$$

and $\mu_B(x)$ is defined as

$$\mu_B(x) = \begin{cases} 1, & x \in B \\ 0, & x \notin B \end{cases} \qquad (8.3)$$

Using the above definitions, the following operations are defined [3].

Union The union between two sets, i.e., $C = A \cup B$, where $\cup$ is the union operator, represents all those elements in the universe which reside in either the set A or set B or both [4], (see Figure 8.2). The characteristic function μ_C is defined in Equation 8.4.

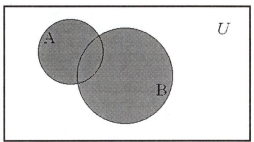

Figure 8.2: Union.

$$\forall x \in U : \mu_C = \max[\mu_A(x), \mu_B(x)] \qquad (8.4)$$

The operator in Equation 8.4 is referred to as the *max-operator*.

Intersection The intersection of two sets, i.e., $C = A \cap B$, where $\cap$ is the intersection operator, represents all those elements in the universe U which reside in both sets A and B simultaneously (see Figure 8.3). Equation 8.5 shows how to obtain the characteristic function μ_C.

$$\forall x \in U : \mu_C = \min[\mu_A(x), \mu_B(x)] \qquad (8.5)$$

The operator in Equation 8.5 is referred to as the *min-operator*.

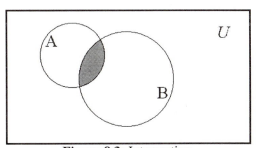

Figure 8.3: Intersection.

Complement The complement of a set A, denoted $\overline{A}$, is defined as the collection of all elements in the universe which do not reside in the set A (see Figure 8.4). The characteristic function $\mu_{\overline{A}}$ is defined by Equation 8.6.

$$\forall x \in U : \mu_{\overline{A}} = 1 - \mu_A(x) \qquad\qquad (8.6)$$

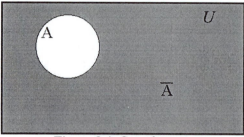

Figure 8.4: Complement.

8.4   PROPERTIES OF CLASSICAL SET

Properties of classical sets are very important to consider because of their influence on the mathematical manipulation. Some of these properties are listed below [5].

Commutativity:

$$A \cup B = B \cup A \qquad\qquad (8.7)$$
$$A \cap B = B \cap A \qquad\qquad (8.8)$$

Associativity:

$$A \cup (B \cup C) = (A \cup B) \cup C \qquad\qquad (8.9)$$
$$A \cap (B \cap C) = (A \cap B) \cap C \qquad\qquad (8.10)$$

Distributivity:

$$A \cup (B \cap C) = (A \cup B) \cap (A \cup C) \qquad\qquad (8.11)$$
$$A \cap (B \cup C) = (A \cap B) \cup (A \cap C) \qquad\qquad (8.12)$$

Idempotency:

$$A \cup A = A \qquad\qquad (8.13)$$
$$A \cap A = A \qquad\qquad (8.14)$$

Identity:

$$A \cup \phi = A \qquad\qquad (8.15)$$
$$A \cap X = A \qquad\qquad (8.16)$$
$$A \cap \phi = \phi \qquad\qquad (8.17)$$
$$A \cup X = X \qquad\qquad (8.18)$$

Excluded middle laws are very important since they are the only set operations that are not valid for both classical and fuzzy sets. Excluded middle laws consist of two laws. The first, known as *Law of Excluded Middle*, deals with the union of a set *A* and its complement. The second law, known as *Law of Contradiction*, represents the intersection of a set *A* and its complement. The following equations describe these laws:

Law of Excluded Middle

$$A \cup \overline{A} = X \tag{8.19}$$

Law of Contradiction

$$A \cap \overline{A} = \phi \tag{8.20}$$

8.5   FUZZY SETS

The definition of a fuzzy set [1] is given by the characteristic function

$$\mu_F : U \to [0,1] \tag{8.21}$$

In this case the elements of the universe of discourse can belong to the fuzzy set with any value between 0 and 1. This value is called the *degree of membership*. If an element has a value close to 1, the degree of membership, or truth value is high. The characteristic function of a fuzzy set is called the *membership function*, for it gives the degree of membership for each element of the universe of discourse. If now the characteristic function μ_{Fhot} is considered, one can express the human opinion, for example, that 37°C is still fairly hot, and that 38°C is hot, but not as hot as 40°C and higher. This result in a gradual transition from membership (completely true) to non-membership (not true at all). Figure 8.5 shows the membership function μ_{Fhot} for the fuzzy set F_{hot}.

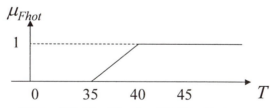

Figure 8.5: The Membership Function μ_{Fhot}

8.5.1   Fuzzy Membership Functions

The membership functions for fuzzy sets can have many different shapes, depending on definition. Figure 8.6 provides a description of the various features of membership functions. Some of the possible membership functions are shown in Figure 8.7.

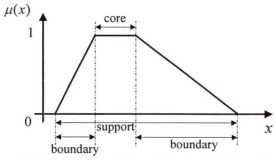

Figure 8.6: Description of Fuzzy Membership Functions [4].

Figure 8.7 illustrates some of the possible membership functions, we have: (a) the Γ-function: an increasing membership function with straight lines; (b) the L-function: a decreasing function with straight lines; (c) Λ-function: a triangular function with straight lines; (d) the singleton: a membership function with a membership function value 1 for only one value and the rest is zero. There are many other possible functions such as trapezoidal, Gaussian, sigmoidal or even arbitrary.

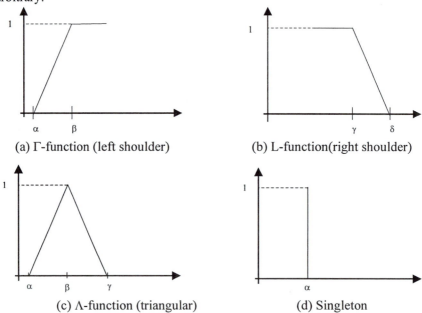

(a) Γ-function (left shoulder)   (b) L-function(right shoulder)

(c) Λ-function (triangular)   (d) Singleton

Figure 8.7: Examples of Membership Functions.

A notation convention for fuzzy sets that is popular in the literature when the universe of discourse U, is discrete and finite, is given below for a fuzzy set A by

$$A = \frac{\mu_A(x_1)}{x_1} + \frac{\mu_A(x_2)}{x_2} + \ldots = \sum_i \frac{\mu_A(x_i)}{x_i} \qquad (8.22)$$

and, when the universe of discourse U is continuous and infinite, the fuzzy set A is denoted by

$$A = \int \frac{\mu_A(x)}{x} \qquad (8.23)$$

8.6 FUZZY SET OPERATIONS

As in the traditional crisp sets, logical operations, e.g., union, intersection, and complement, can be applied to fuzzy sets [1].

Union The union operation (and the intersection operation as well) can be defined in many different ways. Here, the definition that is used in most cases is discussed. The union of two fuzzy sets A and B with the membership functions $\mu_A(x)$ and $\mu_B(x)$ is a fuzzy set C, written as $C = A \cup B$, whose membership function is related to those of A and B as follows:

$$\qquad (8.24)$$
$$\forall x \in U : \mu_C = \max\left[\mu_A(x), \mu_B(x)\right]$$

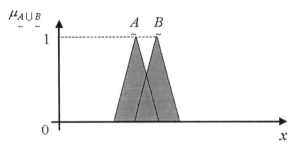

Figure 8.8: Union of Two Fuzzy Sets.

Intersection According to the *min-operator* the intersection of two fuzzy sets A and B with the membership functions $\mu_A(x)$ and $\mu_B(x)$, respectively, is a fuzzy set C, written as $C = A \cap B$, whose membership function is related to those of A and B as follows:

$$\qquad (8.25)$$
$$\forall x \in U : \mu_C = \min\left[\mu_A(x), \mu_B(x)\right]$$

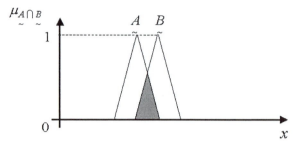

Figure 8.9: Intersection of Two Fuzzy Sets.

Complement The complement of a set A, denoted $\overline{A}$, is defined as the collection of all elements in the universe which do not reside in the set A.

$$\forall x \in U : \mu_{\overline{A}} = 1 - \mu_A(x) \tag{8.26}$$

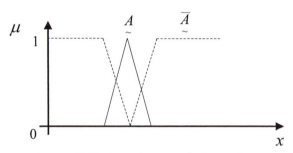

Figure 8.10: Complement of a Fuzzy Set.

Keep in mind that even though the equations of the union, intersection, and complement appear to be the same for classical and fuzzy sets, they differ in the fact that $\mu_A(x)$ and $\mu_B(x)$ can take only a value of zero or one in the case of classical set, while in fuzzy sets they include the whole interval from zero to one.

8.7   PROPERTIES OF FUZZY SETS

Similar to classical sets, fuzzy sets also have some properties that are important for mathematical manipulations [5,6]. Some of these properties are listed below.

Commutativity:

$$A \cup B = B \cup A \tag{8.27}$$

$$A \cap B = B \cap A \tag{8.28}$$

Associativity:

$$A \underset{\sim}{\cup} (B \underset{\sim}{\cup} C) = (A \underset{\sim}{\cup} B) \underset{\sim}{\cup} C \tag{8.29}$$

$$A \underset{\sim}{\cap} (B \underset{\sim}{\cap} C) = (A \underset{\sim}{\cap} B) \underset{\sim}{\cap} C \tag{8.30}$$

Distributivity:

$$A \underset{\sim}{\cup} (B \underset{\sim}{\cap} C) = (A \underset{\sim}{\cup} B) \underset{\sim}{\cap} (A \underset{\sim}{\cup} C) \tag{8.31}$$

$$A \underset{\sim}{\cap} (B \underset{\sim}{\cup} C) = (A \underset{\sim}{\cap} B) \underset{\sim}{\cup} (A \underset{\sim}{\cap} C) \tag{8.32}$$

Idempotency:

$$A \underset{\sim}{\cup} A = A \tag{8.33}$$

$$A \underset{\sim}{\cap} A = A \tag{8.34}$$

Identity:

$$A \underset{\sim}{\cup} \phi = A \tag{8.35}$$

$$A \underset{\sim}{\cap} X = A \tag{8.36}$$

$$A \underset{\sim}{\cap} \phi = \phi \tag{8.37}$$

$$A \underset{\sim}{\cup} X = X \tag{8.38}$$

Most of the properties that hold for classical sets (e.g., commutativity, associativity, and idempotence) hold also for fuzzy sets except for following two properties [5]:

1. *Law of contradiction* ($A \underset{\sim}{\cap} \overline{A} \neq \phi$): One can easily notice that the intersection of a fuzzy set and its complement results in a fuzzy set with membership values of up to ½ and thus does not equal the empty set (as in the case of classical sets) as shown in Figure 8.11.

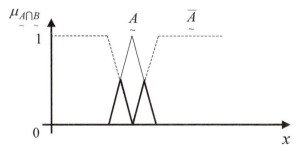

Figure 8.11: Law of Contradiction.

2. *Law of excluded middle* ($A \cup \overline{A} \neq U$): The union of a fuzzy set and its complement does not give the universe of discourse (see Figure 8.12).

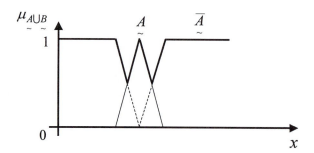

Figure 12: Law of Excluded Middle.

8.7.1   Alpha-Cut Fuzzy Sets

It is the crisp domain in which we perform all computations with today's computers. The conversion from fuzzy to crisp sets can be done by two means, one of which is

alpha-cut sets.

Given a fuzzy set A , the alpha-cut (or lambda cut) set of A is defined by

$$A_\alpha = \left\{ x \middle| \mu_A(x) \geq \alpha \right\}$$

(8.39)

Note that by virtue of the condition on $\mu_A(x)$ in Equation 8.39, i.e., a common property, the set A_α in Equation 8.39 is now a crisp set. In fact, any fuzzy set can be converted to an infinite number of cut sets.

8.7.2   Extension Principle

In fuzzy sets, just as in crisp sets, one needs to find means to extend the domain of a function, i.e., given a fuzzy set A and a function $f(\cdot)$, then what is the value of function $f(A)$? This notion is called the *extension principle* which was first proposed by Zadeh.

Let the function f be defined by

$$f : U \rightarrow V$$

(8.40)

where U and V are domain and range sets, respectively. Define a fuzzy set $A \subset U$ as,

$$A = \left\{ \frac{\mu_1}{u_1} + \frac{\mu_2}{u_2} + \ldots + \frac{\mu_n}{u_n} \right\} \tag{8.41}$$

Then the extension principle asserts that the function f is a fuzzy set, as well, which is defined below:

$$B = f(A) = \left\{ \frac{\mu_1}{f(u_1)} + \frac{\mu_2}{f(u_2)} + \ldots + \frac{\mu_n}{f(u_n)} \right\} \tag{8.42}$$

The complexity of the extension principle would increase when more than one member of $u_1 \times u_2$ is mapped to only one member of v; one would take the maximum membership grades of these members in the fuzzy set A.

Example 8.1

Given two universes of discourse $U_1 = U_2 = \{1,2,\ldots,10\}$ and two fuzzy sets (numbers) defined by

$$\text{"Approximately 2"} = \frac{0.5}{1} + \frac{1}{2} + \frac{0.8}{3}$$

and

$$\text{"Approximately 5"} = \frac{0.6}{3} + \frac{0.8}{4} + \frac{1}{5}$$

It is desired to find "approximately 10"

SOLUTION:

The function $f = u_1 \times u_2 : \rightarrow v$ represents the arithmetic product of these two fuzzy numbers and is given by

$$\text{"approximately 10"} = \left(\frac{0.5}{1} + \frac{1}{2} + \frac{0.8}{3} \right) \times \left(\frac{0.6}{3} + \frac{0.8}{4} + \frac{1}{5} \right) = \frac{\min(0.5,0.6)}{3} +$$

$$\frac{\min(0.5,0.8)}{4} + \frac{\min(0.5,1)}{5} + \frac{\min(1,0.6)}{6} + \frac{\min(1,0.8)}{8} +$$

$$\frac{\min(1,1)}{10} + \frac{\min(0.8,0.6)}{9} + \frac{\min(0.8,0.8)}{12} + \frac{\min(0.8,1)}{15} =$$

$$= \frac{0.5}{3} + \frac{0.5}{4} + \frac{0.5}{5} + \frac{0.6}{6} + \frac{0.8}{8} + \frac{0.6}{9} + \frac{1}{10} + \frac{0.8}{12} + \frac{0.8}{15}$$

The above resulting fuzzy number has its *prototype*, i.e., value 10 with a membership function 1 and the other 8 pairs are spread around the point (1, 10).

Example 8.2

Consider two fuzzy sets (numbers) defined by

$$\text{``Approximately 2''} = \frac{0.5}{1} + \frac{1}{2} + \frac{0.5}{3}$$

and

$$\text{``Approximately 4''} = \frac{0.8}{2} + \frac{0.9}{3} + \frac{1}{4}$$

It is desired to find "approximately 8"

SOLUTION:

The function $f = u_1 \times u_2 :\rightarrow v$ represents the arithmetic product of these two fuzzy numbers and is given by

$$\text{``approximately 8''} = \left(\frac{0.5}{1} + \frac{1}{2} + \frac{0.5}{3}\right) \times \left(\frac{0.8}{2} + \frac{0.9}{3} + \frac{1}{4}\right) = \frac{\min(0.5,0.8)}{2} +$$

$$\frac{\min(0.5,0.9)}{3} + \frac{\max[\min(0.5,1),\min(1,0.8)]}{4} +$$

$$\frac{\max[\min(1,0.9),\min(0.5,0.8)]}{6} + \frac{\min(1,1)}{8} + \frac{\min(0.5,0.9)}{9} +$$

$$\frac{\min(0.5,1)}{12} = \frac{0.5}{2} + \frac{0.5}{3} + \frac{0.8}{4} + \frac{0.9}{6} + \frac{1}{8} + \frac{0.5}{9} + \frac{0.5}{12}$$

8.8   CLASSICAL RELATIONS vs. FUZZY RELATIONS

Classical relations are structures that represent the presence or absence of correlation or interaction among elements of various sets. There are only two degrees of relationship between elements of the sets in a crisp relation, namely, the relationships "completely related" or "not related". Fuzzy relations, on the other hand, are developed by allowing the relationship between elements of two or more sets to take an infinite number of degrees of relationship between the extremes of "completely related" and "not related" [6,7].

The classical relation of two universes U and V is defined as

$$U \times V = \{(u,v) | u \in U, v \in V\} \tag{8.43}$$

which combines $\forall u \in U$ and $\forall v \in V$ in an ordered pair and forms unconstrained matches between u and v. That is, every element in universe U is related completely to every element in universe V. The *strength* of this relationship between ordered pairs of elements in each universe is measured by the characteristic function, where a value of unity is associated with *complete relationship* and a value of zero is associated with *no relationship*, i.e., the binary values 1 and 0.

As an example, if $U=\{1,2\}$ and $V=\{a,b,c\}$, then $U\times V=\{(1,a), (1,b), (1,c), (2,a), (2,b), (2,c)\}$. The above product is said to be *crisp relation*, which can be expressed by either a matrix expression

$$R = U \times V = \begin{matrix} & \begin{matrix} a & b & c \end{matrix} \\ \begin{matrix} 1 \\ 2 \end{matrix} & \begin{bmatrix} 1 & 1 & 1 \\ 1 & 1 & 1 \end{bmatrix} \end{matrix} \tag{8.44}$$

Or in a so-called *Sagittal* diagram (see Figure 8.13)

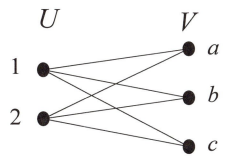

Figure 8.13: Sagittal Diagram.

Fuzzy relations map elements of one universe to those of another universe, through Cartesian product of the two universes. Unlike crisp relations, the *strength* of the relation between ordered pairs of the two universes is not measured with the characteristic function, but rather with a membership function expressing various *degrees* of the strength of the relation on the unit interval [0,1]. In other words, a fuzzy relation R is a mapping:

$$\underset{\sim}{R}:U \times V \rightarrow [0,1] \tag{8.45}$$

The following example illustrates this relationship, i.e.,

$$\mu_{\underset{\sim}{R}}(u,v) = \mu_{\underset{\sim}{A}\times\underset{\sim}{B}}(u,v) = \min(\mu_{\underset{\sim}{A}}(u), \mu_{\underset{\sim}{B}}(v)) \tag{8.46}$$

Example 8.3

Consider two fuzzy sets $A_1 = \dfrac{0.2}{x_1} + \dfrac{0.9}{x_2}$ and $A_2 = \dfrac{0.3}{y_1} + \dfrac{0.5}{y_2} + \dfrac{1}{y_3}$.

Determine the fuzzy relation between these sets.

SOLUTION:

The fuzzy relation R is

$$R = A_1 \times A_2 = \begin{bmatrix} 0.2 \\ 0.9 \end{bmatrix} \times \begin{bmatrix} 0.3 & 0.5 & 1 \end{bmatrix} = \begin{bmatrix} \min(0.2,0.3) & \min(0.2,0.5) & \min(0.2,1) \\ \min(0.9,0.3) & \min(0.9,0.5) & \min(0.9,1) \end{bmatrix} =$$

$$= \begin{bmatrix} 0.2 & 0.2 & 0.2 \\ 0.3 & 0.5 & 0.9 \end{bmatrix}$$

Let R be a relation that relates elements from universe U to universe V, and let S be a relation that relates elements from universe V to universe W. Is it possible to find the relation T that relates the same elements in universe U that R contains to elements in universe W that S contains? The answer is yes, using an operation known as *composition*.

In crisp or fuzzy relations, the composition of two relations, using the max-min rule, is given below. Given two fuzzy relations $R(u,v)$ and $S(v,w)$, then the composition of these is

$$T = R \circ S = \max_{v \in V} \left\{ \min(\mu_R(u,v), \mu_S(v,w) \right\}$$

(8.47)

or using the max-product rule, the characteristic function is given by

$$\mu_T(u,w) = \max_{v \in V} \left\{ \mu_R(u,v) \cdot \mu_S(v,w) \right\}$$

(8.48)

The same composition rules hold for crisp relations.

Example 8.4

Consider two fuzzy relations

$$R = \begin{bmatrix} 0.6 & 0.8 \\ 0.7 & 0.9 \end{bmatrix} \text{ and } S = \begin{bmatrix} 0.3 & 0.1 \\ 0.2 & 0.8 \end{bmatrix}$$

It is desired to evaluate $R \circ S$ and $S \circ R$

SOLUTION:

Using the max-min composition for $R \circ S$ we have

$$R \circ S = \begin{bmatrix} 0.3 & 0.8 \\ 0.3 & 0.8 \end{bmatrix}$$

where, for example, the element (1,1) is obtained by $\max\{\min(0.6,0.3), \min(0.8,0.2)\}=0.3$.

For $S \circ R$ we get the following result

$$S \circ R = \begin{bmatrix} 0.3 & 0.3 \\ 0.7 & 0.8 \end{bmatrix} \neq R \circ S$$

Using the max-product rule, we have

$$R \circ S = \begin{bmatrix} 0.18 & 0.64 \\ 0.21 & 0.72 \end{bmatrix}$$

where, for example, the element (2,2) is obtained by $\max\{(0.7)(0.1), (0.9)(0.8)\}=0.72$.

For $S \circ R$ we get the following result

$$S \circ R = \begin{bmatrix} 0.18 & 0.24 \\ 0.56 & 0.72 \end{bmatrix} \neq R \circ S$$

8.9   CONCLUSION

In this chapter a quick overview of classical and fuzzy sets was given. Main similarities and differences between classical and fuzzy sets were introduced. In general, set operations are the same for classical and fuzzy sets. The exceptions were excluded middle laws. Alpha-cut sets and extension principle were presented followed by a brief introduction to classical vs. fuzzy relations. This chapter presented issues that are important in understanding fuzzy sets and their advantages over classical sets. A set of problems at the end of the book will further enhance the reader's understanding of these concepts.

REFERENCES

1. Zadeh, L. A, Fuzzy sets, *Information and Control,* Vol. 8, 338–353, 1965.
2. Mamdani, E. H., Applications of fuzzy algorithms for simple dynamic plant, *Proc. IEE, 121,* No. 12, 1585–1588, 1974.

3. Jamshidi, M., Titli, A., Zadeh, L.A. and Bverie, S. (eds.), *Applications of Fuzzy Logic - Toward High Machine Intelligence Quotient Systems*, Vol. 9, Prentice Hall series on Environmental and Intelligent Manufacturing Systems (M. Jamshidi, ed.), Prentice Hall, Upper Saddle River, NJ, 1997.
4. Ross, T. J., *Fuzzy Logic with Engineering Application,* McGraw-Hill, New York, 1995.
5. Jamshidi, M., Vadiee, N. and Ross, T. J. (eds.), *Fuzzy Logic and Control: Software and Hardware Applications*. Vol 2. Prentice Hall Series on Environmental and Intelligent Manufacturing Systems, (M. Jamshidi, ed.). Prentice Hall, Englewood Cliffs, NJ, 1993.
6. Dubois, D. and Prade, H., *Fuzzy Sets and Systems, Theory and Applications*, Academic, New York, 1980.
7. Zimmermann, H., *Fuzzy Set Theory and Its Applications*, 2nd ed., Kluwer Academic Publishers, Dordrecht, Germany, 1991.

9 INTRODUCTION TO FUZZY LOGIC

Mo Jamshidi, Aly El-Osery, and Timothy J. Ross

9.1   INTRODUCTION

The need and use of multilevel logic can be traced from the ancient works of Aristotle, who is quoted as saying, "There will be a sea battle tomorrow." Such a statement is not yet true or false, but is potentially either. Much later, around AD 1285-1340, William of Occam supported two-valued logic but speculated on what the truth value of "if *p* then *q*" might be if one of the two components, *p* or *q*, as neither true nor false. During the time period of 1878-1956, Lukasiewicz proposed a three-level logic as a "true" (l), a "false" (0), and a "neuter" (1/2), which represented half true or half false. In subsequent times, logicians in China and other parts of the world continued on the notion of multi-level logic. Zadeh, in his seminal 1965 paper [1], finished the task by following through with the speculation of previous logicians and showing that what he called "fuzzy sets" were the foundation of any logic, regardless of the number of truth levels assumed. He chose the innocent word "fuzz" for the continuum of logical values between 0 (completely false) and 1 (completely true). The theory of fuzzy logic deals with two problems 1) the fuzzy set theory, which deals with the vagueness found in semantics, and 2) the fuzzy measure theory, which deals with the ambiguous nature of judgments and evaluations.

The primary motivation and "banner" of fuzzy logic is the possibility of exploiting tolerance for some inexactness and imprecision. Precision is often very costly, so if a problem does not require precision, one should not have to pay for it. The traditional example of parking a car is a noteworthy illustration. If the driver is not required to park the car within an exact distance from the curb, why spend any more time than necessary on the task as long as it is a legal parking operation? Fuzzy logic and classical logic differ in the sense that the former can handle both symbolic and numerical manipulation, while the latter can handle symbolic manipulation only. In a broad sense, fuzzy logic is a union of fuzzy (fuzzified) crisp logics [2]. To quote Zadeh, "Fuzzy logic's primary aim is to provide a formal, computationally-oriented system of concepts and techniques for dealing with modes of reasoning which are approximate rather than exact." Thus, in fuzzy logic, exact (crisp) reasoning is considered to be the limiting case of approximate reasoning. In fuzzy logic one can see that everything is a matter of degrees.

This chapter is organized as follows. In section 9.2, a brief introduction to predicate logic is given. In section 9.3, fuzzy logic is presented, followed by approximate reasoning in section 9.4.

9.2   PREDICATE LOGIC

Let a predicate logic proposition P be a linguistic statement contained within a universe of propositions that are either completely true or false. The truth value of the proposition P can be assigned a binary truth value, called $T(P)$, just as an element in a universe is assigned a binary quantity to measure its membership in a particular set. For binary (Boolean) predicate logic, $T(P)$ is assigned a value of 1 (truth) or 0 (false). If U is the universe of all propositions, then T is a mapping of these propositions to the binary quantities (0,1), or

$$T : U \to \{0,1\} \tag{9.1}$$

Now let P and Q be two simple propositions on the same universe of discourse that can be combined using the following five logical connectives

(i)   disjunction ($\vee$)
(ii) conjunction ($\wedge$)
(iii) negation ($-$)
(iv) implication ($\to$)
(v)   equality ($\leftrightarrow$ or $\equiv$)

to form logical expressions involving two simple propositions. These connectives can be used to form new propositions from simple propositions.

Now define sets A and B from universe X where these sets might represent linguistic ideas or thoughts. Then a propositional calculus will exist for the case where proposition P measures the truth of the statement that an element, x, from the universe X is contained in set A and the truth of the statement that this element, x, is contained in set B, or more conventionally

P: truth that $x \in A$
Q: truth that $x \in B$, where truth is measured in terms of the truth value, i.e.,
If $x \in A$, $T(P)= 1$; otherwise $T(P)= 0$.
If $x \in B$, $T(Q) = 1$; otherwise $T(Q) = 0$, or using the characteristic function to represent truth (1) and false (0):

$$\chi_A(x) = \begin{cases} 1, & x \in A \\ 0, & x \notin A \end{cases} \tag{9.2}$$

The above five logical connectives can be used to create compound propositions, where a compound proposition is defined as a logical proposition formed by logically connecting two or more simple propositions. Just as one is interested in the truth of a simple proposition, predicate logic also involves the assessment of the truth of compound propositions. Given a proposition $P : x \in A, \overline{P} : x \notin A$, the resulting compound propositions are defined below in terms of their binary truth values:

Disjunction:
$$P \vee Q \Rightarrow x \in A \text{ or } B$$
$$\text{Hence}, T(P \vee Q) = \max(T(P), T(Q))$$
(9.3)

Conjunction:
$$P \wedge Q \Rightarrow x \in A \text{ and } B$$
$$\text{Hence}, T(P \wedge Q) = \min(T(P), T(Q))$$
(9.4)

Negation:
$$\text{If } T(P) = 1, \text{ then } T(\overline{P}) = 0; \text{ If } T(P) = 0, \text{ then } T(\overline{P}) = 1$$
(9.5)

Equivalence:
$$P \leftrightarrow Q \Rightarrow x \in A, B$$
$$\text{Hence}, T(P \leftrightarrow Q) \Rightarrow T(P) = T(Q)$$
(9.6)

Implication:
$$P \rightarrow Q \Rightarrow x \notin A \text{ or } x \in B$$
$$\text{Hence}, T(P \rightarrow Q) = T(\overline{P} \cup Q)$$
(9.7)

The logical connective implication presented here is also known as the classical implication, to distinguish it from an alternative form due to Lukasiewicz, a Polish mathematician in the 1930s, who was first credited with exploring logic other than Aristotelian (classical or binary) logic. This classical form of the implication operation requires some explanation.

For a proposition P defined on set A and a proposition Q defined on set B, the implication "P implies Q" is equivalent to taking the union of elements in the complement of set A with the elements in the set B. That is, the logical implication is analogous to the set-theoretic form.

$$P \rightarrow Q \equiv \overline{A} \cup B \text{ is true} \equiv \text{ either "not in } A\text{" or "in } B\text{"}$$
(9.8)

So that $(P \rightarrow Q) \leftrightarrow (\overline{P} \vee Q)$

$$T(P \rightarrow Q) = T(\overline{P} \vee Q) = \max(T(\overline{P}), T(Q))$$
(9.9)

This is linguistically equivalent to the statement, "P implies Q is true" when either "*not A*" or "*B*" is true [6]. Graphically, this implication and the analogous set operation are represented by the Venn diagram in Figure 9.1. As noted, the region represented by the difference $A \setminus B$ is the set region where the implication "P implies Q" is false (the implication *fails*). The shaded region in Figure 9.1 represents the collection of elements in the universe where the implication is true, i.e., the shaded area is the set:

$$\overline{A \setminus B} = \overline{A} \cup B = \overline{(A \cap \overline{B})}$$

(9.10)

If x is in A and x is not in B then

$$A \to B \equiv \textit{fails A\textbackslash B (difference)}$$

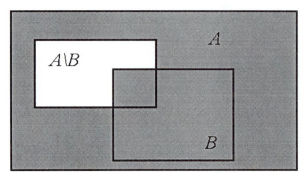

Figure 9.1: Classical Implication Operation (Shaded Area is Where Implication Holds) [2].

Now, with two propositions (*P* and *Q*) each being able to take on one of two truth values (*true* or *false*, 1 or 0), there will be a total of $2^2 = 4$ propositional situations. These situations are illustrated in Table 9.1, along with the appropriate truth values for the propositions *P* and *Q* and the various logical connectives between them in the truth table.

To help understand this concept, assume you have two propositions *P* and *Q*. *P*: you are a graduate student and *Q*: you are a university student. Let us examine the implication "*P* implies *Q*". If you are a student in general, and a graduate student in particular, then the implication is true. On the other hand, the implication would be false if you are a graduate student without being a student. Now, let us assume that you are an undergraduate student; regardless whether you are graduate or not, then the implication is true (since in the case you are not a graduate student does not negate the fact that you are an undergraduate). Then, we come to the final case: you are neither a graduate nor undergraduate student. In this case the implication is true, because the fact that you are not a graduate or undergraduate student does not negate the implication that for you to be a graduate student you have to be a student at the university.

Table 9.1

P	Q	$\overline{P}$	$P \vee Q$	$P \wedge Q$	$P \to Q$	$P \leftrightarrow Q$
True	True	False	True	True	True	True
True	False	False	True	False	False	False
False	True	True	True	False	True	False
False	False	True	False	False	True	True

Suppose the implication operation involves two different universes of discourse, *P* is a proposition described by set *A*, which is defined on universe *X*, and Q is a proposition described by set B, which is defined on universe Y. Then the implication "P implies Q" can be represented in set theory terms by the relation R, where R is defined by

$$R = (A \times B) \cup (\overline{A} \times Y) \equiv \text{IF } A, \text{THEN } B \qquad (9.11)$$

If $x \in A$ (where $x \in X$, $A \subset X$)

Then $y \in B$ (where $y \in Y$, $B \subset Y$)

where $A \times B$ and $A \times Y$ are Cartesian products [3].

This implication is also equivalent to the linguistic rule form: IF A, THEN B. The graphic shown in Figure 9.2 represents the Cartesian space of the product $X \times Y$, showing typical sets A and B, and superimposed on this space is the set theory equivalent of the implication. That is,

$$P \to Q \Rightarrow \text{IF } x \in A, \text{then } y \in B, \text{ or } P \to Q \equiv \overline{A} \cup B \qquad (9.12)$$

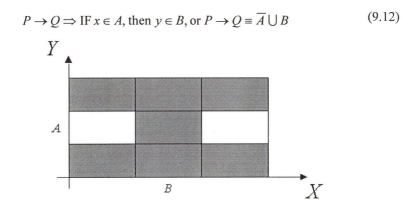

Figure 9.2: Cartesian Space Demonstrating IF A THEN B [3].

The shaded regions of the compound Venn diagram in Figure 9.2 represent the truth domain of the implication, IF A, THEN B (P implies Q).

9.2.1 Tautologies

In predicate logic it is useful to consider compound propositions that are always true, irrespective of the truth values of the individual simple propositions. Classical logic compound propositions with this property are called *tautologies*. Tautologies are useful for deductive reasoning and for making deductive inferences. So, if a compound proposition can be expressed in the form of a tautology, the truth-value of that compound proposition is known to be true. Inference schemes in expert systems often employ tautologies. The reason for this is that tautologies are logical formulas that are true on logical grounds alone [3].

One of these, known as *Modus Ponens* deduction, is a very common inference scheme used in forward chaining rule-based expert systems. It is an operation whose task is to find the truth-value of a consequent in a production rule, given the truth-value of the antecedent in the rule. Modus Ponens deduction concludes that, given two propositions, *a* and *a*-implies-*b*, both of which are true, then the truth of the simple proposition *b* is automatically inferred. Another useful tautology is the *Modus Tollens* inference, which is used in backward-chaining expert systems. In Modus Tollens an implication between two propositions is combined with a second proposition and both are used to imply a third proposition. Some common tautologies are listed below.

$$\overline{B} \cup B \leftrightarrow X \tag{9.13}$$

$$A \cup X \leftrightarrow X \tag{9.14}$$

$$\overline{A} \cup X \leftrightarrow X \tag{9.15}$$

$$(A \wedge (A \to B)) \to B \qquad (\textit{Modus Ponens}) \tag{9.16}$$

$$(\overline{B} \wedge (A \to B)) \to \overline{A} \qquad (\textit{Modus Tollens}) \tag{9.17}$$

9.2.2   Contradictions

Compound propositions that are always false, regardless of the truth-value of the individual simple propositions comprising the compound proposition, are called contradictions. Some simple contradictions are listed below.

$$\overline{B} \cap B \leftrightarrow \phi \tag{9.18}$$

$$A \cap \phi \leftrightarrow \phi \tag{9.19}$$

$$\overline{A} \cap \phi \leftrightarrow \phi \tag{9.20}$$

9.2.3   Deductive Inferences

The Modus Ponens deduction is used as a tool for inferencing in rule-based systems. A typical IF–THEN rule is used to determine whether an antecedent (cause or action) infers a consequent (effect or action). Suppose we have a rule of the form,

$$\text{IF } A, \text{ THEN } B \tag{9.21}$$

This rule could be translated into a relation using the Cartesian product sets *A* and *B*, that is

$$R = (A \times B) \cup (\overline{A} \times Y) \tag{9.22}$$

Now suppose a new antecedent, say *A'*, is known. Can we use Modus Ponens deduction to infer a new consequent, say *B'*, resulting from the new antecedent? That is, in rule form

$$\text{IF } A', \text{THEN } B'? \tag{9.23}$$

The answer, of course, is yes, through the use of the composition relation. Since "A implies B" is defined on the Cartesian space $X \times Y$, B' can be found through the following set-theoretic formulation,

$$B' = A' \circ R = A' \circ ((A \times B) \cup (\overline{A} \times Y)) \tag{9.24}$$

Modus Ponens deduction can also be used for the compound rule,

$$\text{IF } A, \text{THEN } B, \text{ELSE } C \tag{9.25}$$

Using the relation defined as,

$$R = (A \times B) \cup (\overline{A} \times C) \tag{9.26}$$

and hence $B' = A' \circ R$.

Example 9.1

Let two universes of discourse be described by $X=\{1,2,3,4,5,6\}$ and $Y=\{1,2,3,4\}$ and define the crisp set $A=\{2,3\}$ on X and $B=\{3,4\}$ on Y. Determine the deductive inference IF A, THEN B.

SOLUTION

Expressing the crisp sets in Zadeh's notation,

$$A = \frac{0}{1} + \frac{1}{2} + \frac{1}{3} + \frac{0}{4}$$
$$B = \frac{0}{1} + \frac{0}{2} + \frac{1}{3} + \frac{1}{4} + \frac{0}{5} + \frac{0}{6}$$

Taking the Cartesian product $A \times B$ which involves taking the pairwise min of each pair from the sets A and B [3]

$$
A \times B = \begin{array}{c}
 \\ 1 \\ 2 \\ 3 \\ 4
\end{array}
\begin{array}{cccccc}
1 & 2 & 3 & 4 & 5 & 6 \\
\left[\begin{array}{cccccc}
0 & 0 & 0 & 0 & 0 & 0 \\
0 & 0 & 1 & 1 & 0 & 0 \\
0 & 0 & 1 & 1 & 0 & 0 \\
0 & 0 & 0 & 0 & 0 & 0
\end{array}\right]
\end{array}
$$

Then computing $\overline{A} \times Y$

$$\overline{A} = \frac{1}{1} + \frac{0}{2} + \frac{0}{3} + \frac{1}{4}$$

$$Y = \frac{1}{1} + \frac{1}{2} + \frac{1}{3} + \frac{1}{4} + \frac{1}{5} + \frac{1}{6}$$

$$\overline{A} \times Y = \begin{matrix} & \begin{matrix} 1 & 2 & 3 & 4 & 5 & 6 \end{matrix} \\ \begin{matrix} 1 \\ 2 \\ 3 \\ 4 \end{matrix} & \begin{bmatrix} 1 & 1 & 1 & 1 & 1 & 1 \\ 0 & 0 & 0 & 0 & 0 & 0 \\ 0 & 0 & 0 & 0 & 0 & 0 \\ 1 & 1 & 1 & 1 & 1 & 1 \end{bmatrix} \end{matrix}$$

again using pairwise min for the Cartesian product.

The deductive inference yields the following characteristic function in matrix form, following the relation,

$$R = (A \times B) \cup (\overline{A} \times Y) = \begin{matrix} & \begin{matrix} 1 & 2 & 3 & 4 & 5 & 6 \end{matrix} \\ \begin{matrix} 1 \\ 2 \\ 3 \\ 4 \end{matrix} & \begin{bmatrix} 1 & 1 & 1 & 1 & 1 & 1 \\ 0 & 0 & 1 & 1 & 0 & 0 \\ 0 & 0 & 1 & 1 & 0 & 0 \\ 1 & 1 & 1 & 1 & 1 & 1 \end{bmatrix} \end{matrix}$$

9.3   FUZZY LOGIC

The extension of the above discussions to fuzzy deductive inference is straightforward. The fuzzy proposition P has a value on the closed interval $[0,1]$. The truth-value of a proposition P is given by

$$T(P) = \mu_A(x) \quad \text{where } 0 \le \mu_A \le 1 \tag{9.27}$$

Thus, the degree of truth for $P : x \in A$ is the membership grade of x in A. The logical connectives of negation, disjunction, conjunction, and implication are similarly defined for fuzzy logic, e.g., disjunction.

Negation:

$$T(\overline{P}) = 1 - T(P) \tag{9.28}$$

Disjunction:

$$P \vee Q \Rightarrow x \in A \text{ or } B$$

$$\text{Hence, } T(P \vee Q) = \max(T(P), T(Q)) \tag{9.29}$$

Conjunction:
$$P \wedge Q \Rightarrow x \in A \text{ and } B$$
$$\text{Hence, } T(P \wedge Q) = \min(T(P), T(Q)) \tag{9.30}$$

Implication:
$$P \to Q \Rightarrow x \text{ is } A, \text{ then } x \text{ is } B$$
$$T(P \to Q) = T(\overline{P} \vee Q) = \max(T(\overline{P}), T(Q)) \tag{9.31}$$

Thus, a fuzzy logic implication would result in a fuzzy rule

$$P \to Q \Rightarrow \text{If } x \text{ is } A, \text{ then } y \text{ is } B \tag{9.32}$$

and the equivalent to the following fuzzy relation

$$R = (A \times B) \cup (\overline{A} \times Y) \tag{9.33}$$

with a grade membership function,

$$\mu_R = \max\left\{(\mu_A(x) \wedge \mu_B(y)), (1 - \mu_A(x))\right\} \tag{9.34}$$

Example 9.2
Consider two universes of discourse described by $X=\{1,2,3,4\}$ and $Y=\{1,2,3,4,5,6\}$. Let two fuzzy sets A and B be given by

$$A = \frac{0.8}{2} + \frac{1}{3} + \frac{0.3}{4}$$

$$B = \frac{0.4}{2} + \frac{1}{3} + \frac{0.6}{4} + \frac{0.2}{5}$$

It is desired to find a fuzzy relation R corresponding to IF A', THEN B'.

SOLUTION
Using the relation in Equation 9.33 would give

$$A \times B = \begin{matrix} & \begin{matrix} 1 & 2 & 3 & 4 & 5 & 6 \end{matrix} \\ \begin{matrix} 1 \\ 2 \\ 3 \\ 4 \end{matrix} & \begin{bmatrix} 0 & 0 & 0 & 0 & 0 & 0 \\ 0 & 0.4 & 0.8 & 0.6 & 0.2 & 0 \\ 0 & 0.4 & 1 & 0.6 & 0.2 & 0 \\ 0 & 0.3 & 0.3 & 0.3 & 0.2 & 0 \end{bmatrix} \end{matrix}$$

$$\overline{A} \times Y = \begin{array}{c} \\ 1 \\ 2 \\ 3 \\ 4 \end{array} \begin{array}{cccccc} 1 & 2 & 3 & 4 & 5 & 6 \\ \begin{bmatrix} 1 & 1 & 1 & 1 & 1 & 1 \\ 0.2 & 0.2 & 0.2 & 0.2 & 0.2 & 0.2 \\ 0 & 0 & 0 & 0 & 0 & 0 \\ 0.7 & 0.7 & 0.7 & 0.7 & 0.7 & 0.7 \end{bmatrix} \end{array}$$

and hence $R = \max\{A \times B, \overline{A} \times Y\}$

$$R = \begin{array}{c} \\ 1 \\ 2 \\ 3 \\ 4 \end{array} \begin{array}{cccccc} 1 & 2 & 3 & 4 & 5 & 6 \\ \begin{bmatrix} 1 & 1 & 1 & 1 & 1 & 1 \\ 0.2 & 0.4 & 0.8 & 0.6 & 0.2 & 0.2 \\ 0 & 0.4 & 1 & 0.6 & 0.2 & 0 \\ 0.7 & 0.7 & 0.7 & 0.7 & 0.7 & 0.7 \end{bmatrix} \end{array}$$

9.4   APPROXIMATE REASONING

The primary goal of fuzzy systems is to formulate a theoretical foundation for reasoning about imprecise propositions, which is termed *approximate reasoning* in fuzzy logic technological systems [4,5].

Let us have a rule-based format to represent fuzzy information. These rules are expressed in conventional antecedent-consequent form, such as

Rule 1: IF x is A, THEN y is B (9.35)

where A and B represent fuzzy propositions (sets).

Now let us introduce a new antecedent, say A', and we consider the following rule:

Rule 2: IF x is A', THEN y is B' (9.36)

From the information derived from Rule 1, is it possible to derive the consequent Rule 2, B'? The answer is yes, and the procedure is a fuzzy composition. The consequent B' can be found from the composition operation

$$B' = A' \circ R   (9.37)$$

Example 9.3

Reconsider the fuzzy system of Example 9.2. Let a new fuzzy set A' be given by $A' = \dfrac{0.5}{1} + \dfrac{1}{2} + \dfrac{0.2}{3}$. It is desired to find an approximate reason (consequent) for the rule IF A', THEN B'.

SOLUTION

The relations 9.33 and 9.37 are used to determine B'.

$$B' = A' \circ R = [0.5 \quad 0.5 \quad 0.8 \quad 0.6 \quad 0.5 \quad 0.5]$$

or

$$B' = \frac{0.5}{1} + \frac{0.5}{2} + \frac{0.8}{3} + \frac{0.6}{4} + \frac{0.5}{5} + \frac{0.5}{6}$$

where the composition is of the max-min form.

Note the inverse relation between fuzzy antecedents and fuzzy consequences arising from the composition operation. More exactly, if we have a fuzzy relation $R : A \to B$, then will the value of the composition $A \circ R = B$? The answer is no, and one should not expect an inverse to exist for fuzzy composition. This is not, however, the case in crisp logic, i.e., $B' = A' \circ R = A \circ R = B$, where all these latter sets and relations are crisp [5,6]. The following example illustrates the nonexistence of the inverse.

Example 9.4

Let us reconsider the fuzzy system of Example 9.2 and 9.3. Let $A' = A$ and evaluate B'.

SOLUTION

we have

$$B' = A' \circ R = A \circ R = \frac{0.3}{1} + \frac{0.4}{2} + \frac{0.8}{3} + \frac{0.6}{4} + \frac{0.3}{5} + \frac{0.3}{6} \neq B$$

which yields a new consequent, since the inverse is not guaranteed. The reason for this situation is the fact that fuzzy inference is imprecise, but approximate. The inference, in this situation, represents approximate linguistic characteristics of the relation between two universes of discourse.

9.5 CONCLUSION

This chapter introduced, very briefly, classical and fuzzy logic. For more in depth details, readers are encouraged to read Ross [3]. Most of the tools needed to form an idea about fuzzy logic and its operation have been introduced. These tools are essential in understanding the next chapter addressing fuzzy control and stability.

REFERENCES

1. Zadeh, L. A. , Fuzzy Sets, *Information and Control*, Vol. 8, 338–353, 1965.
2. Jamshidi, M., Vadiee, N., and Ross, T. J. (eds.), *Fuzzy Logic and Control: Software and Hardware Applications*, Vol. 2. Prentice Hall

Series on Environmental and Intelligent Manufacturing Systems, (M. Jamshidi, (ed.). Prentice Hall, Englewood Cliffs, NJ, 1993.

3. Ross, T. J. *Fuzzy Logic with Engineering Application,* McGraw-Hill, New York, 1995.

4. Zadeh, L. A., A Theory of Approximate Reasoning, in J. Hayes, D. Michie, and L. Mikulich (eds.), *Machine Intelligence*, Halstead Press, New York, 149–194, 1979.

5. Gaines, B., Foundation of Fuzzy Reasoning, *Int. J. Man Mach. Stud.*, vol. 8, 623–688, 1976.

6. Yager, R. R., On the Implication Operator in Fuzzy Logic, *Inf. Sci.*, Vol. 31, 141–164, 1983.

10 FUZZY CONTROL AND STABILITY

Mo Jamshidi and Aly El-Osery

10.1 INTRODUCTION

The aim of this chapter is to define fuzzy control systems and cover relevant results and development. Traditionally, an *intelligent control* system is defined as one in which classical control theory is combined with artificial intelligence (AI) and possibly OR (Operations Research). Stemming from this definition, two approaches to intelligent control have been in use. One approach combines expert systems in AI with differential equations to create the so called *expert control*, while the other integrates *discrete event systems* (Markov chains) and differential equations [1]. The first approach, although practically useful, is rather difficult to analyze because of the different natures of differential equations (based on mathematical relations) and AI expert systems (based on symbolic manipulations). The second approach, on the other hand, has well developed and solid theory, but is too complex for many practical applications. It is clear, therefore, that a new approach and a change of course are called for here. We begin with another definition of an intelligent control system. An intelligent control system is one in which a physical system or a mathematical model of it is being controlled by a combination of a knowledge-base, approximate (humanlike) reasoning, and/or a learning process structured in a hierarchical fashion. Under this simple definition, any control system which involves fuzzy logic, neural networks, expert learning schemes, genetic algorithms, genetic programming or any combination of these would be designated as intelligent control.

Among the many applications of fuzzy sets and fuzzy logic, fuzzy control is perhaps the most common. Most industrial fuzzy logic applications in Japan, the U.S., and Europe fall under fuzzy control. The reasons for the success of fuzzy control are both theoretical and practical [1].

From a theoretical point of view, a fuzzy logic rule-base, can be used to identify both a model, as a "universal approximation," as well as a nonlinear controller. The most relevant information about any system comes in one of three ways—a mathematical model, sensory input/output data, and human expert knowledge. The common factor in all these three sources is knowledge. For many years, classical control designers began their effort with a mathematical model and did not go any further in acquiring more knowledge about the system, i.e., designers put their entire trust in a mathematical model whose accuracy may sometimes be in question. Today, control engineers can use all of the above sources of information. Aside from a mathematical model whose utilization is clear, numerical (input/output) data can be used to develop an approximate model (input/output nonlinear mapping) as well as a controller, based on the acquired fuzzy IF-THEN rules.

Some researchers and teachers of fuzzy control systems subscribe to the notion that fuzzy controls should always use a model free design approach and, hence, give the impression that a mathematical model is irrelevant. As indicated before, the authors, however, believe strongly that if a mathematical model does exist, it would be the first source of knowledge used in building the entire knowledge base. From a mathematical model, through simulation, for example, one can further build the knowledge base. Through utilization of the expert operator's knowledge which comes in the form of a set of linguistic or semi-linguistic IF-THEN rules, the fuzzy controller designer would get a big advantage in using every bit of information about the system during the design process.

On the other hand, it is quite possible that a system, such as high dimensional large-scale systems, is so complex that a reliable mathematical tool either does not exist or is very costly to attain. This is where fuzzy control (or intelligent control) comes in. Fuzzy control approaches these problems through a set of local humanistic (expert-like) controllers governed by linguistic fuzzy IF-THEN rules. In short, fuzzy control falls into the category of intelligent controllers, which are not solely model-based, but also, knowledge-based.

From a practical point of view, fuzzy controllers, which have appeared in industry and in manufactured consumer products, are easy to understand, simple to implement, and inexpensive to develop. Because fuzzy controllers emulate human control strategies, they are easily understood even by those who have no formal background in control. These controllers are also very simple to implement.

This chapter is organized as follows. Section 10.2 is a basic definition of fuzzy control systems and their components. Section 10.3 introduces different methods to fuzzy control design and provides an example. Section 10.4 is an analysis of fuzzy control systems. Section 10.5 addresses the stability of fuzzy control systems, and the conclusion is given in Section 10.6.

10.2   BASIC DEFINITIONS

A common definition of a fuzzy control system is that it is a system which emulates a human expert. In this situation, the knowledge of the human operator would be put in the form of a set of fuzzy linguistic rules. These rules would produce an approximate decision, just as a human would. Consider Figure 10.1, where a block diagram of this definition is shown. As shown, the human operator observes quantities by observing the inputs, i.e., reading a meter or measuring a chart, and performs a definite action (e.g., pushes a knob, turns on a switch, closes a gate, or replaces a fuse) thus leading to a crisp action, shown here by the output variable $y(t)$. The human operator can be replaced by a combination of a fuzzy rule-based system (FRBS) and a block called *defuzzifier*. The input sensory (crisp or numerical) data are fed into FRBS where physical quantities are represented or compressed into linguistic variables with appropriate membership functions. These linguistic variables are then used in the *antecedents* (IF-Part) of a set of fuzzy rules within an inference engine to result in a new set of fuzzy linguistic variables or *consequent* (THEN-Part). Variables are then denoted in this figure by z, and are combined and changed to a crisp (numerical) output $y^*(t)$ which represents an approximation to actual output $y(t)$.

It is therefore noted that a fuzzy controller consists of three operations: (1) fuzzification, (2) inference engine, and (3) defuzzification.

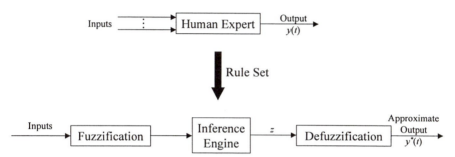

Figure 10.1: Conceptual Definition of a Fuzzy Control System.

Before a formal description of fuzzification and defuzzification processes is made, let us consider a typical structure of a fuzzy control system which is presented in Figure 10.2. As shown, the sensory data go through two levels of interface, i.e., the analog to digital and the crisp to fuzzy and at the other end in reverse order, i.e. fuzzy to crisp and digital to analog.

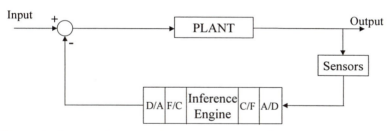

Figure 10.2: Block Diagram for a Laboratory Implementation of a Fuzzy Controller.

Another structure for a fuzzy control system is a fuzzy inference, connected to a knowledge base, in a supervisory or adaptive mode. The structure is shown in Figure 10.3. As shown, a classical crisp controller (often an existing one) is left unchanged, but through a fuzzy inference engine or a fuzzy adaptation algorithm the crisp controller is altered to cope with the system's unmodeled dynamics, disturbances, or plant parameter changes much like a standard adaptive control system. Here the function $h(\cdot)$ represents the unknown nonlinear controller or mapping function $h\!:\!e \rightarrow u$ which along with any two input components e_1 and e_2 of e represents a nonlinear surface, sometimes known as the *control surface* [2].

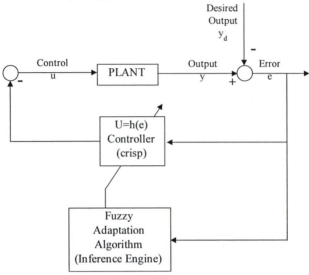

Figure 10.3: An Adaptive (Tuner) Fuzzy Control System, Fuzzification.

The fuzzification operation, or the *fuzzifier* unit, represents a mapping from a crisp point $x = (x_1 \ x_2 \ \dots \ x_n)^T \in X$ into a fuzzy set $A \in X$, where X is the universe of discourse and T denotes vector or matrix transposition[*]. There are normally two categories of fuzzifiers in use. The first is singleton and the second is nonsingleton. A singleton fuzzifier has one point (value) x_p as its fuzzy set support, i.e., the membership function is governed by the following relation:

$$\mu_A(x) = \begin{cases} 1, & x = x_p \in X \\ 0, & x \neq x_p \in X \end{cases} \tag{10.1}$$

The nonsingleton fuzzifiers are those in which the support is more than a point. Examples of these fuzzifiers are triangular, trapezoidal, Gaussian, etc. In these fuzzifiers, $\mu_A(x) = 1$ at $x=x_p$, where x_p may be one or more than one point, and then $\mu_A(x)$ decreases from 1 as x moves away from x_p or the "core" region to which x_p belongs such that $\mu_A(x_p)$ remains 1 (see Section 10.5). For example, the following relation represents a Gaussian-type fuzzifier:

$$\mu_A(x) = \exp\left\{ -\frac{(x - x_p)^T (x - x_p)}{\sigma^2} \right\} \tag{10.2}$$

---

[*] For convenience, in this chapter, the tilde (~) sign that was used earlier to express fuzzy sets will be omitted.

where the variance, σ^2, is a parameter characterizing the shape of $\mu_A(x)$.

10.2.1 Inference Engine

The cornerstone of any expert controller is its inference engine, which consists of a set of expert rules, which reflect the knowledge base and reasoning structure of the solution of any problem. A fuzzy (expert) control system is no exception and its rule base is the heart of the nonlinear fuzzy controller. A typical fuzzy rule can be composed as [3]

$$\text{IF } A \text{ is } A_1 \text{ AND } B \text{ is } B_1 \text{ OR } C \text{ is } C_1 \qquad (10.3)$$
$$\text{THEN } U \text{ is } U_1$$

where A, B, C and U are fuzzy variables, A_1, B_1, C_1 and U_1 are fuzzy linquistic values (membership functions or fuzzy linguistic labels), "AND", "OR", and "NOT" are connectives of the rule. The rule in Equation 10.3 has three antecedents and one consequent. Typical fuzzy variables may in fact, represent physical or system quantities such as: "temperature," "pressure," "output," "elevation," etc. and typical fuzzy linguistic values (labels) may be "hot", "very high," "low," etc. The portion "very" in a label "very high" is called a *linquistic hedge*. Other examples of a hedge are "much," "slightly," "more," or "less," etc. The above rule is known as Mamdani type rule. In Mamdani rules the antecedents and the consequent parts of the rule are expressed using linguistic labels. In general in fuzzy system theory, there are many forms and variations of fuzzy rules, some of which will be introduced here and throughout the chapter. Another form is *Takagi-Sugeno* rules in which the consequent part is expressed as an analytical expression or equation.

Two cases will be used here to illustrate the process of inferencing graphically. In the first case the inputs to the system are crisp values and we use max-min inference method. In the second case, the inputs to the system are also crisp, but we use the max-product inference method. Please keep in mind that there could also be cases where the inputs are fuzzy variables.

Consider the following rule whose consequent is not a fuzzy implication

$$\text{IF } x_1 \text{ is } A_1^i \text{ AND } x_2 \text{ is } A_2^i \text{ THEN } y^i \text{ is } B^i, \text{ for } i = 1,2,...,l \qquad (10.4)$$

where A_1^i and A_2^i are the fuzzy sets representing the ith-antecedent pairs, and B^i are the fuzzy sets representing the ith-consequent, and l is the number of rules.

Case 10.1: Inputs x_1 and x_2 are crisp values, and max-min inference method is used. Based on the Mamdani implication method of inference, and for a set of *disjunctive rules*, i.e, rules connected by the *OR* connective, the aggregated output for the l rules presented in Equation 10.4 will be given by

$$\mu_{B^i}(y) = \max_i[\min[\mu_{A_1^i}(x_1), \mu_{A_2^i}(x_2)]], \text{ for } i = 1,2,\dots,l \qquad (10.5)$$

Figure 10.4 is a graphical illustration of Equation 10.5, for $l=2$, where A_1^1 and A_2^1 refer to the first and second fuzzy antecedents of the first rule, respectively, and B^1 refers to the fuzzy consequent of the first rule. Similarly, A_1^2 and A_2^2 refer to the first and second fuzzy antecedents of the second rule, respectively, and B^2 refers to the fuzzy consequent of the second rule. Because the antecedent pairs used in general form presented in Equation 10.4 are connected by a logical *AND*, the minimum function is used. For each rule, minimum value of the antecedent propagates through and truncates the membership function for the consequent. This is done graphically for each rule. Assuming that the rules are disjunctive, the aggregation operation *max* results in an aggregated membership function comprised of the outer envelope of the individual truncated membership forms from each rule. To compute the final crisp value of the aggregated output, defuzzification is used, which will be explained in the next section.

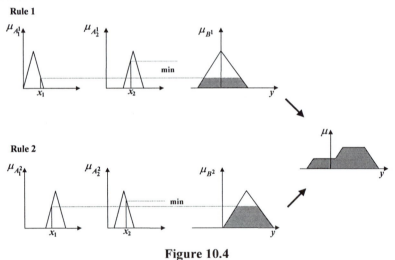

Figure 10.4

Case 10.2: Inputs x_1 and x_2 are crisp values, and max-product inference method is used. Based on the Mamdani implication method of inference, and for a set of *disjunctive rules*, the aggregated output for the l rules presented in Equation 10.4 will be given by

$$\mu_{B^i}(y) = \max_i[\mu_{A_1^i}(x_1) \cdot \mu_{A_2^i}(x_2)], \text{ for } i = 1,2,\dots,l \qquad (10.6)$$

Figure 10.5 is a graphical illustration of Equation 10.6, for *l*=2, where A_1^1 and A_2^1 refer to the first and second fuzzy antecedents of the first rule, respectively, and B^1 refers to the fuzzy consequent of the first rule. Similarly, A_1^2 and A_2^2 refer to the first and second fuzzy antecedents of the second rule, respectively, and B^2 refers to the fuzzy consequent of the second rule. Since the antecedent pairs used in general form presented in Equation 10.4 are connected by a logical *AND*, the minimum function is used again. For each rule, minimum value of the antecedent propagates through and scales the membership function for the consequent. This is done graphically for each rule. Similar to the first case, the aggregation operation *max* results in an aggregated membership function comprised of the outer envelope of the individual truncated membership forms from each rule. To compute the final crisp value of the aggregated output, defuzzification is used.

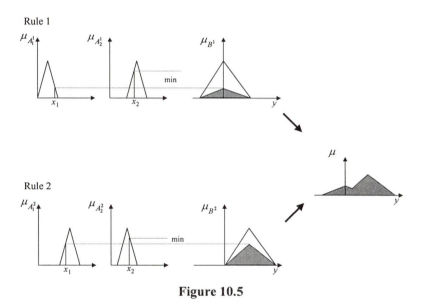

Figure 10.5

10.2.2 Defuzzification

Defuzzification is the third important element of any fuzzy controller. In this section, only the *center of gravity defuzzifier*, which is the most common one, is discussed. In this method the weighted average of the membership function or the center of gravity of the area bounded by the membership function curve is computed as the most typical crisp value of the union of all output fuzzy sets:

$$y_c = \frac{\int y \cdot \mu_A(y) dy}{\int \mu_A(y) dy} \qquad (10.7)$$

10.3   FUZZY CONTROL DESIGN

One of the first steps in the design of any fuzzy controller is to develop a knowledge base for the system to eventually lead to an initial set of rules. There are at least five different methods to generate a fuzzy rule base [4]:

1.   Simulate the closed-loop system through its mathematical model,
2.   Interview an operator who has had many years of experience controlling the system,
3.   Generate rules through an algorithm using numerical input/output data of the system,
4.   Use learning or optimization methods such as neural networks (NN) or genetic algorithms (GA) to create the rules, and
5.   In the absence of all of the above, if a system does exist, experiment with it in the laboratory or factory setting and gradually gain enough experience to create the initial set of rules.

Example 10.1

Consider the linearized model of the inverted pendulum Figure 10.6, described by the equation given below,

$$\dot{x} = \begin{pmatrix} 0 & 1 \\ 15.79 & 0 \end{pmatrix} x + \begin{pmatrix} 0 \\ 1.46 \end{pmatrix} u$$

with l=0.5m, m=100g, and initial conditions $x^T(0) = [\theta(0) \quad \dot{\theta}(0)]^T = [1 \quad 0]^T$. It is desired to stabilize the system using fuzzy rules.

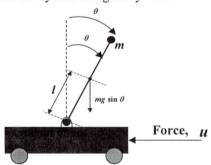

Figure 10.6: Inverted Pendulum.

Clearly this system is unstable and a controller is needed to stabilize it. To generate the rules for this problem only common sense is needed, i.e., if the pole is falling in one direction then push the cart in the same direction to counter the movement of the pole. To put this into rules of the form Equation 10.4 we get the following:

IF θ is θ_Positive AND $\dot{\theta}$ is $\dot{\theta}$_Positive THEN u is u_Negative

IF θ is θ_Negative AND $\dot{\theta}$ is $\dot{\theta}$_Negative THEN u is u_Positive

where the membership functions described above are defined in Figure 10.7.

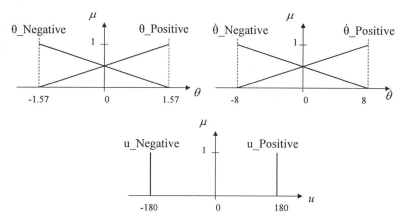

Figure 10.7: Membership Functions for the Inverted Pendulum Problem.

As shown in Figure 10.7, the membership functions for the inputs are half-triangular, while the membership function of the output is singleton. By simulating the system with fuzzy controller we get the response shown in Figure 10.8. It is clear that the system is stable. In this example only two rules were used, but more rules could be added in order to get a better response, i.e., less undershoot.

10.4   ANALYSIS OF FUZZY CONTROL SYSTEMS

In this section, some results of Tanaka and Sugeno [5] with respect to analysis of feedback fuzzy control systems will be briefly discussed. This section would use Takagi-Sugeno models to develop fuzzy block diagrams and fuzzy closed-loop models.

Consider a typical Takagi-Sugeno fuzzy plant model represented by implication P^i in Figure 10.9.

$$P^i : \text{IF } x(k) \text{ is } A_1^i \text{ AND} \ldots x(k-n+1) \text{ is } A_n^i \text{ AND}$$

$$u(k) \text{ is } B_1^i \text{ AND} \ldots \text{AND } u(k-m+1) \text{ is } B_n^i \qquad (10.8)$$

$$\text{THEN } x^i(k+1) = a_0^i + a_1^i x(k) + \ldots + a_n^i x(k-n+1) +$$

$$b_1^i u(k) + \ldots + b_n^i u(k-m+1)$$

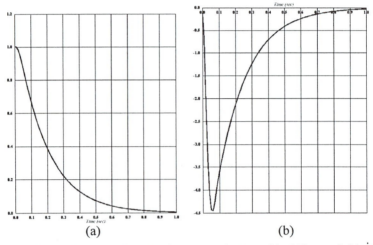

(a)   (b)

Figure 10.8: Simulation Results for Example 10.1: (a) $\theta(t)$, and (b) $\dot{\theta}(t)$.

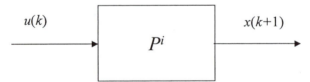

Figure 10.9:   Single-Input, Single-Output Fuzzy Block Represented by ith Implication P^l .

where P^i , $(i = 1,2,\ldots,l)$ is the ith implication, l , is the total number of implications, a_p^i , $(p = 1,2,\ldots,n)$ and b_q^i , $(q = 1,2,\ldots,m)$ are constant consequent parameters, k is time sample, $x(k),\ldots,x(k-n+1)$ are input variables, n and m are the number of antecedents for states and inputs, respectively. The terms A_p^i and B_p^i are fuzzy sets with piecewise-continuous polynomial (PCP) membership functions. PCP is defined as follows.

Definition 10.1

A fuzzy set A satisfying the following properties is said to be a *piecewise-continuous polynomial* (PCP) membership function $A(x)$ [4]:

(a) $A(x) = \begin{cases} \mu_1(x), & x \in [p_0, p_1] \\ \quad \vdots \\ \mu_s(x), & x \in [p_{s-1}, p_s] \end{cases}$

$\hspace{8cm}$ (10.9)

where $\mu_i(x) \in [0,1]$ for $x \in [p_{i-1}, p_i]$, $i = 1, 2, \ldots, s$, and
$-\infty < p_0 < p_1 < \ldots < p_{s-1} < p_s < \infty$.

(b) $\mu_i(x) = \sum_{j=0}^{n_i} c_j^i x^j$

$\hspace{8cm}$ (10.10)

where c_j^i are known parameters of polynomials $\mu_i(x)$.

Given the inputs

$$\mathbf{x}(k) \equiv [x(k) \quad x(k-1) \ldots x(k-n+1)]^T$$
$$\mathbf{u}(k) \equiv [u(k) \quad u(k-1) \ldots u(k-m+1)]^T$$

$\hspace{8cm}$ (10.11)

Using the above vector notation, Equation 10.11 can be represented in the following form,

P^i : IF $\mathbf{x}(k)$ is $\mathbf{A}^i$ AND $\mathbf{u}(k)$ is $\mathbf{B}^i$

$\hspace{7cm}$ (10.12)

$\hspace{1cm}$ THEN $x^i(k+1) = a_0^i + \sum_{p=1}^{n} a_p^i x(k-p+1) + \sum_{q=1}^{m} b_q^i u(k-q+1)$

where $\mathbf{A}^i \equiv [A_1^i \ A_2^i \ldots A_n^i]^T$, $\mathbf{B}^i \equiv [B_1^i \ B_2^i \ldots B_m^i]^T$, and "$\mathbf{x}(k)$ is $\mathbf{A}^i$" are equivalent to antecedent "$x(k)$ is A_1^i AND $\ldots x(k-n+1)$ is A_n^i".

The final defuzzified output of the inference is given by a weighted average of $x^i(k+1)$ values:

$$x(k+1) = \frac{\sum_{i=1}^{l} w^i x^i(k+1)}{\sum_{i=1}^{l} w^i}$$

$\hspace{8cm}$ (10.13)

where it is assumed that the denominator of Equation 10.13 is positive, and $x^i(k+1)$ is calculated from the ith implication, and the weight w^i refers to the overall truth value of the ith implication premise for the inputs in Equation 10.12.

Since the product of two PCP fuzzy sets can be considered as a series connection of two fuzzy blocks of the type in Figure 10.9, it is concluded that the convexity of fuzzy sets in succession is not preserved in general. Now let us consider a fuzzy control system whose plant model and controller are represented by fuzzy implications as depicted in Figure 10.10. In this figure, $r(k)$ represents a reference input. The plant implication P^i is already defined by Equation 10.12, while the controller's jth implication is given by

$$C^j : \text{IF } \mathbf{x}(k) \text{ is } \mathbf{D}^j \text{ AND } \mathbf{u}(k) \text{ is } \mathbf{F}^j \tag{10.14}$$

$$\text{THEN } f^j(k+1) = c_0^j + \sum_{p=1}^{n} c_p^j x(k-p+1)$$

where $\mathbf{D}^j \equiv \left[D_1^j \ D_2^j \dots D_n^j \right]^T$, $\mathbf{F}^i \equiv \left[F_1^i \ F_2^i \dots F_m^i \right]^T$, and of course $u(k)=r(k)-f(k)$. The equivalent implication S^{ij} is given by

$$S^{ij} : \text{IF } \mathbf{x}(k) \text{ is } (\mathbf{A}^i \text{ AND } \mathbf{D}^j) \text{ AND } \mathbf{v}^*(k) \text{ is } (\mathbf{B}^i \text{ AND } \mathbf{F}^j)$$

$$\text{THEN } x^{ij}(k+1) = a_0^j - b^i c_0^j + b^i r(k) + \tag{10.15}$$

$$\sum_{p=1}^{n} (a_p^j - b^i c_p^j) x(k-p+1)$$

where $i=1,\dots,l_1$, $j=1,\dots,l_2$, and l_1 and l_2 are the total number of implications for the plant and the controller, respectively. The term $v^*(k)$ is defined by

$$v^*(k) = \left[r(k) - e^*(x(k)), r(k-1) - e^*(x(k-1)), \tag{10.16} \right.$$
$$\left. \dots, r(k-m+1) - e^*(x(k-m+1)) \right]^T$$

where $e^*(\cdot)$ is the input-output mapping function of block C^j in Figure 10.10, i.e., $f(k)=e^*(x(k))$.

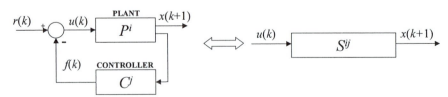

Figure 10.10: A Fuzzy Control System Depicted by Two Implications and its Equivalent Implication [4].

Example 10.2
Consider a fuzzy feedback control system of the type shown in Figure 10.10 with the following implications:

$$P^1 : \text{IF } x(k) \text{ is } A^1 \text{ THEN } x^1(k+1) = 1.85x(k) - 0.65x(k-1) + 0.35u(k)$$

$$P^2 : \text{IF } x(k) \text{ is } A^2 \text{ THEN } x^2(k+1) = 2.56x(k) - 0.135x(k-1) + 2.22u(k)$$

$$C^1 : \text{IF } x(k) \text{ is } D^1 \text{ THEN } f^1(k+1) = k_1^1 x(k) - k_2^1 x(k-1)$$

$$C^2 : \text{IF } x(k) \text{ is } D^2 \text{ THEN } f^2(k+1) = k_1^2 x(k) - k_2^2 x(k-1)$$

It is desired to find the closed-loop implications S^{ij}, $i=1,2$, and $j=1,2$.

SOLUTION
Noting that $u(k)=r(k)-f(k)$ in Figure 10.10 and the implications in Equation 10.15, we have

$$S^{11} : \text{IF } x(k) \text{ is } (A^1 \text{ AND } D^1) \text{ THEN } x^{11}(k+1) = (1.85 - 0.35k_1^1)x(k) +$$
$$(-0.65 - 0.35k_2^1)x(k-1) + 0.35r(k)$$

$$S^{12} : \text{IF } x(k) \text{ is } (A^1 \text{ AND } D^2) \text{ THEN } x^{12}(k+1) = (1.85 - 0.35k_1^2)x(k) +$$
$$(-0.65 - 0.35k_2^2)x(k-1) + 0.35r(k)$$

$$S^{21} : \text{IF } x(k) \text{ is } (A^2 \text{ AND } D^1) \text{ THEN } x^{21}(k+1) = (2.56 - 2.22k_1^1)x(k) +$$
$$(-0.135 - 2.22k_2^1)x(k-1) + 2.22r(k)$$

$$S^{22} : \text{IF } x(k) \text{ is } (A^2 \text{ AND } D^2) \text{ THEN } x^{22}(k+1) = (2.56 - 2.22k_1^2)x(k) +$$
$$(-0.135 - 2.22k_2^2)x(k-1) + 2.22r(k)$$

10.5 STABILITY OF FUZZY CONTROL SYSTEMS

One of the most important issues in any control system fuzzy or otherwise is stability. Briefly, a system is said to be *stable* if it would come to its equilibrium state after any external input, initial conditions, and/or disturbances have impressed the system. The issue of stability is of even greater relevance when questions of safety, lives, and environment are at stake as in such systems as nuclear reactors, traffic systems, and airplane autopilots. The stability test for fuzzy control systems, or lack of it, has been a subject of criticism by many control engineers in some control engineering literature [6].

Almost any linear or nonlinear system under the influence of a closed-loop crisp controller has one type of stability test or as other. For example, the stability of a linear time-invariant system can be tested by a wide variety of methods such as Routh-Hurwitz, root locus, Bode plots, Nyquist criterion, and even through traditionally nonlinear systems methods of Lyapunov, Popov, and circle criterion. The common requirement in all these tests is the availability of a mathematical model, either in time or frequency domain. A reliable mathematical model for a very complex and large-scale system may, in practice, be unavailable or unfeasible. In such cases, a fuzzy controller may be designed based on expert knowledge or experimental practice. However, the issue of the stability of a fuzzy control system still remains and must be addressed. The aim of this section is to present an up-to-date survey of available techniques and tests for fuzzy control systems stability.

From the viewpoint of stability a fuzzy controller can be either acting as a conventional (low-level) controller or as a supervisory (high-level) controller. Depending on the existence and nature of a system's mathematical model and the level in which fuzzy rules are being utilized for control and robustness, four classes of fuzzy control stability problems can be distinguished. These four classes are:

Class 1: Process model is crisp and linear and fuzzy controller is low level.
Class 2: Process model is crisp and nonlinear and the fuzzy controller is low level.
Class 3: Process model (linear or nonlinear) is crisp and a fuzzy tuner or an adaptive fuzzy controller is present at high level.
Class 4: Process model is fuzzy and fuzzy controller is low level.

Figures 10.11-10.14 show all four classes of fuzzy control systems whose stability is of concern. Here, we are concerned mainly with the first three classes. For the last class, traditional nonlinear control theory could fail and is beyond the scope of this section. It will be discussed very briefly. The techniques for testing the stability of the first two classes of systems (Figures 10.11 and 10.12) are divided into two main groups—time and frequency.

Time-Domain Methods
The state-space approach has been considered by many authors [7]-[15]. The basic approach here is to subdivide the state space into a finite number of cells based on the definitions of the membership functions. Now, if a separate rule is defined for every cell, a cell-to-cell trajectory can be constructed from the system's output induced by the new outputs of the fuzzy controller. If every cell of the modified state space is checked, one can identify all the equilibrium points, including the system's stable region. This method should be used with some care since the inaccuracies in the modified description could cause oscillatory phenomenon around the equilibrium points.

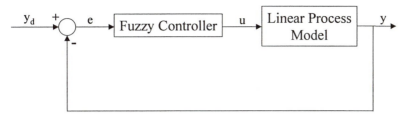

Figure 10.11: Class 1 of Fuzzy Control System Stability Problem.

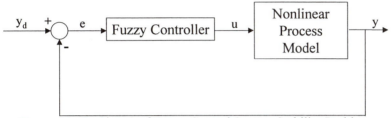

Figure 10.12: Class 2 of Fuzzy Control System Stability Problem.

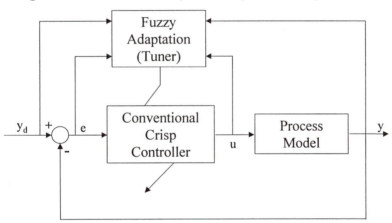

Figure 10.13: Class 3 of Fuzzy Control System Stability Problem.

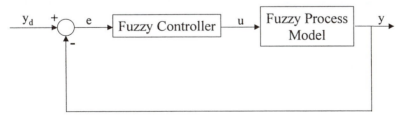

Figure 10.14: Class 4 of Fuzzy Control System Stability Problem.

The second class of methods is based on Lyapunov's method. Several authors, [5], [10], [11], [13] [16]-[23], have used this theory to come up with criterion for stability of fuzzy control systems. The approach shows that the time derivative of the Lyapunov function at the equilibrium point is negative semi definite. Many approaches have been proposed. One approach is to define a Lyapunov function and then derive the fuzzy controller's architecture out of the stability conditions. Another approach uses Aiserman's method [7] to find an adopted Lyapunov function, while representing the fuzzy controller by a nonlinear algebraic function $u=f(y)$, when y is the system's output. A third method calls for the use of so called *facet functions*, where the fuzzy controller is realized by boxwise multilinear facet functions with the system being described by a state space model. To test stability, a numerical parameter optimization scheme is needed.

The *hyperstability* approach, considered by other authors [24]-[26] has been used to check stability of systems depicted in Figure 10.11. The basic approach here is to restrict the input-output behavior of the nonlinear fuzzy controller by inequality and to derive conditions for the linear part of the closed-loop system to be satisfied for stability.

Bifurcation theory [13] can be used to check stability of fuzzy control systems of the class described in Figure 10.12. This approach represents a tool in deriving stability conditions and robustness indices for stability from small gain theory. The fuzzy controller, in this case, is described by a nonlinear vector function. The stability in this scheme could only be lost if one of the following conditions becomes true: (1) the origin becomes unstable if a pole crosses the imaginary axis into the right half-plane—static bifurcation, (2) the origin becomes unstable if a pair of poles would cross over the imaginary axis and assumes positive real parts—Hopf bifurcation—or (3) new additional equilibrium points are produced.

The last time-domain method is the use of graph theory [13]. In this approach conditions for special nonlinearities are derived to test the BIBO stability.

Frequency-Domain Methods

There are three primary groups of methods which have been considered here. The harmonic balance approach, considered in references [27]-[29], among others, has been used to check the stability of the first two classes of fuzzy control systems (see Figures 10.11 and 10.12). The main idea is to check if permanent oscillations occur in the system and whether these oscillations with known amplitude or frequency are stable. The nonlinearity (fuzzy controller) is described by a complex-valued describing function and the condition of harmonic balance is tested. If this condition is satisfied, then a permanent oscillation exists. This approach is equally applicable to MIMO systems.

The *circle criterion* [8],[26],[30],[31] and *Popov criterion* [32],[33] have been used to check stability of the first class of systems (Figure 10.11). In both criteria, certain conditions on the linear process model and static nonlinearity (controller) must be satisfied. It is assumed that the characteristic value of the nonlinearity remains within certain bounds, and the linear process model must be open-loop stable with proper transfer function. Both criteria can be graphically evaluated in simple manners. A summary of many stability approaches for fuzzy control systems has been presented in Jamshidi[4].

10.5.1   Lyapunov Stability

One of the most fundamental criteria of any control system is to ensure stability as part of the design process. In this section, some theoretical results on this important topic are detailed.

We begin with the ith Takagi-Sugeno implication of a fuzzy system:

$$P^i : \text{IF } x(k) \text{ is } A_1^i \text{ AND} \ldots x(k-n+1) \text{ is } A_n^i$$
$$\text{THEN } x^i(k+1) = a_0^i + a_1^i x(k) + \ldots + a_n^i x(k-n+1) \tag{10.17}$$

with $i=1,\ldots,l$. It is noted that this implication is similar to Equation 10.12 except since we are dealing with Lyapunov stability, the inputs $u(k)$ are absent. The stability of a fuzzy control system with the presence of the inputs will be considered shortly. The consequent part of Equation 10.17 represents a set of linear subsystems and can be rewritten as [5]

$$P^i : \text{IF } x(k) \text{ is } A_1^i \text{ AND} \ldots x(k-n+1) \text{ is } A_n^i$$
$$\text{THEN } \mathbf{x}(k+1) = \mathbf{A}_i \mathbf{x}(k) \tag{10.18}$$

where $\mathbf{x}(k)$ is defined by Equation 10.11 and $n{\times}n$ matrix $\mathbf{A}_i$ is

$$\mathbf{A}_i = \begin{bmatrix} a_1^i & a_2^i & \cdots & a_{n-1}^i & a_n^i \\ 1 & 0 & \cdots & 0 & 0 \\ 0 & 1 & \cdots & 0 & 0 \\ \vdots & \vdots & \ddots & \vdots & \vdots \\ 0 & 0 & \cdots & 1 & 0 \end{bmatrix} \tag{10.19}$$

The output of the fuzzy system described by Equations 10.17-10.19 is given by

$$\mathbf{x}(k+1) = \frac{\sum_{i=1}^{l} w^i \mathbf{A}_i \mathbf{x}(k)}{\sum_{i=1}^{l} w^i} \tag{10.20}$$

where w^i is the overall truth value of the ith implication and l is the total number of implications. Using this notation we then present the first stability result of fuzzy control systems [5].

Theorem 10.1

The equilibrium point of a fuzzy system Equation 10.20 is globally asymptotically stable if there exists a common positive definite matrix $\mathbf{P}$ for all subsystems such that

$$A_i^T \mathbf{P} A_i - \mathbf{P} < 0 \quad \text{for} \quad i = 1, \dots, l. \tag{10.21}$$

It is noted that the above theorem can be applied to any nonlinear system which can be approximated by a piecewise linear function if the stability condition (10.21) is satisfied. Moreover, if there exists a common positive definite matrix $\mathbf{P}$, then all the A_i matrices are stable. Since Theorem 10.1 is a sufficient condition for stability, it is possible not to find a $\mathbf{P} > 0$ even if all the A_i matrices are stable. In other words, a fuzzy system may be globally asymptotically stable even if a $\mathbf{P} > 0$ is not found. The fuzzy system is not always stable even if all the A_i's are stable.

Theorem 10.2
Let A_i be stable and nonsingular matrices for $i=1,\dots,l$. Then $A_i A_j$ are stable matrices for $i,j=1,\dots,l$, if there exists a common positive definite matrix $\mathbf{P}$ such that

$$A_i^T \mathbf{P} A_i - \mathbf{P} < 0 \quad \text{for} \quad i = 1, \dots, l. \tag{10.22}$$

Example 10.3
Consider the following fuzzy system:

$$P^1 : \text{IF } x(k) \text{ is } A^1 \text{ THEN } x^1(k+1) = 1.2x(k) - 0.6x(k-1)$$

$$P^2 : \text{IF } x(k) \text{ is } A^2 \text{ THEN } x^2(k+1) = x(k) - 0.4x(k-1)$$

where A^i are fuzzy sets shown in Figure 10.15. It is desired to check the stability of this system.

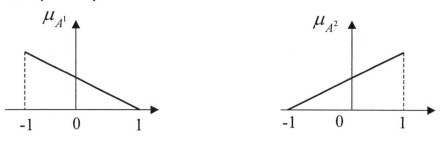

Figure 10.15: Fuzzy Sets for Example 10.3.

SOLUTION
The two subsystems' matrices are

$$A_1 = \begin{pmatrix} 1.2 & -0.6 \\ 1 & 0 \end{pmatrix}, \quad A_2 = \begin{pmatrix} 1 & -0.4 \\ 1 & 0 \end{pmatrix}$$

The product of matrix $A_1 A_2$ is

$$A_1 A_2 = \begin{pmatrix} 0.6 & -0.48 \\ 1 & -0.4 \end{pmatrix}$$

whose eigenvalues are $\lambda_{1,2}=0.1\pm j0.48$ which indicates that $\mathbf{A}_1\mathbf{A}_2$ is a stable matrix. Thus, by Theorem 10.2 a common $\mathbf{P}$ exists, and if we use $\mathbf{P}$ with the following,

$$\mathbf{P} = \begin{pmatrix} 2 & -1.2 \\ -1.2 & 1 \end{pmatrix}$$

then both equations $\mathbf{A}_i^T \mathbf{P} \mathbf{A}_i - \mathbf{P} < 0$ *for* $i = 1,2$ are simultaneously satisfied. This result was also verified using simulation. Figure 10.16 shows the simulation result, which is clearly stable.

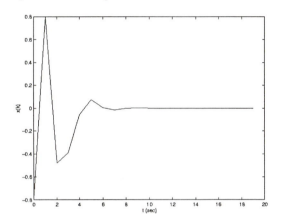

Figure 10.16: Simulation Result for Example 10.3.

Thus far, the criteria which have been presented treat autonomous (either closed-loop or no input) systems. Consider the following non-autonomous fuzzy system:

$$P^i : \text{IF } x(k) \text{ is } A_1^i \text{ AND} \ldots \text{AND } x(k-n+1) \text{ is } A_n^i \text{ AND}$$
$$u(k) \text{ is } B_1^i \text{ AND} \ldots \text{AND } u(k-m+1) \text{ is } B_m^i \quad (10.23)$$
$$\text{THEN } x^i(k+1) = a_0^i + a_1^i x(k) + \ldots + a_n^i x(k-n+1) +$$
$$b_1^i u(k) + \ldots + b_m^i x(k-m+1)$$

Here, we use some results from Tahani and Sheikholeslam [23] to test the stability of the above system. We begin with a definition.

Definition 10.2
The nonlinear system

$$\mathbf{x}(k+1) = \mathbf{f}[\mathbf{x}(k), \mathbf{u}(k), k], \quad \mathbf{y} = \mathbf{g}[\mathbf{x}(k), \mathbf{u}(k), k] \quad (10.24)$$

is *totally stable* if and only if for any bounded input $\mathbf{u}(k)$ and bounded initial state $\mathbf{x}_0$, the state $\mathbf{x}(k)$ and the output $\mathbf{y}(k)$ of the system are bounded, i.e., we have

For all $\|\mathbf{x}_0\| < \infty$ and for all $\|\mathbf{u}(k)\| < \infty \Rightarrow \|\mathbf{x}(k)\| < \infty$ and $\|\mathbf{y}(k)\| < \infty$ (10.25)

Now, we consider the following theorem:

Theorem 10.3
The fuzzy system Equation 10.23 is totally stable if there exists a common positive definite matrix **P** *such that the following inequalities hold*

$$\mathbf{A}_i^T \mathbf{P} \mathbf{A}_i - \mathbf{P} < 0 \quad for \quad i = 1,...,l. \tag{10.26}$$

where $\mathbf{A}_i$ *is defined by Equation 10.19. The proof of this theorem can be found in Sheikholeslam [34].*

Example 10.4
Consider the following fuzzy system:

P^1 : IF $x(k)$ is A^1 THEN $x^1(k+1) = 0.85x(k) - 0.25x(k-1) + 0.35u(k)$

P^2 : IF $x(k)$ is A^2 THEN $x^2(k+1) = 0.56x(k) - 0.25x(k-1) + 2.22u(k)$

where A^i are fuzzy sets shown in Figure 10.17. It is desired to check the stability of this system. Assume that the input $u(k)$ is bounded.

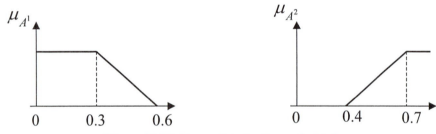

Figure 10.17: Fuzzy Sets for Example 10.4.

SOLUTION
    The two subsystems' matrices are

$$\mathbf{A}_1 = \begin{pmatrix} 0.85 & -0.25 \\ 1 & 0 \end{pmatrix}, \quad \mathbf{A}_2 = \begin{pmatrix} 0.56 & -0.25 \\ 1 & 0 \end{pmatrix}$$

If we choose the positive definite matrix **P**

$$\mathbf{P} = \begin{pmatrix} 3 & -1 \\ -1 & 1 \end{pmatrix}$$

then it can be easily verified that the systems is totally stable.
The product of matrix $\mathbf{A}_1\mathbf{A}_2$ is

$$\mathbf{A}_1\mathbf{A}_2 = \begin{pmatrix} 0.23 & -0.21 \\ 0.56 & -0.25 \end{pmatrix}$$

The eigenvalues of product of matrix $\mathbf{A}_1\mathbf{A}_2$ eigenvalues are $\lambda_{1,2}=0.012\pm j0.25$ which indicates that $\mathbf{A}_1\mathbf{A}_2$ is a stable matrix.

10.5.2 Stability via Interval Matrix Method

Some results on the stability of time varying discrete interval matrices by Han and Lee [35] can lead us to some more conservative, but computationally more convenient, stability criteria for fuzzy systems of the Takagi-Sugeno type shown by Equation 10.17. Before we can state these new criteria some preliminary discussion will be necessary.

Consider a linear discrete time system described by a difference equation in state form:

$$x(k+1) = (\mathbf{A} + \mathbf{G}(k))x(k), \quad x(0) = x_0 \tag{10.27}$$

where $\mathbf{A}$ is an $n \times n$ constant asymptotically stable matrix, x is the $n \times 1$ state vector, and $\mathbf{G}(k)$ is an unknown $n \times n$ time varying matrix on the perturbation matrix's maximum modulus, i.e.,

$$\left|\mathbf{G}(k)\right| \le \mathbf{G}_m, \quad \text{for all } k \tag{10.28}$$

where the $\left|\cdot\right|$ represents the matrix with modulus elements and the inequality holds element-wise. Now, consider the following theorem.

Theorem 10.4
The time varying discrete time system Equation 10.27 is asymptotically stable if

$$\rho(\left|\mathbf{A}\right| + \mathbf{G}_m) < 1 \tag{10.29}$$

where $\rho(\cdot)$ stands for spectral radius of the matrix. The proof of this theorem is straightforward, based on the evaluation of the spectral norm $\left\|x(k)\right\|$ or $x(k)$ and showing that if condition Equation 10.29 holds, then $\lim_{k \to \infty} \left\|x(k)\right\| = 0$.

The proof can be found in Han and Lee [35].

Definition 10.3
An interval matrix $\mathbf{A}_I(k)$ is an $n \times n$ matrix whose elements consist of intervals $[b_{ij}, c_{ij}]$ for $i,j=1,\ldots,n$, i.e.,

$$\mathbf{A}_I(k) = \begin{bmatrix} [b_{11},c_{11}] & \cdots & [b_{1n},c_{1n}] \\ \vdots & [b_{ij},c_{ij}] & \vdots \\ [b_{n1},c_{n1}] & \cdots & [b_{nn},c_{nn}] \end{bmatrix} \tag{10.30}$$

Definition 10.4
The *center matrix*, $\mathbf{A}_c$ and the *maximum difference matrix*, $\mathbf{A}_m$ of $\mathbf{A}_I(k)$ in Equation 10.30 are defined by

$$\mathbf{A}_c = \frac{\mathbf{B}+\mathbf{C}}{2}, \quad \mathbf{A}_m = \frac{\mathbf{C}-\mathbf{B}}{2} \tag{10.31}$$

where $\mathbf{B}=\{b_{ij}\}$ and $\mathbf{C}=\{c_{ij}\}$. Thus, the interval matrix $\mathbf{A}_I(k)$ in 10.30 can also be rewritten as

$$\mathbf{A}_I(k) = [\mathbf{A}_c - \mathbf{A}_m, \mathbf{A}_c + \mathbf{A}_m] = \mathbf{A}_c + \Delta\mathbf{A}(k) \tag{10.32}$$

with $|\Delta\mathbf{A}(k)| \le \mathbf{A}_m$.

Lemma 10.1
The interval matrix $\mathbf{A}_I(k)$ is asymptotically stable if matrix $\mathbf{A}_c$ is stable and

$$\rho(|\mathbf{A}_c| + \mathbf{A}_m) < 1 \tag{10.33}$$

The proof can be found in Han and Lee [35]. The above lemma can be used to check the sufficient condition for the stability of fuzzy systems of Takagi-Sugeno type given in Equation 10.18. Consider a set of m fuzzy rules like Equation 10.18,

$$\text{IF } x(k) \text{ is } A_1^1 \text{ AND} \ldots x(k-n+1) \text{ is } A_n^1$$
$$\text{THEN } \mathbf{x}(k+1) = \mathbf{A}_1\mathbf{x}(k)$$
$$\vdots \tag{10.34}$$
$$\text{IF } x(k) \text{ is } A_1^m \text{ AND} \ldots x(k-n+1) \text{ is } A_n^m$$
$$\text{THEN } \mathbf{x}(k+1) = \mathbf{A}_m\mathbf{x}(k)$$

where $\mathbf{A}_i$ matrices for $i=1,\ldots,m$ are defined by Equation 10.19. One can now formulate all the m matrices $\mathbf{A}_i$, $i=1,\ldots,m$ as an interval matrix of the form 10.30 by simply finding the minimum and the maximum of all elements at the top row of all the $\mathbf{A}_i$ matrices. In other words, we have

$$\mathbf{A}_I(k) = \begin{bmatrix} [\underline{a}_1,\overline{a}_1] & [\underline{a}_2,\overline{a}_2] & \cdots & [\underline{a}_{n-1},\overline{a}_{n-1}] & [\underline{a}_n,\overline{a}_n] \\ 1 & 0 & \cdots & 0 & 0 \\ 0 & 1 & \cdots & 0 & 0 \\ \vdots & \vdots & \ddots & \vdots & \vdots \\ 0 & 0 & \cdots & 1 & 0 \end{bmatrix} \tag{10.35}$$

where $\underline{a}_i$ and $\overline{a}_i$, for $i=1,\ldots,n$ are the minimum and maximum of the respective element of the first rows of $\mathbf{A}_i$ in Equation 10.19, taken element by element. Using the above definitions and observations, the fuzzy system Equation 10.34 can be rewritten by

$$\text{IF } x(k) \text{ is } A_1^i \text{ AND} \ldots x(k-n+1) \text{ is } A_n^i \tag{10.36}$$
$$\text{THEN } \mathbf{x}(k+1) = \mathbf{A}_I^i\mathbf{x}(k)$$

where $i=1,\ldots,m$ and $\mathbf{A}_I^i$ is an interval matrix of form Equation 10.35 except that $\underline{a}_i = \overline{a}_i = a_i$. Now, finding the weighted average, one has

$$x(k+1) = \frac{\sum\limits_{i=1}^{l} w^i \mathbf{A}_I^i x(k)}{\sum\limits_{i=1}^{l} w^i}. \tag{10.37}$$

Theorem 10.5

The fuzzy system Equation 10.37 is asymptotically stable if the interval matrix $\mathbf{A}_I(k)$ is asymptotically stable, i.e., the conditions in Lemma 10.1 are satisfied.

Example 10.5

Reconsider Example 10.3. It is desired to check its stability via the matrix interval approach

SOLUTION

The system's two canonical matrices are written in the form of an interval matrix (10.30) as

$$\mathbf{A}_I(k) = \begin{pmatrix} [1,1.2] & [-0.6,-0.4] \\ 1 & 0 \end{pmatrix}$$

The center and maximum difference matrices are

$$\mathbf{A}_c = \begin{pmatrix} 1.1 & -0.5 \\ 1 & 0 \end{pmatrix}, \quad \mathbf{A}_m = \begin{pmatrix} 0.1 & 0.1 \\ 0 & 0 \end{pmatrix}$$

Then, condition 10.33 would become,

$$\rho(|\mathbf{A}_c|+\mathbf{A}_m) = \rho\begin{pmatrix} 1.2 & 0.6 \\ 1 & 0 \end{pmatrix} = 1.58 > 1$$

Thus the stability of the fuzzy system under consideration is inconclusive. In fact, it was shown to be stable.

Consider the following fuzzy system:

$$P^1 : \text{IF } x(k) \text{ is } A^1 \text{ THEN } x^1(k+1) = 0.3x(k)+0.5x(k-1)$$

$$P^2 : \text{IF } x(k) \text{ is } A^2 \text{ THEN } x^2(k+1) = 0.2x(k)+0.2x(k-1)$$

where A^i are fuzzy sets shown in Figure 10.17. It is desired to check the stability of this system using matrix interval method.

SOLUTION

The two subsystems' matrices are

$$\mathbf{A}_1 = \begin{pmatrix} 0.3 & 0.5 \\ 1 & 0 \end{pmatrix}, \quad \mathbf{A}_2 = \begin{pmatrix} 0.2 & 0.2 \\ 1 & 0 \end{pmatrix}$$

The systems' two canonical matrices are written in the form of an interval matrix 10.30 as

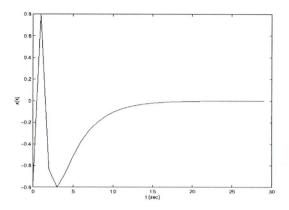

Figure 10.18: Simulation Result for Example 10.6.

$$\mathbf{A}_I(k) = \begin{pmatrix} [0.2,0.3] & [0.2,0.5] \\ 1 & 0 \end{pmatrix}$$

The center and maximum difference matrices are

$$\mathbf{A}_c = \begin{pmatrix} 0.25 & 0.35 \\ 1 & 0 \end{pmatrix}, \quad \mathbf{A}_m = \begin{pmatrix} 0.05 & 0.15 \\ 0 & 0 \end{pmatrix}$$

Then, condition 10.33 would become,

$$\rho(|\mathbf{A}_c| + \mathbf{A}_m) = \rho\begin{pmatrix} 0.3 & 0.5 \\ 1 & 0 \end{pmatrix} = 0.873 < 1$$

Thus the system is stable. This result was also verified by simulation (see Figure 10.18).

10.6   CONCLUSION

This chapter introduced the building blocks of fuzzy control systems. Both Mamdani rules and Takagai-Sugeno rules were presented. Stability analysis of Takagi-Sugeno type fuzzy systems was addressed. Fuzzy control systems are very desirable in situations where precise mathematical models are not available and the human involvement is necessary. In that case fuzzy rules could be used to mimic human behavior and actions.

REFERENCES

1.   Wang, L.-X, *Adaptive Fuzzy Systems and Control,* Prentice Hall, Engelwood Cliffs, NJ, 1994.

2.   Jamshidi, M., Vadiee, N., and Ross, T. J. (eds.), *Fuzzy Logic and Control: Software and Hardware Applications*, Vol 2. Prentice Hall Series on Environmental and Intelligent Manufacturing Systems, (M. Jamshidi, ed.), Prentice Hall, Englewood Cliffs, NJ, 1993.

3. Ross, T. J., *Fuzzy Logic with Engineering Application,* McGraw-Hill, New York, 1995.

4. Jamshidi, M., *Large-Scale Systems—Modeling, Control and Fuzzy Logic,* Prentice Hall Series on Environmental and Intelligent Manufacturing Systems (M. Jamshidi, ed.), Vol. 8, Saddle River, NJ, 1996.

5. Tanaka, K. and Sugeno, M., Stability Analysis and Design of Fuzzy Control Systems, *Fuzzy Sets and Systems,* 45, 135–156, 1992.

6. *IEEE Control Syst. Mag.,* Letters to the Editor, IEEE, Vol. 13, 1993.

7. Bretthauer, G. and Opitz, H.-P, Stability of fuzzy systems, *Proc. EUFIT'94.* Aachen, Germany, Sept., 1994, 283–290, 1994.

8. Aracil, J., Garcia-Cezero, A., Barreiro, A., and Ollero, M., Stability Analysis of Fuzzy Control Systems: A Geometrical Approach, Kulikowski, C.A. and Huber, R.M. (eds.), *AI, Expert Systems and Languages in Modeling and Simulation,* North Holland, Amsterdam, 323–330, 1988.

9. Chen, Y. Y. and Tsao, T. C., A description of the dynamical behavior of fuzzy systems, *IEEE Trans. on Syst., Man and Cyber.,* 19, 745–755, 1989.

10. Wang, P.-Z., Zhang, H. –M, and Xu, W., Pad-Analysis of Fuzzy Control Stability, *Fuzzy Sets and Systems,* 38, 27–42, 1990.

11. Hojo, T., Terano, T., and Masui, S., Stability Analysis of Fuzzy Control Systems, *Proc. IFSA '91,* Engineering, Brussels, 44–49, 1991.

12. Hwang, G.-C and Liu, S. C., A Stability Approach to Fuzzy Control Design for Nonlinear Systems, *Fuzzy Sets and Systems,* 48, 279–287, 1992.

13. Driankov, D., Hellendoorn, H., and Reinfrank, M., *An Introduction to Fuzzy Control,* Springer-Verlag, Berlin, 1993.

14. Kang, H., Stability and Control of Fuzzy Dynamic Systems via Cell-State Transitions in Fuzzy Hypercubes, *IEEE Trans. on Fuzzy Systems,* 1, 267–279, 1993.

15. Demaya, B., Boverie, S., and Titli, A., Stability Analysis of Fuzzy Controllers via Cell-to-cell Root Locus Analysis, *Proc. EVFIT '94,* Aachen, Germany, 1168–1174, 1994.

16. Langari, G. and Tomizuka, M., Stability of Fuzzy Linguistic Control Systems, *Proc. IEEE Conf. Decision and Control,* Hawaii, 2185-2190, 1990.

17. Bouslama, F. and Ichikawa, A., Application to Limit Fuzzy Controllers to Stability Analysis, *Fuzzy Sets and Systems,* 49, 103–120, 1992.

18. Chen, C.-L, Chen, P.-C., and Chen, C.-K, Analysis and Design of a Fuzzy Control System, *Fuzzy Sets and Systems,* 57, 125-140, 1993.

19. Chen, Y. Y., Stability Analysis of Fuzzy Control –a Lyapunov Approach, *IEEE Ann. Conf. Syst., Man, and Cyber.,* 19, 1027-1031, 1987.

20. Franke, D., Fuzzy Control with Lyapunov Stability, *Proc. European Control Conf.,* Groningen, 1993.

21.   Gelter, J. and Chang, H. W., An Instability Indicator for Expert Control, *IEEE Trans. on Control Syst.,* Vol. 31, 14–17, 1986.

22.   Kiszka, J. B., Gupta, M. M., and Nikiforuk, P. N., Energistic Stability of Fuzzy Dynamic Systems, *IEEE Trans. on Syst., Man and Cyber.,* 15, 783-792, 1985.

23.   Tahani, V. and Sheikholeslam, F., Extension of New Results on Nonlinear Systems Stability of Fuzzy Systems, *Proc. EUFIT'94,* Aachen, Germany, 638–686, 1994.

24.   Barreiro, A. and Aracil, J., Stability of Uncertain Dynamical Systems. *Proc., IFAC Symp. on AI in Real-Time Control,* Delft, 177–182, 1992.

25.   Opitz, H. P., Fuzzy Control, Teil 6: Stabilitat von Fuzzy-Regelungen, *Automatisierungstechnik,* 41, A21–24, 1993.

26.   Opitz, H.P., Stability Analysis and Fuzzy Control, *Proc. Fuzzy Duisburg '94, Int. Workshop on Fuzzy Technologies in Automation and Intelligent Systems,* Duisburg, 1994.

27.   Braee, M. and Rutherford, D. A., Selection of Parameters for a Fuzzy Logic Controller, *Fuzzy Set and Syst.,* 49, 103–120, 1978.

28.   Braee, M. and Rutherford, D. A., Theoretical and Linguistic Aspects of the Fuzzy Logic Controller, *Automatica,* 15, 553–577, 1979.

29.   Kickert, W. J. and Mamdani, E.H., Analysis of Fuzzy Logic Controller, *Fuzzy Sets and Syst.,* 1, 29–44, 1978.

30.   Ray, K. S. and Majumder, D. D., Application of Circle Criteria for Stability Analysis Associated with Fuzzy Logic Controller, *IEEE Trans. on Syst., Man and Cyber.,* 14, 345-349, 1984.

31.   Ray, K. S., Ananda, S. G., and Majumder, D. D., L-stability and the Related Design Concept for SISO Linear Systems Associated with Fuzzy Logic Controller, *IEEE Trans. on Syst., Man and Cyber.,* 14, 932–939, 1984.

32.   Böhm, R., Ein Ansatz Zur Stabilitätasalyse von Fuzzy-Reglern. *Forschungsberichte Universitäte Dortmund, Fakultät fur Elektrotechnik, Band Nr. 3,2. Workshop Fuzzy Control des GMA-UA 1.4.2.* am 19/20.11.1992, 24–35, 1992.

33.   Bühler, H., Stabilitatsuntersuchung von Fuzzy-Regelungssystemem, *Proc., 3, Workshop Fuzzy Control des GMA-UA 1.4.1,* Dortmund, 1-12, 1993.

34.   Sheikholeslam, F, Stability Analysis of Nonlinear and Fuzzy Systems, M.Sc. Thesis, Department of EECS Isfahan University of Technology, Isfahan, Iran, 1994.

35.   Han, H. S. and Lee, J. G, Necessary and Sufficient Conditions for Stability of Time-varying Discrete Interval Matrices, *Int. J. Control,* Vol. 59, 1021–1029, 1994.

11

SOFT COMPUTING APPROACH TO SAFE NAVIGATION OF AUTONOMOUS PLANETARY ROVERS

Edward Tunstel, Homayoun Seraji, and Ayanna Howard

11.1   INTRODUCTION

During the past decade, the National Aeronautics and Space Administration (NASA) has been engaged in the conceptualization and implementation of space flight missions to planet Mars. As an integral part of its initiatives to explore the planet's surface, NASA has opted to employ mobile robots that are designed to rove across the surface in search of clues and evidence about the geologic and climatic history of the planet.   These planetary rovers must have mobility characteristics that are sufficient for traversing rough and rugged terrain. Moreover, due to the extreme remoteness of their operating environment, Mars rovers must be capable of operating autonomously and intelligently.

The first autonomous planetary rover, named *Sojourner*, was deployed on Mars in the summer of 1997. This planetary rover was a part of the payload on the NASA Mars Pathfinder lander, which also carried a stereo imaging system, various science instruments, and a telecommunications system that served as a communications relay between Earth and the rover. *Sojourner* was used to demonstrate the viability of exploring planetary surfaces using mobile robot technology; its mission was limited to minimal scientific surface exploration confined to an area in close proximity to the lander. At NASA, the focus of ongoing research for subsequent rover deployments is on enhanced mobility and increased autonomy. In 2003, NASA plans to launch a follow up Mars mission that will use two rovers to explore distinct regions of the planet's surface. These Mars exploration rovers will have greater mobility and autonomy than *Sojourner* since they are expected to traverse up to 100 meters each Martian day and to conduct exploration independent of a surface lander.   The longer-term technology requirements for future Mars missions call for rovers that are capable of traversing distances on the order of kilometers over high risk and challenging terrain.   This chapter describes fundamental research aimed at achieving such long term objectives through application of soft computing techniques for safe and reliable autonomous rover navigation

11.1.1   Practical Issues in Planetary Rover Applications

Autonomous rovers designed for planetary surface exploration must be capable of point-to-point navigation in the presence of varying obstacle distributions (rocks, boulders, etc.), surface characteristics, and hazards. Mobility and navigation hazards include extreme slopes, sand/dust-covered pits, ditches, cliffs and otherwise unstable surfaces. As in the Mars Pathfinder mission scenario, the navigation task can be facilitated by knowledge of a series of waypoints (path sub goals) furnished by mission operations personnel or an automated path planner, which lead to designated intermediate goals. Waypoints can be selected with the aid of images taken at the scene local to the rover. This mode of operation may also prevail on the 2003 rover mission, albeit with significantly longer traverse distances to locations viewable within the images captured by the rovers' onboard cameras. The round trip communication time delay between Earth and Mars, coupled with lack of frequent opportunities for communication with landed resources on Mars, makes direct control of a Mars rover all but impractical. Supervised autonomous control of the rover must therefore be achieved without the luxury of continuous or frequent remote communication between the Earth-based mission operations facility and the Mars rover.

Advanced rovers must have autonomy sufficient to avoid hazards and negotiate (if necessary) challenging terrain if they are to be of practical use for carrying out the goals of scientific exploration in an environment as harsh as the Martian surface. In essence, a capacity for safe navigation and survivability is required for the types of long-duration missions included on the NASA "roadmap" for Mars exploration.   For typical missions, rover autonomy capabilities must be provided under significant constraints on power, computation, weight, and communications bandwidth. To further increase the challenge, many popular and fast state-of-the-art processors that enable advanced capabilities in laboratory research robots are infeasible for planetary rover applications. This is due to the fact that space flight projects require the use of proven, radiation-hardened or otherwise space-flight-qualified electronics that will survive and operate in the harsh temperature and radiation extremes of space. The meager availability of fast and/or powerful space-qualified processors for onboard computation intensifies the need for efficient algorithms for implementing the necessary onboard autonomy.

In order to advance rover navigation capabilities beyond those of *Sojourner*, and even the twin Mars exploration rovers planned for the NASA 2003 Mars mission, advanced algorithms and computational approaches to autonomy and intelligent control must be pursued that comply with the practical constraints. Our research has revealed that the various components of soft computing hold promise as strong candidate technologies that can enable significant advances. The flexibility in applying soft computing techniques, individually or as a hybrid system, facilitates the formulation of efficient solutions to the problems of safe rover navigation in challenging terrain. We have developed a fuzzy-logic-based reasoning and control framework that is complemented by neural networks and visual perception algorithms to realize a practical rover navigation system.

In the following sections, we describe the various components of the safe navigation system and several ways in which soft computing techniques have been applied to solve different aspects of the rover navigation problem. Section 11.2 provides a high level description of the navigation system and its fuzzy logic foundation. In section 11.3, fuzzy logic methods for reasoning about rover vehicle health and safety are described. Next, a methodology for factoring perception of terrain quality into the navigation logic is presented in section 11.4. Section 11.5 describes the fuzzy behavior-based approach and elemental motion behaviors of the system. The soft computing algorithms have been implemented on a commercial mobile robot used as a testbed for outdoor navigation research. In section 11.6, we discuss experimental investigations with this robot that demonstrate the various component technologies. This is followed by a summary and concluding remarks.

11.2   NAVIGATION SYSTEM OVERVIEW

Upon viewing images of the Martian landscape (see Figure 11.1), one would agree that the terrain could be difficult to traverse even for a human driver of an off-road vehicle. The difficulty of the problem increases by orders of magnitude for an autonomous robotic rover. Nonetheless, human driver performance is a worthy goal to strive for in the design of a rover navigation system. In our design, we exploit the fact that fuzzy logic provides a viable means for endowing a computing system with human-like algorithmic reasoning capabilities. In part, we have sought to develop fuzzy inference systems for navigation that emulate human judgment and reasoning as derived from off-road driving heuristics [1] and loose analogies to rating systems used by rock climbers to assess the difficulty of traversing rough terrain [2].

Figure 11.1: Mars Pathfinder Landing Site, 1997.

The safe navigation system is comprised of the various modules and components shown in Figure 11.2. With the exception of the low-level rover motion control system, each component is implemented using soft computing techniques — primarily fuzzy reasoning and control along with artificial neural networks, embedded within a behavior-based structure. The system consists primarily of modules dedicated to rover safety reasoning and strategic navigation control. These are accompanied by associated perception and actuation functionality. The safety reasoning module focuses on vehicle survivability and health, while the strategic navigation module focuses on mission and goal-directed motion from place to place.

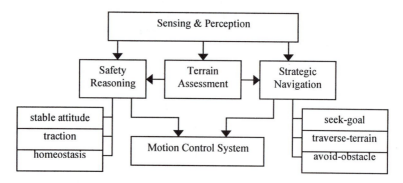

Figure 11.2: Modular System Diagram.

11.2.1   Fuzzy Behavior-Based Structure

We have adopted a fuzzy behavior-based approach [3] for implementation of the knowledge-based reasoning and control components. The architectural design is based on the premise that autonomous navigation functionality can be decomposed into a finite number of special purpose task achieving and decision-making behaviors. The basic building block, then, of the navigation strategy is a *behavior*. A behavior represents a mapping, from perceptions or goals to actions or decisions, aimed at achieving a given desired objective. That is, behaviors may be of two general types: control behaviors and decision behaviors. Fuzzy control behaviors are encoded as fuzzy rule bases with distinct control policies governed by fuzzy inference. The control behaviors are typically simple and self-contained behaviors that serve a single purpose while operating in a reactive (nondeliberative) or reflexive (memoryless) fashion. Within each control module, fuzzy control behaviors perform nonlinear mappings from different subsets of the available sensor suite to set-points for common actuators. If X and U are input and output universes of discourse of a behavior with a rule base of size n, the usual fuzzy IF-THEN rule takes the following form

$$\text{IF } x \text{ is } C_i, \text{ THEN } u \text{ is } A_i \qquad (11.1)$$

where x and u represent input and output fuzzy linguistic variables, respectively, and C_i and A_i ($i = 1...n$) are fuzzy subsets denoting linguistic values of x and u, which represent possible *conditions* and *actions*. In our case, the input x refers to sensory data; u refers to motion control variables that influence rover translation and rotation. The control variables serve as set points for low level classical PID (proportional integral derivative) motor controllers. In general, the rule antecedent consisting of the condition x is C_i could be replaced by a compound fuzzy proposition consisting of conjunctions, disjunctions, or complements of similar propositions. Similarly, the rule consequent consisting of the action u is A_i could be composed of multiple rule base output propositions. Equation 11.1 represents a typical rule that expresses the actions taken by an expert human driver based on the prevailing conditions.

The control behaviors can be executed individually or concurrently to produce intelligent behavior for goal-directed navigation. Concurrent execution of fuzzy behaviors is facilitated by fuzzy decision-making modules, which combine the individual capabilities by implementing a fuzzy set theoretic approach to inferring control gains and computing control inputs for the rover. Within each decision module, fuzzy decision behaviors map perceptual and goal information to appropriate gains based on the current situation or context. Reasoning is governed by rules of the following form

$$\text{IF } x \text{ is } S_k, \text{ THEN } w \text{ is } G_k \tag{11.2}$$

where x and w are fuzzy linguistic variables that represent sensor/goal data and control behavior gains, respectively. Here, S_k and G_k are fuzzy subsets of x and w, which represent possible navigational *situations* and adjustable *gains*.

Implementation details of each component are presented in the following sections. In the next section, we discuss relevant rover health and safety issues. We then describe how fuzzy logic can be applied to provide an intrinsic safety cognizance and a capacity for reactive mitigation of navigation risks. Having described how a nominal level of safety assurance can be achieved, we move on in subsequent sections to discuss higher-level cognitive components of the system that provide the strategic navigation capabilities necessary to perform mission- and goal-directed tasks.

11.3 FUZZY-LOGIC-BASED ROVER HEALTH AND SAFETY

Built-in safe operation and health cognizance are essential for autonomous traversal through challenging terrain over extended time and distance. In many existing systems [4, 5], it is common to consider basic monitoring of individual hardware components for proper operation, but without explicit autonomous reaction or counteraction by the rover. Efficient management of onboard resources, such as power and science data storage capacity and regulation of energy and internal temperature are common concerns for maintaining vehicle health [5-7]. In addition to vehicle health, operational safety is of primary importance. Navigation systems have also been developed which account for some measure of risk mitigation with respect to accidental damage (as due to tipover) and/or vehicle entrapment [8, 9]. However, few field mobile robot systems have been reported in the literature that feature efficient implementation of both active vehicle health *and* safety countermeasures.

11.3.1 Health and Safety Indicators

The ability of a system to provide substantial safety countermeasures depends upon its capacity for assessing vehicle status with respect to the operating environment. Various observable states, events, and terrain features can be considered for online assessment of a rover's operational status. Table 11.1 lists a number of possible health and safety indicators (HSIs) associated with rover on-board subsystems, which convey some aspect(s) of rover operational well being as it relates to safe terrain traversal. At any given moment, the amount of power available to a rover system is perhaps the strongest indicator of its operational health. Solar energy is the primary power source for planetary rovers, although some systems have the luxury of rechargeable batteries. The attitude (pitch and roll) of the vehicle chassis can be monitored in order to avoid instabilities associated with ascent/descent of slopes, traversal of rocky terrain, and turning within vehicle curvature constraints. In addition to surface irregularities, the type and condition of the terrain surface provide clues for safety assessment. Human automobile drivers are able to perceive certain road conditions (e.g., oil slicks, pot holes, and ice patches) as measures of safety, which can be reacted to in order to reduce the risk of potential accidents. In a similar manner, rover potential safety can also be inferred and reacted to based on knowledge of the terrain type or condition. Wheel-soil interactions are important mobility considerations in natural terrain. Excessive wheel slip reduces the effective traction that a rover can achieve and, therefore, its ability to make significant forward progress (not to mention the dramatic effect it can have on the accumulation of errors in estimated position and orientation over distance and time). On soft soils, such as sand, excessive wheel slip can often lead to wheel sinkage and eventual entrapment of the vehicle. Unfortunately, wheel slip and sinkage are often difficult to measure and estimate in a simple manner. Some progress has been made, however, in developing statistical estimation approaches for planetary rovers [10]. One simple approach involves the detection of drive motor stall via current sensing. A detected stall condition for one or more drive motors could be indicative of sinking, trapped, or stuck wheels. However, additional reasoning beyond speculation of the possible causes of a stalled motor would likely be necessary to assess the actual vehicle status. Other HSIs can be considered that are related to critical internal environmental conditions such as temperatures of hardware components that are sensitive to thermal variations. In addition, general dynamic and kinematic states can be monitored for compliance with vehicle mechanical capabilities and constraints.

Table 11.1: Rover Health and Safety Indicators.

Health	Safety
Available power	Chassis attitude
Component failure or anomaly	Terrain type or condition
Component temperature	Wheel slip and sinkage
Drive motor stall	Dynamic/kinematic compliance

Ultimately, a comprehensive autonomous vehicle health and safety system is desired to increase rover survivability. Perhaps consideration of all items in Table 11.1 would make this possible, but such complete observability is rare in practice. To this end, we have concentrated on providing some of the elements necessary to approach the ultimate goal. As a baseline set of HSIs, we have considered chassis attitude, terrain type and condition, and available power.

The safety module will also incorporate a reasoning approach to homeostatic regulation of onboard resources. That is, the addition of automated mechanisms for self-regulation of internal operating condition is planned. A capability such as this is analogous to self regulating functions provided by parts of human or animal physiology. An example of how this can be done is discussed in Arkin [11], where a homeostatic control approach for mobile robots is proposed based on an analogy with the mammalian endocrine system. In that work, internal sensing is used to stimulate behavioral reactions through gain modulation and parameter adjustment, which contribute to regulation of energy and internal temperature. In our navigation system, this can be achieved through rover speed modulation and adjustment of relevant fuzzy set membership function parameters, to contribute to power efficiency and thermal regulation. A related approach applied to planetary rover prototypes is described by Huntsberger and Rose [6]. Reactions to power and internal temperature threshold violations are automatically invoked in response to internal sensing. The reactions consist of halting rover motion to cool down or recharge batteries via solar panels, and activating internal heaters to warm up when necessary.

At this stage of development, the safety module employs concise fuzzy systems that provide autonomous reasoning to facilitate maintenance of stable vehicle attitude (pitch and roll) and wheel traction on rough terrain. The system employs off-road driving heuristics to facilitate avoidance of hazardous vehicle configurations and excessive wheel slip. In each case, our system is designed to produce safe speed recommendations associated with the current perception of the safety status of the rover. In the following section, we discuss the associated soft computing solutions.

11.3.2 Stable Attitude Control

For indoor mobile robots, mobility and navigation problems can often be addressed in two dimensions since the typical operating environments consist of flat and smooth floors. In sharp contrast to this, mobility and navigation problems for outdoor rough terrain vehicles are characterized by significantly higher levels of difficulty. This is due to the fact that complex motions in the third dimension occur quite frequently as the vehicle traverses undulated terrain, encountering multidirectional impulsive and resistive forces throughout. The problem is more pronounced for vehicles with more or less rigid suspensions than it is for vehicles with articulated chassis. In any case, sufficient measures must be taken to maintain upright stability of the vehicle in both static and dynamic configurations.

For safety monitoring, the rover is outfitted with a two-axis inclinometer/tilt-sensor to measure pitch and roll. It is model CXTA02, manufactured by Crossbow Technology, Inc., which features +/- 75° range and 0.05° resolution. In this case, perhaps the simplest approach is to stop rover motion when either axis senses tilt beyond a critical threshold. In a few instances this may be sufficient. More often than not, however, dynamic effects such as momentum will quickly defeat the simplest approach and cause the rover to reach marginal stability (a point at which the vehicle begins to tip over), or worse yet, to actually tip over. Even though planetary rovers are typically driven at low speeds (e.g., maximum average speed of ~0.3 m/s), more sophistication is required beyond binary threshold reactions. We have elected to formulate a strategy in which the recommended safe speed for the rover is proportionately modulated in reaction to changes in attitude (pitch and roll). When the rover travels on a relatively level surface, a maximum safe speed is recommended. As pitch and/or roll approaches extremes near marginal stability, gradual reductions in safe speed are recommended (including the stop condition). At attitudes between these extremes, recommended safe speeds are computed by interpolation via fuzzy sets and logical inference.

By considering various off-road driving heuristics for traversing rock fields, ravines, and hills (up-, down-, and side-hill), a set of fuzzy logic rules is formulated to maintain stable rover attitudes for safe navigation. Fuzzy subset partitions and membership function definitions for pitch and roll are derived based on subjective assessment of the problem. Pitch is represented by five fuzzy sets with linguistic labels {NEG-HIGH, NEG-LOW, ZERO, POS_LOW, POS-HIGH}, while roll is partitioned using three fuzzy sets with linguistic labels {NEG, ZERO, POS}. Here, positive and negative are abbreviated by "pos" and "neg," respectively. Bounds on the universe of discourse for attitude measurements are chosen in accordance with the rover stability constraints and the level of acceptable risk. The rules and input membership functions for the stable attitude control component are shown in Figure 11.3. As is typical in fuzzy control systems, the membership functions, used to express uncertainty in the variables of each system component, take on triangular and/or trapezoidal shapes.

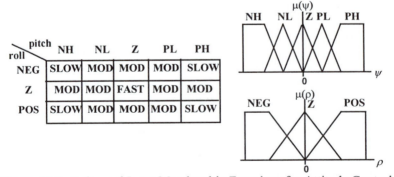

pitch roll	NH	NL	Z	PL	PH
NEG	SLOW	MOD	MOD	MOD	SLOW
Z	MOD	MOD	FAST	MOD	MOD
POS	SLOW	MOD	MOD	MOD	SLOW

Figure 11.3: Rules and Input Membership Functions for Attitude Control.

Fifteen fuzzy logic rules are employed to map the range of stable attitudes to safe driving speed recommendations. In addition to these rules, a crisp rule is applied to handle the extreme cases when marginal stability is reached and the safest reaction is to stop the motion. However, in contrast to the binary threshold scheme, as marginal stability is approached the rover speed is smoothly decreased to near zero due to the interpolation provided by the fuzzy logic rules.

11.3.3 Traction Management

In the absence of some measure of control, wheeled vehicles are prone to loss of traction under certain conditions. On dry paved roads, traction performance is perhaps maximal for most wheeled vehicles due to the high coefficient of friction/adhesion between the road and tread (whether rubber or metal as in the case of some rover wheels). On off-road terrain, however, a variety of surface types are typically encountered including sand, gravel, densely packed soil, ice, mud, and so on. Based on current knowledge about the surface of Mars, rovers may encounter additional types of hard and soft surfaces on which rover wheels are susceptible to slippage. As mentioned above, loss of traction due to excessive wheel slip can lead to wheel sinkage and ultimately, vehicle entrapment. Frequent loss of traction during a traverse from one place to another will also detract significantly from the ability to maintain good position estimates. To improve mobility and navigation performance of rovers, a mechanism for regulating or minimizing wheel slip is highly desirable.

The problem of traction control is not new. It is a common problem in automobile and general transportation vehicle design with a variety of effective solutions. In many cases, solutions are derived from analyses based on the following equation for wheel slip ratio, λ, which is defined nondimensionally as a percentage of vehicle forward speed, v:

$$\lambda = \left(1 - \frac{v}{r_w \omega_w}\right) * 100 \tag{11.3}$$

Here, r_w is the wheel radius and ω_w is the wheel rotational speed. Equation 11.3 expresses the normalized difference between vehicle and wheel speed. Therefore, when this difference is nonzero, wheel slip occurs. The objective of traction control is to regulate λ to maximize traction. This is a relatively straightforward regulation task if v *and* ω_w are observable. Wheel speed is typically available from shaft encoders or tachometers. However, it is often difficult to measure the actual over-the-ground speed for off-road wheeled vehicles. The problem is further complicated by nonlinearities and time varying uncertainties due to wheel-ground interactions. Despite this, effective solutions have been found for automotive applications. In fact, fuzzy logic is a common tool for antilock (deceleration) and antislip (acceleration) control [12-15]. In these cases, measurement of v is facilitated by the even surface on which the vehicle travels. For example, in Arkin [11] an accelerometer is used to measure vehicle speed and the slip ratio is estimated based on deceleration of the four wheels. In Bauer and Tomizuka [13], the measurement of vehicle speed is facilitated by the use of magnetic markers alongside the road in an intelligent highway automation system. In this case, the vehicle speed is measured according to travel time between markers. For application to an electrically driven locomotive, the solution in Palm and Storjohann [14] makes use of a model of the friction-slip relationship, which is fixed for the wheel-rail interaction. On outdoor terrain, the friction-slip relationship varies with surface type. In large part, the available solutions are not directly transferable to off-road vehicle applications for which the terrain is uneven as opposed to being relatively flat, as is the case for automobiles and locomotives.

The use of an accelerometer to measure off-road vehicle speed is problematic since the gravity effects of traversing longitudinal and lateral slopes will interfere with the measurement. For an accelerometer used to measure horizontal acceleration, any off horizontal vehicle tilt will be sensed as a change in acceleration; as a result, the integrated velocity will be in error. This is realized in Van der Burg and Blazevic [16] where an alternative traction control concept for rovers is considered. In that case, a non-driven "free wheel" is proposed for measuring actual vehicle speed. Another promising solution was proposed for rovers with an articulated chassis, which enables active control of the vehicle center of gravity. For those vehicles, the use of accelerometers in concert with rate gyroscopes is suggested [17].

In our work, we have elected to take a simple linguistic approach that does not rely on accurate sensing of vehicle speed. Instead, visual perception of terrain surface type is used to infer an appropriate speed of traversal. Results from traction tests performed on the actual rover are used to determine appropriate speeds for a range of potential surface types. In particular, the rover is tested on different terrain surfaces (e.g., sand, gravel, densely packed soil, etc.) to determine the maximum speeds achieved before the onset of wheel slippage. Given this information, commanded vehicle speed can be modulated during traversal based on visual classification of the terrain surface type just ahead of the rover. This is analogous to the perception-action process that takes place when a human driver notices an icy road surface ahead and decelerates to maintain traction. For the rover, such speed modulation allows management of traction by mitigating the risk of wheel slippage.

Given the results of actual traction tests, the formulation of fuzzy rules to achieve speed modulation is relatively straightforward. The success of the traction management approach depends more heavily on the ability to perceive and classify the various terrain surface types. The problem of off-road surface type identification would be quite formidable for systems equipped with only proximity sensors, range finders, and/or tactile probes. However, visual image-based classification has been found to be particularly promising [18]. We will now describe an artificial neural network solution to this problem that provides qualitative information about the expected surface traction ahead of the rover. This information is used to infer tractive rover speeds via fuzzy inference.

11.3.3.1 Neuro-Fuzzy Solution

Distinct terrain surfaces reflect different textures in visual imagery. The ability to associate image textures to terrain surface properties such as traction, hardness, or bearing strength has clear benefits for safe autonomous navigation. To provide this capability, we make use of an onboard camera pointed such that its field of view (FOV) covers an area on the ground in front of the rover. In this way, the projected image provides a downward looking view of the surface as illustrated in Figure 11.4a. Using a neural network (Figure 11.4b), texture analysis is performed on image data acquired by the camera. That is, a neural network classifier, trained to associate texture with several surface types, provides the information needed to make any necessary adjustments to wheel speed in order to maintain traction on the classified surface. Based on typical surfaces that the rover may encounter, three texture prototypes are selected: sand, gravel, and compacted soil (Figure 11.5).

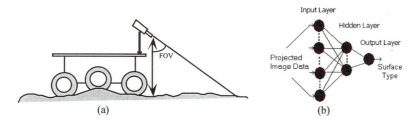

Figure 11.4: (a) Camera Mounted on Rover; (b) Neural Network for Surface Classification.

The method proceeds as follows. Assuming the section of the image just ahead of the front wheels is free of obstacles, a set of 40x40 pixel image blocks is randomly selected from a camera image of size 320x280 pixels. To reduce the large data dimensionality inherent in typical vision-based applications, a filtering step is performed. This permits effective extraction of features embedded in the surface image data set in real time. The image blocks are normalized to compensate for lighting variations and the data is used to train the neural network classifier. After training the network on typical image data representing different surface prototypes, we utilize it to classify the surface types during run time.

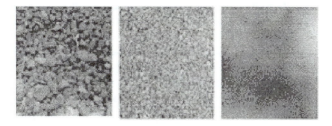

Figure 11.5: Terrain Surface Texture Images: Gravel, Sand, Compacted Soil.

The neural network is trained to provide texture prototype outputs in the unit interval [0, 1], with 0 corresponding to surfaces of very low traction (e.g., ice) and 1 corresponding to surfaces of high traction (e.g., dry cement). This is a design decision motivated by a desire to establish some correlation to actual wheel-terrain coefficients of friction. In this way, we can make a qualitative association between neural network output and expected traction in front of the rover. In the sequel, we will refer to the texture prototype output as the *traction coefficient*, denoted by C_t.

Wheel-terrain friction coefficients for a variety of tread and surface types are widely published in the literature on vehicle mechanics. However, published friction coefficients for identical tread and surface types vary from source to source. This is due to the fact that measured values depend heavily on the variety of tests and conditions from which they were generated. Nevertheless, common ranges of friction coefficients for given tread and surface types are widely agreed upon. The following are typical estimates of the friction coefficient for rubber tires on various surfaces: icy road/snow (0.1), sand (0.3), slippery/wet road (0.4), hard unpaved road (0.65), grass (0.7), and dry paved road (0.8-1.0).

Given the uncertainty in associating exact friction coefficients with certain terrain surface types, and the loose correlation provided by the traction coefficient, we elect to reason about traction using fuzzy logic. The range of traction coefficients, [0,1], obtained from the neural network classifier is partitioned using three fuzzy sets with linguistic labels {LOW, MEDIUM, HIGH}. Triangular membership functions are used which are equally distributed throughout the universe of discourse. Based on these definitions, the following simple fuzzy logic rules are applied to manage rover traction on varied terrain:

- IF C_t is LOW, THEN v is SLOW.
- IF C_t is MEDIUM, THEN v is MODERATE.
- IF C_t is HIGH, THEN v is FAST.

Here, membership functions for the rover speed v are defined over the range of tractive speeds that result from traction tests on various surface types. Note that the neural network can be trained to map its inputs directly to the actual range of tractive speeds (rather than the range of C_t). However, in this neuro-fuzzy approach, fuzzy inference serves to accommodate uncertainties in both the surface classification and the subsequent specification of tractive speed.

In summary, the stable attitude and traction management components of the safety module combine to provide active countermeasures to potential vehicle tip over and excessive wheel slip. The minimum of the rover speeds inferred by the two components is issued as the safe speed recommendation v_{safe}. The interface between the safety module and the strategic navigation module is depicted in Figure 11.6. As indicated by the diagram, safe speeds recommended by the safety module are compared to the strategic speed recommendations, and the safest speed is issued as the commanded set point for the motion control system. The determination of safe rover speed is independent of the behavior fusion process (discussed later) that produces the strategic navigation speed. This allows recommended safe speeds to override strategic speeds, if necessary, to ensure vehicle safety. This is also the approach taken in Murphy and Dawkins [19] where it is asserted that distributing speed control across all behaviors makes it difficult to ensure that the interactions will yield a safe speed.

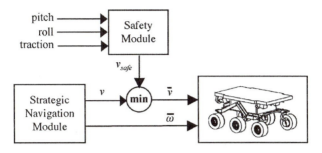

Figure 11.6: Safety and Strategic Navigation Module Interface.

11.4   TERRAIN-BASED FUZZY NAVIGATION

In dealing with day-to-day processes, humans make subjective decisions based on qualitative information. Their perception of processes is based on qualitative, rather than quantitative, assessments obtained from imprecise and approximate measurements. The human control strategy for a process typically consists of simple, intuitive, and heuristic rules based on prior experience that are brought to bear to affect the process. For instance, in the process of driving a car, the human driver turns the steering wheel to the right if the car veers too far to the left, and vice versa. The driver intuitively determines the degree of course correction based on driving experience, rather than resorting to mathematical modeling and formulation of the process. His actions are based on how far from the lane the car has moved and on how fast the car is moving. Similarly, the driver adjusts the speed of the car based on his subjective judgment of the road conditions, e.g., the car speed is decreased in off-road driving on a bumpy and rough terrain but is increased on a smooth and flat surface. This human control strategy exhibits characteristics of reactivity, set point tracking, and regulation, all perceptually guided by qualitative situational assessments. As mentioned earlier, it is highly desirable to capture the essence of the tight perception-action control loop exhibited by human drivers for implementation in autonomous navigation strategies for planetary rovers.

To develop intelligent navigation controllers, we formulate simple and intuitive fuzzy logic rules that capture the attributes of human driver reasoning and decision-making. Robust navigation behavior in practical rover systems can be achieved when perception uncertainty and actuator imprecision is accommodated by the rover control system. Such is the case when fuzzy logic control and decision systems are employed. That is, the linguistic values in the rule antecedents can be chosen to convey the imprecision associated with on-board sensor measurements, while those in the rule consequents can represent the vagueness inherent in the reasoning processes and the imprecision inherent in actuator operation. Having developed a number of desired navigation behaviors in this way, one may rely upon the computational mechanisms of fuzzy logic to provide robust inference and approximate reasoning under practical uncertainties. The operational strategies of the human expert driver then, can be transferred via fuzzy logic tools to the robot navigation strategy in the form of several fuzzy behaviors. The main advantages of such a navigation strategy lie in the ability to extract heuristic rules from human experience and to obviate the need for an analytical model of the process. To complement this methodology, we have developed soft computing solutions for robust qualitative assessment of terrain traversability, which permits further advancement toward achieving human driver performance. Our approach is enabled by intelligent visual perception using terrain imagery captured by cameras onboard the rover.

11.4.1 Visual Terrain Traversability Assessment and Fuzzy Reasoning

Outdoor navigation systems for autonomous field mobile robots must consider terrain characteristics in order to support safe and efficient traversal from place to place. Two important attributes that characterize the difficulty of a terrain for traversal are the slope and roughness of the region. In current methods of terrain assessment [20-24], terrain traversability is defined as an analytical function of the terrain slope and roughness in the region local to the vehicle. The slope is determined by finding the least squares fit of a geometric plane covering the region, while the roughness is calculated as the residual of the best plane fit. Once the traversability of each region is evaluated, a traversable path for the robot to follow is then constructed. These analytical representations of the terrain traversability rely on accurate interpretation of the sensory data, as well as an exact mathematical definition of the traversability function. Here, we present an alternative approach based on fuzzy reasoning. Real time terrain assessment is achieved by computing physical properties of the terrain (such as roughness and slope) using data provided by stereo cameras mounted on the rover. The terrain properties are then used to *infer* traversability according to a recently introduced measure called the *fuzzy traversability index* [25, 26]. The fuzzy traversability index is a simple measure for quantifying the suitability of a natural terrain for traversal by a mobile robot. It can be inferred from knowledge of physical terrain properties, but it also depends on the properties of the robot mobility mechanism, which determines its hill and rock climbing capabilities. In order to quantify the roughness and slope of a region, image processing algorithms are applied for terrain feature extraction as described below.

11.4.1.1 Terrain Roughness Extraction
During navigation, images of the viewable scene are periodically captured by the rover vision system. An algorithm is applied to a pair of stereo camera images that determines the sizes and concentration of rocks/ditches in the viewable scene. These parameters are used to infer terrain roughness, β, which is represented by fuzzy sets with linguistic labels {SMOOTH, ROUGH, ROCKY}. Equally spacing trapezoidal membership functions are used.

The rock size and concentration parameters are represented in terms of a two-parameter vector $r = [r_{small}, r_{large}]$, where r_{small} denotes the concentration of small rocks and r_{large} represents the concentration of large rocks contained within the image. In order to compute these parameters, a horizon line extraction program is run that identifies the peripheral boundary of the ground plane. This, in effect, recognizes the point at which the ground and the landscaped backdrop intersect. The algorithm then identifies target objects located on the ground plane using a region growing method [27]. In effect, target objects that differ from the ground surface are identified and counted as rocks for inclusion in the roughness assessment. The denser the rock concentration, the higher the calculated roughness of the associated region. Figure 11.7 shows an example output of the rock identification algorithm.

Original Image   Horizon Line Extraction   Rock Detection

Figure 11.7: Visual Terrain Roughness Extraction.

To determine the number of small and large rocks contained within the image, the number of pixels that comprise a target object are first enumerated. Those targets with a pixel count less than a user defined threshold are labeled as belonging to the class of small rocks and those with a count above the threshold are classified as large rocks. The threshold is determined based on the mechanical characteristics of the rover, such as wheel size, wheel base, body height, and so on. This defines fuzzy sets with linguistic labels {SMALL, LARGE}, which represent the rock sizes, R_s. All such labeled target objects are then grouped according to their sizes in order to determine the small and large rock concentration parameters. These values are then used to populate the two-parameter vector r, which is characterized by fuzzy sets with linguistic labels {FEW, MANY} and used as input for the following fuzzy logic rules, where R_c represents the rock concentration:

- IF R_c is FEW AND R_s is SMALL, THEN β is SMOOTH.
- IF R_c is FEW AND R_s is LARGE, THEN β is ROUGH.
- IF R_c is MANY AND R_s is SMALL, THEN β is ROUGH.
- IF R_c is MANY AND R_s is LARGE, THEN β is ROCKY.

The terrain roughness is thus derived directly from the rock size and concentration parameters of the associated image scene.

11.4.1.2 Terrain Slope Extraction

Slope characterizes the average incline/decline of the ground surface to be traversed. To obtain the surface slope, an innovative approach is utilized to obtain depth information from two uncalibrated cameras. The process involves training a neural network to learn the relationship between slope and correlated image points that lie along the horizon line.

Given a pair of camera images, the algorithm first locates correlated points by determining the position of the largest rocks located along the horizon line and centered within both images (Figure 11.8). Once these points are extracted, the pixel locations in the two images are used as inputs to a trained neural network for slope extraction.

Figure 11.8: Determination of Correlated Image Points.

Using our algorithm, we wish to find a relationship between corresponding image points located along the horizon line and the slope of the viewable terrain. Initially, we train the network by finding a set of weights that will give us the desired slope output. We utilize a three-layer feedforward neural network with error backpropagation. In this process, we present a set of correlated image points and the corresponding slope value to the network. Given this input, the network will calculate the output, which is then compared with the desired slope parameter. The difference between the network slope output and the desired slope value is then used to change the network weights, thus minimizing network error. In this way, the network can learn the desired relationship between correlated image points and slope.

Our network has four input nodes corresponding to the image positions of the correlated points in the two images, and one output node corresponding to the terrain slope parameter. The hidden layer has two processing elements. After training the network on typical imagery data representing different positive and negative sloped examples, we utilize it to extract the slope during run time. The network output provides the terrain slope parameter, α, whose magnitude is then converted into the linguistic representation {FLAT, SLOPED, STEEP}, with membership functions similar to those defined for β.

11.4.1.3 Fuzzy Inference of Terrain Traversability

Once the slope and roughness parameters of the region are determined from the camera images, the fuzzy traversability index, τ, is inferred and used to classify the ease of terrain traversal. The index is represented by three trapezoidal fuzzy sets with linguistic labels {LOW, MEDIUM, HIGH}. The fuzzy traversability index is defined in terms of the terrain slope α and the terrain roughness β by a set of simple fuzzy logic relations summarized in Figure 11.9. Observe that this approach to terrain assessment gives an intuitive, linguistic definition of terrain roughness and traversability as used by a human observer, in contrast to the mathematical definitions (as the residual of the least squares plane fit and as analytical functions of slope and roughness) used previously [20-24]. This representation has the advantage of being robust and tolerant to uncertainty and imprecision in measurements and in the interpretation of sensor data. It conveys sufficient qualitative information about the terrain to permit intelligent assessment of traversability. In addition, it can be easily extended to include consideration of additional terrain features in the reasoning process.

β \ α	FLAT	SLOPED	STEEP
SMOOTH	HIGH	MED	LOW
ROUGH	MED	LOW	LOW
ROCKY	LOW	LOW	LOW

Figure 11.9: Fuzzy Rule Table for Traversability Index.

11.5 STRATEGIC FUZZY NAVIGATION BEHAVIORS

The robot navigation strategy presented in this section is comprised of three simple motion behaviors: seek-goal, traverse-terrain, and avoid-obstacle. These behaviors operate at different perceptual resolutions. The fuzzy logic rules for the seek-goal behavior make use of *global* information about the goal position to make recommendations for rover speed and steering. The fuzzy logic rules for the traverse-terrain behavior incorporate the *regional* information about the terrain quality to produce recommendations for rover speed and steering. The fuzzy logic rules for the avoid-obstacle behavior utilize *local* information about en route obstacles to generate the appropriate speed and steering recommendations. The output of each behavior is a recommendation over all possible control actions from the perspective of achieving that behavior's objective. Each control recommendation is represented by a fuzzy possibility distribution over the space of speed and steering commands. To facilitate behavioral rule formulation, the rule set for each motion behavior has been de-coupled into turn rules and move rules. In the final stage before commanding rover actuators, the individual fuzzy recommendations from the three behaviors are aggregated and defuzzified to yield crisp control inputs. This process of behavior fusion is facilitated by the use of weighting factors inferred from navigational contexts. The approach yields an autonomous navigation strategy for the rover that requires no *a priori* information (e.g., maps) about the environment. We will now describe in detail the individual fuzzy control behaviors and the behavior fusion approach to realizing goal-directed navigation.

11.5.1   Seek-Goal Behavior

The problem addressed in this section is to navigate a rover on a natural terrain from a known initial position to a user-specified goal position. The rover control variables for this behavior are the translational speed v and the rotational speed ω_r. The vehicle speed v is represented by four fuzzy sets with linguistic labels {STOP, SLOW, MODERATE, FAST}. Triangular membership functions are defined which are equally distributed throughout a range of allowable rover speeds. Similarly, the rover turn rate ω is represented by five fuzzy sets with linguistic labels {FAST-LEFT, SLOW-LEFT, ON-COURSE, SLOW-RIGHT, FAST-RIGHT}, defined by equally spaced triangular membership functions over a range of allowable turn rates.

The fuzzy navigation rules for the seek-goal behavior direct the rover to initially perform an in place rotation toward the goal to nullify the heading error, ϕ, which is the relative angle by which the rover needs to turn to face the goal directly. Once the rover is aligned with the goal direction, it then proceeds toward the goal position. A similar rule set can also be formulated for robots that cannot perform in place rotation.

The fuzzy rules for rover rotational motion are listed below, where the heading error input ϕ is represented by five fuzzy sets with linguistic labels {GOAL-FAR LEFT, GOAL-LEFT, GOAL-HEAD ON, GOAL-RIGHT, GOAL-FAR RIGHT}. The turn rules are followed by a list of fuzzy rules used for rover translational motion, where the position error input (goal distance) d is represented by four fuzzy sets with linguistic labels {VERY NEAR, NEAR, FAR, VERY FAR}. The universe of discourse for both ϕ and d is partitioned by an equal distribution of triangular membership functions.

- IF ϕ is GOAL-FAR LEFT, THEN ω is FAST-LEFT.
- IF ϕ is GOAL-LEFT, THEN ω is SLOW-LEFT.
- IF ϕ is GOAL-HEAD ON, THEN ω is ON-COURSE.
- IF ϕ is GOAL-RIGHT, THEN ω is SLOW-RIGHT.
- IF ϕ is GOAL-FAR RIGHT, THEN ω is FAST-RIGHT.

- IF d is VERY NEAR OR ϕ is NOT GOAL-HEAD ON, THEN v is STOP.
- IF d is NEAR AND ϕ is GOAL-HEAD ON, THEN v is SLOW.
- IF d is FAR AND ϕ is GOAL-HEAD ON, THEN v is MODERATE.
- IF d is VERY FAR AND ϕ is GOAL-HEAD ON, THEN v is FAST.

The first rule for translational motion keeps the rover stationary while it is correcting its heading. In the remaining translational motion rules, the rover is aligned with the goal direction and moves with a speed proportional to its distance from the goal.

11.5.2 Traverse-Terrain Behavior

This section presents fuzzy logic rules that use the fuzzy traversability index to infer the vehicle turn rate and speed while moving on natural terrain. It is assumed that the robot can only move in the forward direction (i.e., reverse motion is not allowed). The visual sensor coverage area of the terrain region in front of the rover spans 180°. This sensor horizon is partitioned into three 60° sectors, namely: front, right, and left of the rover position, each extending outward to a distance of up to five meters. The indices for the three sectors, τ_f, τ_r, τ_l, are inferred in real time from the values of terrain slope and roughness extracted by the onboard vision system. The fuzzy rules for determining rover steering direction based on the terrain traversability data are summarized in Figure 11.10a (R:RIGHT, L:LEFT, O:No Turn). The rule table in Figure 11.10b corresponds to steering behavior for obstacle avoidance (discussed below). These rules emulate the steering actions of the human driver during an off-road driving session.

Examining Figure 11.10a, we see that a turn maneuver is not initiated when either the front region is the most traversable, or the right and left regions have the same traversability indices as the front region. Also, observe that the preferred direction of turn is chosen arbitrarily to be LEFT, i.e., when the rover needs to turn to face a more traversable region, it tends to turn left. The choice of LEFT instead of RIGHT is arbitrary, but selection of a preferred turn direction is essential to avoid the possibility that simultaneous left and right rotations can result in a no-turn recommendation even though there may be an impassable region directly ahead of the rover.

τ_l \ τ_r	high	med	low	τ_f
high	0	0	0	
med	0	0	0	high
low	0	0	0	
high	L	L	L	
med	R	0	0	med
low	R	0	0	
high	L	L	L	
med	R	L	L	low
low	R	R	0	

(a)

d_l \ d_r	F	N	VN	d_f
F	0	0	0	
N	0	0	0	F
VN	0	0	0	
F	L	L	L	
N	R	0	0	N
VN	R	0	0	
F	L	L	L	
N	R	L	L	VN
VN	R	R	0	

(b)

Figure 11.10: Turn Rules for (a) Traverse-Terrain and (b) Avoid-Obstacle.

Once the direction of traverse is chosen based on the relative values of τ, the rover speed v can be determined based on the value $\tau*$ of the traversability index τ in the *chosen region*. This determination is formulated as a set of two simple fuzzy logic rules for speed of traverse: IF $\tau*$ is LOW, THEN v is STOP, and IF $\tau*$ is MEDIUM, THEN v is SLOW. The effect of these rules is analogous to that of the human driver adjusting the car speed based on the surface conditions.

11.5.3. Avoid-Obstacle Behavior

In this section, fuzzy logic rules are presented which govern rover behavior based on the local information about en route obstacles, such as large rocks. In general, obstacles may belong to any variety of mobility and navigation hazards such as extreme slopes, sand/dust-covered pits, crevasses, cliffs and otherwise unstable terrain. Also included are so called negative obstacles such as ditches and craters, and their complements such as ridges and boulders. Rocks that are considered obstacles are those with sizes that exceed the obstacle climbing threshold for which the rover is designed. In the case of the Mars rover *Sojourner*, the threshold was 1.5 wheel diameters. Without loss of generality, we may refer to the general category of untraversable patches of terrain as navigation obstacles. This local obstacle information is acquired online and in real time by the proximity sensors mounted on the rover. For space robotics applications, different types of proximity sensors can be used, ranging from low-resolution infrared sensors to high-resolution and longer-range laser detectors [28]. A wider range of options is available for use in more general mobile robot applications [29]. The range of reliable operation of proximity sensors is typically 20 to 50 cm, which is about an order of magnitude shorter than that of regional sensor coverage. Note, however, that precise measurement of the obstacle distance is *not* needed, because of the multivalued nature of the fuzzy sets used to describe it.

In the present implementation, it is assumed that there are three groups of proximity sensors mounted on the robot facing the three different directions of front, right, and left. These sensors report the distances between the robot and the closest front obstacle d_f, the closest right obstacle d_r, and the closest left obstacle d_l within their ranges of operation. The three obstacle distances are continuously measured and updated during rover motion. The steering and speed rules for avoiding obstacles use this local information to maneuver the robot around the obstacles and to avoid potential collisions. Each obstacle distance d_f, d_r, or d_l is represented by the three fuzzy sets with linguistic labels {VERY NEAR, NEAR, FAR}. Equally distributed trapezoidal membership functions are defined for each obstacle distance. Typically, different fuzzy set bounds are defined on the universe of discourse for the front obstacle distance and side (left and right) obstacle distances so that front and side collision detection will have different sensitivities.

The behavioral objectives of the obstacle avoidance rules are to direct the rover to: (a) turn to face a region with the least nearby obstacles, and (b) adjust its speed of motion depending on the distance to the closest front obstacle. The goal of the steering rule set is to steer the robot clear of all obstacles. This goal is accomplished by sensing the three obstacle distances and reacting according to the fuzzy logic rule sets summarized above in the Figure 11.10b. The following points are noted about the above steering rules. First, when d_f is FAR, i.e., the front of the rover is clear of obstacles, the rover will not collide with any obstacles and no corrective action needs to be taken. Therefore, the collision avoidance steering rules are activated only when the situation is otherwise. Second, observe that the preferred direction of turn is chosen to be LEFT, i.e., when the rover needs to turn to avoid an impending collision, it tends to turn left. The choice of LEFT instead of RIGHT is arbitrary, but selection of a preferred turn direction is essential to avoid the possibility that simultaneous left and right obstacles can result in a no-turn recommendation even though there may be an obstacle in front of the vehicle.

The speed rules for collision avoidance are very simple. Basically, the robot is required to slow down as it approaches the closest front obstacle. Again, note that when the front obstacle distance is FAR, collision avoidance is not activated and no corrective action needs to be taken. There are two fuzzy logic rules as follows: IF d_f is VERY NEAR, THEN v is STOP, and IF d_f is NEAR, THEN v is SLOW.

11.5.4. Fuzzy-Behavior Fusion

The decision-making process used to combine recommendations from multiple behaviors is commonly referred to as *behavior coordination* [3]. The most common approach is *behavior arbitration*, which employs a prioritization scheme wherein the control recommendation of only one behavior among several competing behaviors is taken while recommendations from the remaining behaviors with lower priorities are ignored. In contrast to this switching type of arbitration, we advocate using a more comprehensive blending scheme. The preferred coordination scheme permits more than one behavior to influence the resultant control action to the extent governed by variable gains or weighting factors assigned dynamically according to the prevailing context — a scheme referred to as *behavior fusion*. Behavior fusion is facilitated by fuzzy set theoretic computations; however, nonfuzzy implementations are also possible [8]. Thus, in the proposed approach, weight rules combine elemental behaviors, not through fixed priority arbitration, but rather through a generalization of dynamic gains that are determined based on consideration of the situational status of the rover. The weight rules continuously update the behavior weighting factors during rover motion based on the prevailing conditions.

The gains or weighting factors s^w, t^w, and a^w represent the strengths by which the seek-goal, traverse-terrain, and avoid-obstacle recommendations are taken into account to compute the final control actions $\bar{v}$ and $\bar{\omega}$. These weights are represented by two fuzzy sets with linguistic labels {NOMINAL, HIGH}. Three sets of decision rules for the respective motion behavior gains are listed below.

- IF d is VERY NEAR, THEN s^w is HIGH.
- IF d is NOT VERY NEAR, THEN s^w is NOMINAL.

- IF d is NOT VERY NEAR AND d_f is NOT VERY NEAR, THEN t^w is HIGH.
- IF d is VERY NEAR OR d_f is VERY NEAR, THEN t^w is NOMINAL.

- IF d is NOT VERY NEAR, THEN a^w is HIGH.
- IF d is VERY NEAR, THEN a^w is NOMINAL.

At each control cycle, the above sets of gain rules are used to calculate the three crisp weighting factors using the center-of-gravity (centroid) defuzzification method. Note that with this defuzzification method, overlapping areas between adjacent truncated membership functions in the aggregated fuzzy set are counted twice. The resulting crisp weights are then used to compute the final control actions for the rover speed and turn rate.

Fuzzy recommendations from the seek-goal, traverse-terrain, and avoid-obstacle behaviors are weighted by the corresponding behavior gains prior to defuzzification, as shown in Figure 11.11. The weighted fuzzy outputs for the individual behaviors are aggregated into single fuzzy possibility distributions for both rover speed and turn rate. The final control actions for each cycle are computed using the center-of-gravity defuzzification method.

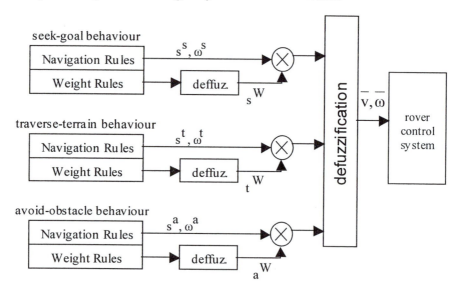

Figure 11.11: Fuzzy-Behavior Fusion.

11.6   ROVER TESTBED AND EXPERIMENTAL RESULTS

Field tests using the Pioneer AT (All-Terrain) rover are conducted on rough terrain near JPL (Pasadena, California) to test the reasoning and decision-making capabilities provided by the fuzzy logic navigation strategy. This commercially available rover is kinematically quite different from planetary rovers designed for Mars. Nonetheless, with certain enhancements it is suitable as a testbed for developing advanced technology and algorithms for infusion into flight rover navigation systems. The Pioneer AT rover, shown on the left of Figure 11.12, is enhanced with additional onboard processing capability, 8-input image multiplexer, a vision system for real time terrain assessment, and a tilt sensor (mentioned in Section 11.3.2). The vision system consists of eight CMOS NTSC video cameras. Six cameras are mounted on a raised platform and used for terrain-based navigation. The right side of Figure 11.12 shows the physical layout of the camera platform used specifically to provide terrain imagery data. These six cameras are placed such that the lens centers are 740 mm above the ground and the optical axis of each camera is tilted down by 8°. The intersecting origin of all cameras is centered above the support polygon formed by the rover wheel ground contact points. In addition, the stereo baseline length is set to 500 mm. This camera placement scheme provides the rover a viewable distance of ~5 m spanning a field of view of ~180°. The remaining two cameras are mounted on a mast below the raised platform and pointed towards the ground for obstacle detection and surface type classification.

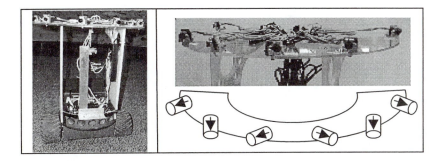

Figure 11.12: Enhanced Pioneer AT with Terrain Assessment Vision System.

The processing power onboard the rover consists of a 333 MHz Pentium II processor housed in a CompactPCI chassis running the Linux Operating System. The system has also been tested using a laptop computer running Windows 95. Resident on the computer are the image processing algorithms and the fuzzy logic computation engine (written in the C language) used to calculate the translational and rotational speed commands issued to control the wheel motors. Using this hardware platform, rover field tests are performed outdoors in natural terrain. We shall now present field test and experimental results for the safety module and the strategic navigation module.

11.6.1 Safe Mobility

In this section, we describe two field tests and associated laboratory experiments performed to evaluate the effect of the safe attitude and traction components. The first test considers reactions to rover pitch and roll during traversal. The second test is concerned with mitigation of wheel slippage.

For the stable attitude test, an obstacle-free swath of undulated terrain is chosen. The rover is commanded to traverse the swath with and without the stable attitude component activated. Without active stable attitude management, the rover traverses the terrain at a nominally fast speed recommended by the strategic navigation system based on the fact that no significant obstacles are present. With active attitude management, the rover traverses the terrain at various reduced speeds in response to changes in its pitch and roll according to the fuzzy logic rules in Figure 11.3. This reactivity reduces the risk of approaching marginal tilt stability, which leads to tip over. It also enhances the ability of rigid suspension vehicles (such as the Pioneer AT) to maintain wheel contact with the ground. A comparative effect of the stable attitude component is shown in Figure 11.13. The left picture corresponds to the test without active attitude management; it shows a case where the rear right wheel loses contact with the ground. The right picture shows the rover at the same approximate location with all wheels making ground contact while actively modulating its speed to maintain stable attitude.

Figure 11.13: Comparative Effect of Stable Attitude Management.

To further illustrate the effect of safe attitude management, we exercise the component in a laboratory experiment where the rover traverses a swath of terrain for ten meters. Synthetic attitude measurements are generated by sinusoidal functions of random amplitude to emulate changes in pitch and roll experienced on a hypothetical undulated and rough terrain. The amplitudes are uniformly distributed random numbers bounded by the maximum stable pitch and roll of the rover. It is assumed that the strategic navigation module recommends a constant normalized speed of 75 percent (of maximum allowable speed) throughout the traverse. The results of this experiment are shown in Figure 11.14 in plots of pitch, roll and v_{safe} (normalized) vs. distance. The strategic speed is shown in the speed distance plot as a dashed line. Observe that v_{safe} is modulated low in response to near-extreme attitudes. This is most apparent when both pitch and roll are simultaneously large in magnitude.

To test safe traction management, a benign portion of terrain comprising two distinct surface types (hard compact soil and gravel) is chosen on which the rover will be susceptible to wheel slippage when traversing the surface transition at nominally fast speeds. The scenario is depicted in Figure 11.15 where the rover is about to transition from a hard compact soil to gravel surface. The rover is commanded to traverse the transition with and without the safe traction management component activated. Again, without active traction management, the rover traverses the terrain at a nominally fast speed. With active traction management, the rover reduces its speed upon encountering a surface of lower perceived traction (as classified by the vision-based neural network classifier described earlier) according to the fuzzy logic rules presented in section 11.3.3.1. This reactivity mitigates the risk of excessive wheel slippage during transitions between and traversal on surfaces of different traction characteristics.

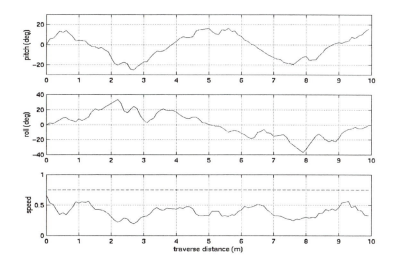

Figure 11.14: Speed Modulation for Attitude Management.

Figure 11.15: Rover Approaching Surface Type Transition.

To further illustrate the effect of the safe traction management, we exercise the component in a laboratory experiment where the rover traverses a 12 meter swath of terrain consisting of different surface types for which the traction coefficient C_t is 0.5 for 5 m, 0.2 for 3 m, and 0.9 for 4 m. We assume, for the sake of discussion, that these values correspond to sand, gravel, and concrete, and that the surface texture camera has a ground surface view horizon out to 0.3 m in front of the rover wheels. In this experiment, the strategic navigation module recommends a constant normalized speed of 80 percent throughout the 12 m traverse. The result is shown in Figure 11.16 where the recommended rover speeds are plotted vs. distance; the strategic speed is shown as a dashed line. Images of the three terrain surface types corresponding to distance are inset in the figure as well. As expected, changes in perceived traction result in reactive management of the safe speed recommended by the safe traction component to avoid the risk of excessive wheel slippage. Note that our laboratory experiment accounts for a reaction delay between classification of the surface type and the actual change in set points for v_{safe}. Thus far, our tests have revealed that v_{safe} is consistently lower than the strategic speed, thus exhibiting the caution of the safety module in reaction to cognizance of vehicle safety and changing "road" conditions.

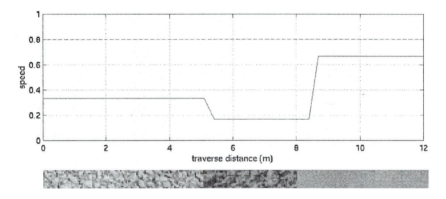

Figure 11.16: Speed Modulation for Traction Management.

11.6.2 Safe Navigation

The strategic navigation module was also tested in the field. In this section, we present results of a point-to-point navigation run in natural terrain. To navigate from a starting position to a user-specified goal position, the rover employs three navigation behaviors — seek-goal, traverse-terrain, and avoid-obstacle. The goal position is located approximately 20 m in front of the rover. Directly in between the starting and the goal positions are two regions having low traversability — one region contains a highly sloped hill and the other contains a large cluster of rocks. Figure 11.17 shows the path traversed by the rover from its original starting position until it has autonomously reached the specified goal position using its onboard fuzzy logic navigation rules. The rover begins by first analyzing the traversability of the three partitioned 60° sectors (left, front, right) of the terrain located in front of the rover. The front and left sectors (which contain the large sloped hill) are found to have low traversability. The rover therefore turns toward the right sector, which is found to be highly traversable, and proceeds to enter the safe region. Once in the safe region, the rover turns and navigates toward the goal, while ensuring that it is still physically located in the highly traversable sector; this corresponds to the last scene in the top row of images in Figure 11.17. Note that the viewpoint of the camera recording the path in Figure 11.17 is different for the top and bottom rows of images. Images on the top row are captured from a location behind the rover; the bottom row of images is captured from a location ahead of the rover.

After traversing a distance of about 10 m from start, the rover stops, turns toward the goal, and re-analyzes the traversability of the terrain ahead of it This time the front sector is found to have low traversability due to the large cluster of rocks located in this area. The left region is found to have low traversability due to the large sloped hill, and the right region is once again found to have high traversability. The rover thus turns to the right and proceeds into the safe region. At the point when the rover is within 1.5 m of the goal, the weight on the traverse-terrain recommendation is reduced automatically, and the seek-goal behavior becomes dominant. At this point, the rover heads directly toward the goal and stops when it is reached.

Figure 11.17: Navigation Path using Strategic Navigation Behaviors. Top-left Image Shows the Initial Position; Bottom-right Image Indicates Goal Achievement.

As shown in the sequence of test images, the navigation system directs the rover through the *safest* traversable regions. The combination of terrain assessment, safety, and strategic navigation modules in the safe navigation system thus demonstrates the viability of soft computing algorithms for enabling safe traversal of the rover on challenging terrain.

11.7   SUMMARY AND CONCLUSIONS

Safe and autonomous long range navigation of a rover on hazardous natural terrain offers significant technical challenges. An autonomous planetary rover must be able to operate intelligently with minimal interaction with mission operators on Earth. To accomplish this goal, the rover must have the onboard intelligence needed to traverse highly unstructured, poorly modeled terrain with a high level of robustness and reliability. For operation over extended time and distance, some capacity for built-in safe operation and health cognizance is required.   The rover onboard software intelligence must be capable of supporting real time navigation and motion planning based on poor and noisy sensor data. At the same time, it must be realizable in practical rover computing hardware. As such, efficient algorithms are essential for intelligent control.

As a goal, we have focused on achieving human driver performance through the application of soft computing techniques. This chapter presents the current state of development of a safe rover navigation system designed with this goal in mind. Various components of the safe navigation system are described in detail. Several soft computing solutions to different aspects of the rover navigation problem are also presented. Through this research and application experience, we have found that fuzzy logic provides a natural framework for expressing the human reasoning and decision-making processes for driving a rover on hazardous terrain. The human driving strategy can be transferred easily to the onboard rover navigation system and executed in real time.

Robot navigation strategies based on fuzzy logic offer major advantages over analytical methods. First, the fuzzy rules that govern the robot motion are easily understandable, intuitive, and emulate the human driver's experience. Second, the tolerance of fuzzy logic to imprecision and uncertainties in sensory data is particularly appealing for outdoor navigation because of the inevitable inaccuracies in measuring and interpreting the terrain quality data, such as slope and roughness. And third, the fuzzy logic strategy has a modular structure that can be extended very easily to incorporate new capabilities, whereas this requires complete reformulation for analytical methods. Multiple fuzzy behaviors can be blended readily into a unified navigation strategy that permits smooth interpolation between behaviors, thereby avoiding abrupt and discontinuous behavioral transitions.

The addition of the onboard terrain sensing and traversability analysis, coupled with the traverse-terrain behavior that takes advantage of this information, is a significant and novel contribution. These capabilities allow the navigation system to take preventive measures by looking ahead and preventing the rover from potential entrapment in rock clusters and other impassable regions and thus, guiding the vehicle to circumnavigate such regions. The technology described herein will lead to survivable rover systems that are of practical use for performing long duration missions involving long range traversal over challenging and high risk terrain.

ACKNOWLEDGMENT

The research described in this chapter was performed at the Jet Propulsion Laboratory, California Institute of Technology, under contract with the National Aeronautics and Space Administration.

REFERENCES

1.   DeLong, B., *4-Wheel Freedom: The Art of Off-Road Driving*, Paladin Press, Boulder, CO, 1996.

2.   Graydon, D. and Hanson, K. (eds.), *Mountaineering: The Freedom of the Hills*, The Mountaineers, Seattle, WA, 1997.

3.   Tunstel, E., Fuzzy-behavior Synthesis, Coordination, and Evolution in an Adaptive Behavior Hierarchy, in Saffiotti, A. and Driankov, D. (eds.) *Fuzzy Logic Techniques for Autonomous Vehicle Navigation*, Springer-Verlag Studies in Fuzziness and Soft Computing Series, Berlin/Heidelberg, 2000.

4.   Morrison, J. C. and Nguyen, T. T., On-Board Software for the Mars Pathfinder Microrover, *Proc. 2nd IAA Intl. Conf. on Low-Cost Planetary Missions*, IAA-L-0504, 1996.

5.   Washington, R., Golden, K., Bresina, J., Smith, D. E., Anderson, C., and Smith, T., Autonomous Rovers for Mars Exploration, *Proc. IEEE Aerospace Conf.*, 1999.

6.   Huntsberger, T.L. and Rose, J., BISMARC: A Biologically Inspired System for Map-Based Autonomous Rover Control, *IEEE Trans. on Neural Networks*, 11(7/8), 1497, 1998.

7.   DeCoste, D., Adaptive Resource Profiling, *Proc. Intl. Symp. on AI, Robotics and Automation in Space*, Noordwijk, Netherlands, 285, 1999.

8.   Rosenblatt, J. K., DAMN: A Distributed Architecture for Mobile Navigation, *J. Experimental and Theoretical AI*, 9(2/3), 339, 1997.

9.   Kelly A. and Stentz A., An Approach to Rough Terrain Autonomous Mobility, *Proc. Intl. Conf. on Mobile Planetary Robots*, 1, 1997.

10.   Wilcox, B.H., Non-Geometric Hazard Detection for a Mars Microrover, *Proc. NASA/AIAA Conf. on Intelligent Robotics in Field, Factory, Service, and Space*, 675, 1994.

11.   Arkin, R.C., Homeostatic Control for a Mobile Robot: Dynamic Replanning in Hazardous Environments, *J. Robotic Syst.*, 9(2), 197, 1992.

12.   Mauer, G. F., A Fuzzy Logic Controller for an ABS Braking System, *IEEE Trans. Fuzzy Syst.*, 3(4), 381, 1995.

13.   Bauer, M. and Tomizuka, M., Fuzzy Logic Traction Controllers and Their Effect on Longitudinal Vehicle Platoon Systems, California PATH Research Report, UCB-ITS-PRR-95-14, Univ. of California, Berkeley, 1995.

14.   Palm, R. and Storjohann, K., Slip Control and Torque Optimization using Fuzzy Logic, in Jamshidi, et al., (eds.), *Applications of Fuzzy Logic: Towards High Machine Intelligent Quotient Systems*, 8, Prentice-Hall PTR, Upper Saddle River, NJ, 1997.

15.   Colyer, R. E. and Economou, J. T., Soft Modeling and Fuzzy Logic Control of Wheeled Skid-Steer Electric Vehicles with Steering Prioritization, *J. Approx. Reasoning*, 22, 31, 1999.

16.   Van der Burg, J. and Blazevic, P., Anti-Lock Braking and Traction Control Concept for All-Terrain Robotic Vehicles, *Proc. IEEE Intl. Conf. on Robotics and Automation*, 1400, 1997.

17.   Sreenivasan, S. V. and Wilcox, B. H., Stability and Traction Control of an Actively Actuated Micro-Rover, *J. Robotic Syst.*, 11(6), 487, 1994.

18.   Marra, M., Dunlay, R. T., and Mathis, D., Terrain Classification using Texture for the ALV, *Proc. Mobile Robots III*, SPIE 1007, 64, 1988.

19.   Murphy, R. R. and Dawkins, D. K., Behavioral Speed Control Based
       on Tactical Information, *Proc. IEEE Intl. Conf. on Intell. Robots and
       Syst.*, 1715, 1996.

20.   Gennery, D. B., Traversability Analysis and Path Planning for a
       Planetary Rover, *J. Autonomous Robots*, 6, 131, 1999.

21.   Krotkov, E., Hebert, M., Henriksen, L., Levin, P., Maimone, M.,
       Simmons, R., and Teza, J.,   Field Trials of a Prototype Lunar Rover
       under Multi-Sensor Safeguarded Teleoperation Control, *Proc. Am.
       Nuclear Society 7th Topical Meeting on Robotics and Remote Syst.*, 1,
       575, Augusta, 1997.

22.   Langer, D., Rosenblatt, J. K., and Hebert, M.,   A Behavior-Based
       System for Off-Road Navigation, *IEEE Trans. on Robotics and
       Automation*, 10(6), 776, 1994.

23.   Shiller, Z. and Gwo, Y. R., Dynamic Motion Planning of Autonomous
       Vehicles, *IEEE Trans. on Robotics and Automation*, 7(2), 241, 1991.

24.   Iagnemma, K., Genot, F., and Dubowsky, S. Rapid Physics-Based
       Rough-Terrain Rover Planning with Sensor and Control Uncertainty,
       Proc. IEEE Intl. Conf. on Robotics and Automation, 3, 2286, Detroit,
       1999.

25.   Seraji, H., Traversability Index: A New Concept for Planetary Rovers,
       Proc. IEEE Intl. Conf. on Robotics and Automation, 3, 2006, Detroit,
       1999.

26.   Seraji, H., Fuzzy Traversability Index: A New Concept for Terrain-
       Based Navigation, *J. Robotic Syst.*, 17(2), 75, 2000.

27.   Horn, B., *Robot Vision*, MIT Press, MA, 1986.

28.   Volpe, R. and Ivlev, R. A Survey and Experimental Evaluation of
       Proximity Sensors for Space Robotics, *Proc. IEEE Intl. Conf. on
       Robotics and Automation*, 4, 3466, San Diego, 1994.

29.   Everett, H. R., *Sensors for Mobile Robots: Theory and Application*, A
       K Peters, Ltd., Wellesley, MA, 1995.

12 AUTONOMOUS UNDERWATER VEHICLE CONTROL USING FUZZY LOGIC

Feijun Song and Samuel M. Smith

12.1 INTRODUCTION

In this chapter, we will discuss the applications of fuzzy logic in autonomous underwater vehicle (AUV) control. In particular, we will discuss a special type of fuzzy logic controller named sliding mode fuzzy controller (SMFC) that combines the advantages of sliding mode control and fuzzy logic control. We will show how to design and tune a sliding mode fuzzy controller. The application of such controller structure to AUV control will also be shown.

As an emerging technique for oceanography measurement and littoral survey, AUVs have drawn much attention from researchers with different backgrounds. The pitch and heading control of an AUV forms the basis of any successful mission. However, the environment that a mission has to face gives many difficulties in the controller design. Wave and current are two basic environmental factors that are generally treated as external disturbance by a controller designer. Sensor measurement in such an environment also lends itself to noise. Furthermore, an AUV system is a real time distributed system that consists of many different components; therefore, sampling rate becomes another problem for the controller designer. Thus, robustness must be considered in the controller design yet the controller still needs to perform well.

Most current robust controller design methodologies require a system model. An AUV system is highly nonlinear and difficult to model, this adds more difficulties to the controller design. During the past several years of research and practice, we found that SMFC is a plausible control scheme for AUVs in tough environments. To discuss the application of SMFC in AUVs, we first briefly introduce the dynamics of an AUV, then the sliding mode control. The structure of SMFC is presented in detail, followed by the presentation of at sea experimental results. We conclude this chapter with a discussion of the advantages of SMFC.

12.2 BACKGROUND

Pitch and heading control are low level controls in AUV control architecture. Many control strategies have been adopted; among them are neural network control [1], sliding mode control [2,3], supervisory control [4], linear quadratic gaussian/loop transfer recovery method [5], self-tuning control [6], fuzzy logic control [7], etc. However, it is difficult to design time optimal controllers for

AUVs due to the fact that most optimal control design methodologies require analytical system models (equations of motion) of the AUVs, which is highly nonlinear and difficult to obtain.

Generally, for time optimal control, there exists a nonlinear switching curve where the bang-bang control should switch its sign. The curve also represents the maximal vehicle maneuvering capabilities in terms of time. A time optimal controller should be able to control the AUVs so that the same switching curve is always followed. A fuzzy logic controller can approximate this nonlinear switching curve since fuzzy systems are universal function approximators [8,9,10,11].

Sliding mode control is known for its robustness to the external disturbance and system modeling error. In order to have a controller that is not only time optimal, but also robust, a combination of sliding mode control and fuzzy logic control is needed. This results in the SMFC, in which each fuzzy rule output function is exactly a sliding mode controller. The slope of the sliding mode controller in each rule is determined by the approximate slope of the nonlinear switching curve in that partition of the phase plane that the rule covers. The nonlinearity of the switching curve thus is approximated by the fuzzy rules.

The approximation property of fuzzy logic control and robustness property of sliding mode control make the SMFC idea for AUV time optimal and robust control under rough sea state. However, as in fuzzy logic cotnroller design, the parameters for a SMFC are difficult to determine. An experimental method is presented in this chapter to determine those parameters. The method makes the design of controllers for complex highly nonlinear systems possible without any analytical representations of the system.

The method is based on Pontryagin's maximum principle [12]. Starting from a steady state under maximal rudder or stern plane deflection, an AUV's open loop pitch or heading response generally is a nonlinear curve in a phase plane. This curve represents the maximal vehicle maneuvering capabilities in terms of time. It also represents the switching line where the bang-bang control should switch its sign. A time optimal controller should be able to control the AUV so that the same switching curve is always followed. The parameters of a SMFC should be selected such that the experimental switching curve is approximated.

A pitch and a heading controller have been designed with at sea open loop experimental data generated by the Ocean Explorer (OEX) series AUVs. The at sea closed loop experimental data justified the methodology used to determine the controller parameters.

12.3   AUTONOMOUS UNDERWATER VEHICLES (AUVs)

OEX series AUVs were developed at the Ocean Engineering Department of Florida Atlantic University. The vehicle is depicted in Figure 12.1. It is 7.14 feet long with basic payload and 21 inches in diameter. The maximum cross sectional area is 2.4053 ft^2. Weight in air is 714.2 lbs and displaced weight is 716.7 lbs. Hull volume is 11.1931 ft^3. Following is a brief description of the basic vehicle configuration.

- A tear-drop shaped fiber glass hull based on a modified version of the Gertler Series 58 Model 4154 body shape;
- Aft-mounted cruciform control surfaces;
- A 3-bladed propeller 18 inches in diameter;
- Intelligent Ni-Cd battery packs. The battery packs can supply up to 12 hours of continuous missions at 3-knot cruising speed;
- Main computer and electronics board (MC68030 at 50 MHz on the VME bus). Each of the components is embedded with a LonWorks Neuron node, and the control communication is achieved via LonTalk protocol;
- Sensors include Watson AHRS-C302RS (3-axis acceleration, angles and rates), SIMRAD mesotech 809 (altitude), Druck PTX 1649 (water depth), Sonic Speed (water speed), Differential Global Positioning System, LBL and USBL positioning system.

In OEX series AUVs, the control of heading and pitch is achieved through the adjustment of rudder and stern plane. We will design two time optimal controllers for pitch and heading control of OEX series AUVs. The inputs for the pitch controller are the pitch error and pitch error rate. The output of the pitch controller is the deflection of the stern plane. The inputs for the heading controller are heading error and heading error rate. The output of the heading controller is the deflection of the rudder.

For a detailed system dynamics of the AUV, please refer to reference [13,14].

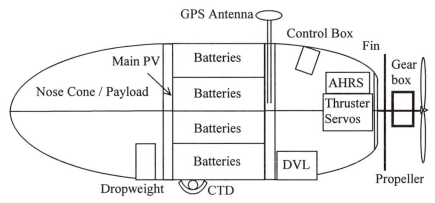

Figure 12.1: OEX Series AUVs.

12.4 SLIDING MODE CONTROL

A sliding mode controller (SMC) is a variable structure controller (VSC). Basically, a VSC includes several different continuous functions that can map plant state to a control surface; the switching among different functions is determined by plant state that is represented by a switching function.

Without loss of generality, consider the design of a sliding mode controller for the following second order system:

$$\ddot{x} = f(x, \dot{x}, t) + bu(t) \tag{12.1}$$

Here we assume $b > 0$. $u(t)$ is the input to the system. The following is a possible choice of the structure of a sliding mode controller [15,16]:

$$u = k\,\text{sgn}(s) + u_{eq} \tag{12.2}$$

where u_{eq} is called equivalent control which is used when the system state is in the sliding mode [16,17]. k is a constant. k is the maximal value of the controller output. s is called switching function because the control action switches its sign on the two sides of the switching surface $s = 0$. s is defined as [15,18]:

$$s = \dot{e} + \lambda e \tag{12.3}$$

where $e = x - x_d$ and x_d is the desired state. λ is a constant. $\text{sgn}(s)$ is a sign function, which is defined as:

$$\text{sgn}(s) = \begin{cases} -1 & \text{if } s < 0 \\ 1 & \text{if } s > 1 \end{cases} \tag{12.4}$$

The control strategy adopted here will guarantee the system trajectories move toward and stay on the sliding surface $s = 0$ from any initial condition if the following condition is met:

$$s\dot{s} \leq -\eta|s| \tag{12.5}$$

where η is a positive constant that guarantees the system trajectories hit the sliding surface in finite time [15].

Using a sign function often causes chattering in practice. One solution is to introduce a boundary layer around the switch surface [17]:

$$u = k\,sat\left(\frac{s}{\phi}\right) + u_{eq} \tag{12.6}$$

where constant factor ϕ defines the thickness of the boundary layer. $sat\left(\frac{s}{\phi}\right)$ is a saturation function that is defined as:

$$sat\left(\frac{s}{\phi}\right) = \begin{cases} \dfrac{s}{\phi} & \text{if } \left|\dfrac{s}{\phi}\right| \leq 1 \\ \text{sgn}\left(\dfrac{s}{\phi}\right) & \text{if } \left|\dfrac{s}{\phi}\right| > 1 \end{cases} \tag{12.7}$$

This controller is actually a continuous approximation of the ideal relay control [15,16]. The consequence of this control scheme is that invariance of sliding mode control is lost. The system robustness is a function of the width of the boundary layer.

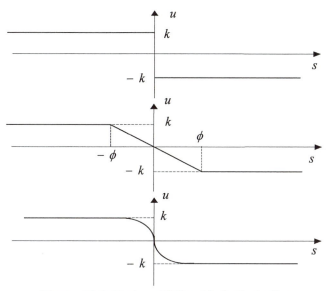

Figure 12.2: Various Sliding Mode Controllers.

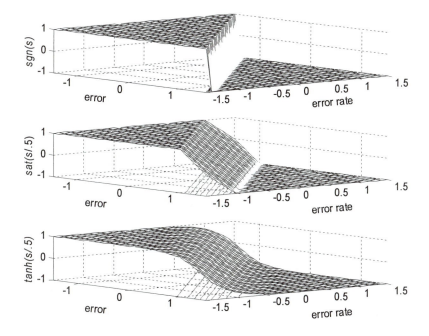

Figure 12.3: Control Surfaces for Various Sliding Mode Controllers.

A variation of the above controller is to use a hyperbolic tangent function instead of a saturation function [3,19]:

$$u = k \tanh(\frac{s}{\phi}) + u_{eq} \qquad (12.8)$$

Figure 12.2 shows different sliding control laws for a two-dimensional system, Their corresponding control surfaces are shown in Figure 12.3.

It is proven that if k is large enough, the sliding model controllers in equations 12.2, 12.6 and 12.8 are guaranteed to be asymptotically stable [16,19].

12.5   SLIDING MODE FUZZY CONTROL (SMFC)

Pontryagin's maximum principle states that for two-dimensional time optimal controller design, there exists a nonlinear switch curve so that the control can have maximal value on one side of the switching curve and minimal value on the other side of the curve. The nonlinear switching curve often has the form depicted in Figure 12.4. Figure 12.4 also shows there can be a switching band around the switching line to alleviate chattering.

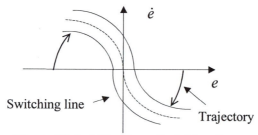

Figure 12.4: A Nonlinear Switching Curve.

There are two problems associated with nonlinear time optimal controller design: First, how to get the true switching curve because for nonlinear systems, this switching curve is very difficult to get analytically. The second problem is how to approximate the nonlinear curve.

To solve the first problem, we could use system open loop experimental data. Under the maximal control command, the system output should be saturated after a period of time. The nonlinear open loop response can be used as a switching curve since it represents the system's fastest response. For example, if the maximal rudder angle is delivered to an AUV constantly, then the yaw rate of the AUV will be gradually saturated as segment AB in Figure 12.5. The delivering of minimal rudder angle in the opposite direction generates the segment CD shown in Figure 12.5.

This open loop response actually represents the maximal maneuvering capability of the AUV. In other words, the curves AB and CD are the quickest way the AUV can move. These curves can be used as switching curves in time optimal controller design.

Another approach to obtain this nonlinear switching line is to use computer aided controller automatic design and optimization methods. We have developed a very efficient and effective cell state space based search algorithm to automatically optimize a general type of controller. For a detailed description of this algorithm and other related materials, please refer to references [20,21,22,23,24].

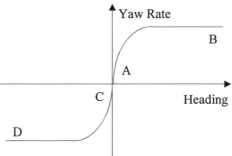

Figure 12.5: Open Loop Step Response for Maximal and Minimal Rudder.

The advantage of the open loop experimental approach for obtaining a nonlinear switching line is that a mathematical model of the system is not necessary. However, due to sensor noise and other factors, the experimental switching line might not be exact, which leads to a less optimal controller. Computer aided automatic method can achieve higher accuracy, but a computational model is always needed.

Once the switching line is obtained, the next step is to approximate this line. Obviously, traditional linear controller design can only linearly approximate this nonlinear curve. We could very well use a Takagi-Sugeno (TS) type fuzzy logic controller to approximate this nonlinear curve. Fuzzy logic controller has been proven to be able to approximate any nonlinear curve with arbitrary accuracy [9].

Generally, a time optimal controller is not robust. The performance of a time optimal controller degrades severely with the external disturbance, measurement noise, or system dynamics changes. To add robustness to a time optimal controller, we need to combine sliding mode control and fuzzy logic control.

In a TS type FLC, the rule output function typically is a linear function of controller inputs. The mathematical expression of this function is similar to a switching function. This similarity indicates that the information from a sliding mode controller can be used to design a fuzzy logic controller, resulting in a sliding mode fuzzy controller. Wu proposed such an approach in which parameters in the output functions for different rules that cover different partitions of the state space are determined by different sliding mode controllers that also cover the corresponding partitions of the state space [25]. The resulting controller is still a typical TS type FLC. In fact, since a fuzzy system can seamlessly connect different control strategies into one system, one can take an even more direct approach to incorporate sliding mode controllers into a fuzzy logic controller [26]. In Xu's approach, each rule is a sliding mode controller. The SMC in each rule can have various forms. The boundary layer and the coefficients of the sliding surface become the coefficients of the rule output function.

The i th rule for an SMFC is expressed as follows:

$$\text{IF } e \text{ is } A_i \text{ and } \dot{e} \text{ is } B_i \text{, THEN } u_i = ksat(\frac{\dot{e} + \lambda_i e + c_i}{\phi_i})$$

Notice that the rule output function is not necessarily a saturation function. It could be a sign function or hyperbolic tangential function too. The fuzzification of e and $\dot{e}$ are illustrated in Figure 12.6.

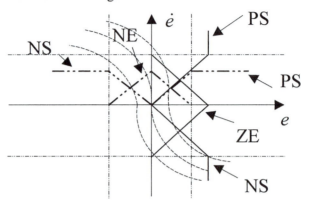

Figure 12.6: Fuzzification of e and $\dot{e}$.

The constant coefficients of λ_i and c_i are determined by the open loop experimental data. They are determined in such a way that the slope of the nonlinear switching curve is followed. Usually, the at sea data has oscillation that reflects the environmental disturbance and measurement noise. The magnitude of the oscillation can be used to determine the coefficient ϕ_i.

Notice that in Figure 12.4 the switching curve can be either a function of e, or a function of $\dot{e}$. Fewer rules are needed to approximate this one-dimensional function. This is how an SMFC reduces the rule base size. A typical rule for the simplified rule base is the following:

$$\text{IF } e \text{ is } A_i \text{, THEN } u_i = ksat(\frac{\dot{e} + \lambda_i e + c_i}{\phi_i})$$

We will use this simplified rule to construct a pitch and a heading controller for an AUV.

12.6   SMFC DESIGN EXAMPLES

A sliding mode fuzzy pitch controller and a sliding mode fuzzy heading controller have been designed for the OEX series AUVs. The inputs to the sliding mode fuzzy heading controller are heading error and heading error rate.

The output is rudder deflection. Figure 12.7 shows the fuzzy sets for the heading errors. There are no fuzzy sets for heading error rate. Table 12.1 shows the rule base of the heading controller.

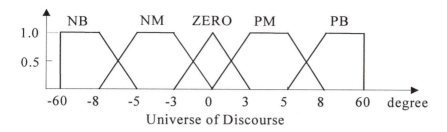

Figure 12.7: Fuzzy Sets for Heading Error.

Table12.1: Rule Base for the Sliding Mode Fuzzy Heading Controller

5 rules for the sliding mode fuzzy heading controller, $k = 20$ degree			
Antecedents	Output Functions		
Heading error e	Thickness ϕ_i	Slope λ_i	Offset c_i
PB	2.5	0.01	13.5
PM	3.0	1.50	1.0
ZERO	3.0	2.00	0.0
NM	3.0	1.50	-1.0
NB	2.5	0.01	-13.5

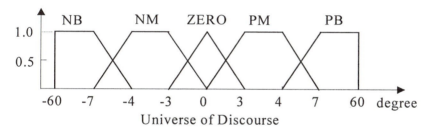

Figure 12.8: Fuzzy Sets for Pitch Error.

The inputs to the pitch controller are pitch error and pitch error rate. The output is stern plane deflection. Figure 12.8 shows the fuzzy sets for the pitch errors. There are no fuzzy sets for pitch error rate. Table 12.2 shows the rule base of the pitch controller.

Figure 12.9 and Figure 12.10 show the control surface and its contour of the sliding mode fuzzy heading controller. Figures 12.11 and 12.12 show the control surface and its contour of the sliding mode fuzzy pitch controller.

Again, we need to emphasize that although the parameters in Table 12.1 and Table 12.2 are from AUV experimental data, there should be trials and errors before we determine the final values. The offset c_i helps to adjust the contour of the resultant control surface. Each time we come to a set of parameters, the corresponding control surface and contour will be plotted. The parameters will be ajdusted slightly to generate a better shaped control surface and contour.

Table 2: Rule Base for the Sliding Mode Fuzzy Pitch Controller

5 rules for the sliding mode fuzzy pitch controller, $k = 20$ degree			
Antecedents	Output Functions		
Pitch error e	Thickness ϕ_i	Slope λ_i	Offset c_i
PB	1.5	0.01	13.5
PM	3.0	2.00	1.5
ZERO	3.0	3.00	0.0
NM	3.0	2.00	-1.5
NB	1.5	0.01	-13.5

Another issue that must be clarified here is that although the original values for the controller parameters are from experimental data, the tuning (trial and error) was done in a simulation environment where the OEX series AUVs were modeled by a six degree-of-freedom (DOF) nonlinear model. Each time the controller parameters were adjusted, the new controller would be tested in the simulation environment. The final controller was then ported to vehicles for at-sea tuning. The at sea tuning of the controller took about one week. The use of simulation toolbox for controller tuning will be covered in other publications.

The design and tuning of an SMFC are summarized in the following steps:

Step 1: Determine a nonlinear switching line and its necessary boundary as in Figure 12.4. As discussed before, there are two ways to find such a line and its boundary. The first way is to use experimental open loop system repsonse to the maximal physical control command. The difference among repeated experiments will give us a rough idea of how thick the boundary layer should be. This difference generally reflects typical sensor noise level, typical external disturbance level and system parameter changes, etc. A properly selected boundary will conpensate for those changes in real control. The second way to find a switching line is to use a system model. With a system model, the system response to the maximal physical control command can be easily obtained and be used in determining the switching line. However, with a system model, the thickness of a boundary layer is not so easy to obtain since there is no means to reflect sensor noise level and external disturbance by model computation. In this case, the specifications of the sensors used in the system can be utilized to determine the thickness of a boundary layer, that is, to determince ϕ_i.

Step 2: Fuzzify the controller inputs as in Figure 12.7 and Figure 12.8. In this step, we need to determine the number of membership functions for each controller input. We also need to determine what kind of membership function should be used. This step is very important to the successful approximation of the nonlinear switching line found in step 1. Intuitively, the more the membership functions, the better the approximation. However, more membership functions mean more rules and more computation complexity. A rule of thumb is to have at least three and at most nine membership functions for each controller input [27]. Generally, for a number of membership functions less than five, gaussian type functions are prefered. Triangle functions are adequate if the number of membership functions is more than five [28].

Step 3: Construct a rule base. After all the controller inputs have been fuzzified, a rule base can be constructed. Some rule reduction method can be applied here although sliding mode fuzzy control already has the potential to reduce a rule base.

Step 4: Choose a defuzzification method. Since real physical systems often require a crisp control command, we need to defuzzify a controller output. There are many defuzzification methods [29,30,31,32]. The most adopted one is averaged sum.

Step 5: Determine rule output function parameters. That is, determine different ϕ_i, λ_i and c_i. Once the controller inputs are fuzzified and the switching line is obtained, the only way to better approxmiate the switching line is to tune the rule output function parameters. Often, initial values for ϕ_i, λ_i and c_i are chosen based on the switching line and boundary layer found in step 1. After that, the parameters are tuned to have a better approximation. A trial and error method is often adopted in this step. A few guildelines on how to tune these parameters are given later in this chapter. If the slidng mode fuzzy controller can not approximate the switching line satisfactorily by tuning ϕ_i, λ_i and c_i only, the designer may need to increase the membership functions for each controller input. In this case, the design goes back to step 2.

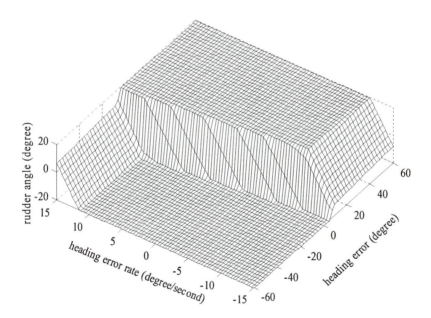

Figure 12.9: Control Surface of the Sliding Mode Fuzzy Heading Controller.

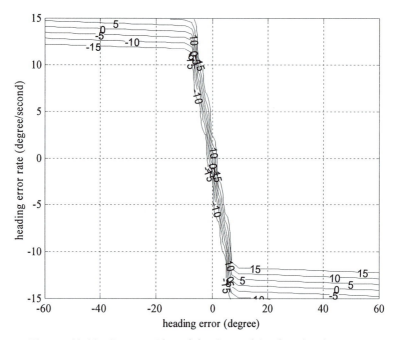

Figure 12.10: Contour Plot of the Control Surface in Figure 12.9.

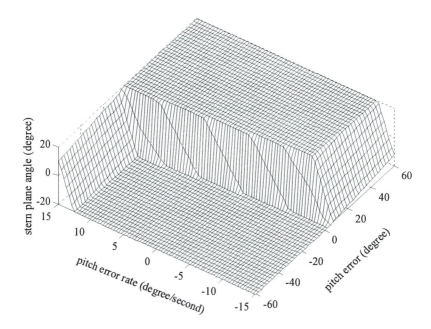

Figure 12.11: Control Surface of the Sliding Mode Fuzzy Pitch Controller.

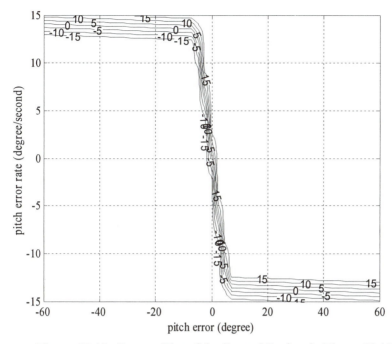

Figure 12.12: Contour Plot of the Control Surface in Figure 12.11.

Step 6: Check the designed sliding mode fuzzy controller performance with a system model. If a system model is not available, this step can be skipped. In this step, the controller rule output function parameters can be further tuned in order to have a better performance instead of a better approximation. This step can save lots of field test efforts if the controllers are properly tuned.

Step 7: Field experiments. This is the last step in the controller tuning. Once the controller design is done and all the rule ouput function parameters are tuned in the steps above, the performance of the controller should be tested with the real system. In this step, controller input membership functions are generally unchaged, but the rule output function parameters can be further tuned to have satisfactory field performance.

12.7 GUIDELINES FOR ONLINE ADJUSTMENT

One of the advantages of SMFC over conventional TS type fuzzy logic control is that all the parameters in a sliding mode fuzzy controller have their own physical meanings, making the online adjustment of a sliding mode fuzzy controller much easier than the online adjustment of a fuzzy logic controller. One can use the experience and knowledge on sliding mode control to adjust the rule output functions of a sliding mode fuzzy controller. Furthermore, the structure of a sliding mode fuzzy controller also opens the door to the online adaptive control. As one may notice, the parameters of a typical TS type FLC have no physical meaning. An online adaptive fuzzy controller scheme often involves complicated adaptation on every piece of rule in each adaptation iteration. However, with a sliding mode fuzzy controller, one can adapt only one rule, or a subset of the rule base. The computation would be much less, which is another attractive feature in real time systems.

Below are some salient features associated with sliding slope λ and thickness of boundary layer ϕ that can be used as guidelines for sliding mode fuzzy controller online adjustment.

12.7.1   Sliding Slope λ Effects

Sliding slope will have much influence on how fast the controller responds. The bigger the sliding slope, the faster the controller and the less stable it tends to be.

When the slope is larger, the controller is more robust, but the chattering might be worse. This is equivalent to a high gain controller; the rise time could be smaller, the overshoot bigger, and the settling time larger; the robustness to varying sample rates is worse, whereas the robustness to parameter variations and disturbance is better.

12.7.2   Thickness of the Boundary Layer ϕ Effects

When thickness of the boundary layer is larger, chattering decreases or disappears; robustness to varying sample rates could be better, but robustness to parameter variations and disturbance could be worse; the steady-state error can be larger.

12.8   AT SEA EXPERIMENTAL RESULTS

Figure 12.13 shows at sea test data. The commanded heading was first set to $20°$ and then to $200°$. The heading controller successfully controlled the vehicle to the desired heading. There is a two degree oscillation in the at sea heading data. This is normal since the sea environment is not clean.

Figure 12.14 shows the performance of the pitch controller. Since in the at sea test the vehicle cannot have a constant pitch with limited sea depth, the performance of the pitch controller was tested through the depth controller. In OEX series AUVs, the depth control is done through the pitch control. The control output from the depth controller is a desired pitch angle. It is the responsibility of the pitch controller to drive the vehicle to the desired pitch. The depth controller used in the test is a linear proportional-derivative controller. In Figure 12.14, the pitch controller successfully drives the vehicle to a five meter depth.

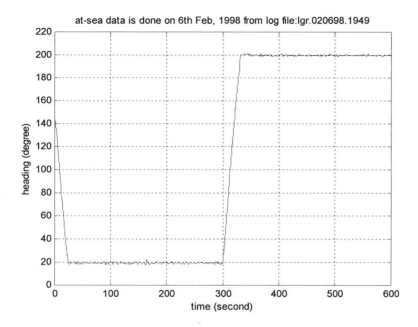

Figure 12.13: At Sea Test for Sliding Mode Fuzzy Heading Controller.

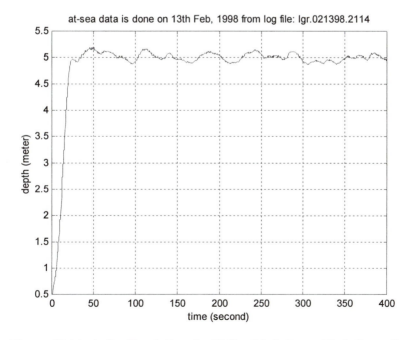

Figure 12.14: At Sea Depth Test for Sliding Mode Fuzzy Pitch Controller.

12.9   SUMMARY

The structure of sliding mode fuzzy control was presented in this chapter. A sliding mode fuzzy controller inherits the interpolation property of fuzzy logic control and robustness property of sliding mode control, therefore making it ideal for time optimal robust control. The at sea experimental data shows that sliding mode fuzzy control can be successfully applied to AUVs, therefore making them another alternative for the robust time optimal control problem. Moreover, since the physical meaning of the rule output function parameters for a sliding mode fuzzy controller is straightforward, the online tuning is made easy. This is exactly why sliding mode fuzzy control is adopted in AUV development at Florida Atlantic University.

REFERENCES

1.   Yuh, J., A Neural Net Controller for Underwater Robotic Vehicles, *IEEE J. of Oceanic Eng.*, vol.15, no.3, pp.161-166, 1990.
2.   Yoerger, D.R. and Slotine, J. E., Robust Trajectory Control of Underwater Vehicles, *IEEE J. of Oceanic Eng.*, vol.10, no.4, pp.462-470, 1985.
3.   Healey, A.J. and Lienard, D., Multivariable Sliding Mode Control for Autonomous Diving and Steering of Unmanned Underwater Vehicles, *IEEE J. of Oceanic Eng.*, vol.18, no.3, pp.327-339, 1993.
4.   Yoerger, D.R., Newman, J.B., and Slotine, J. E., Supervisory Control System for the JASON ROV, *IEEE J. of Oceanic Eng.*, vol.11, no.3, pp.392-399, 1986.
5.   Triantafyllou, M.S. and Grosenbaugh, M.A., Robust Control for Underwater Vehicle Systems with Time Delays, *IEEE J. of Oceanic Eng.*, vol.16, no.1, pp.146-151, 1991.
6.   Goheen, K.R. and Jefferys, E.R., Multivariable Self-Tuning Autopilots for Autonomous and Remotely Operated Underwater Vehicles, *IEEE J. of Oceanic Eng.*, vol.15, no.3, pp.144-151, 1990.
7.   Smith, S.M., Rae, G.J.S., and Anderson, D.T., Applications of Fuzzy Logic to the Control of an Autonomous underwater Vehicle, *IEEE Int. Conf. on Fuzzy Syst.*, pp.1099-1106, San Francisco, CA, 1993.
8.   Zadeh, L.A., Fuzzy Sets, *Inf. and Control*, vol.8, pp.338-353, 1965.
9.   Ying, H., General Takagi-Sugeno Fuzzy Systems Are Universal Approximators, *IEEE Int. Conf. on Fuzzy Syst.*, pp.819-823, Anchorage, AL, 1998.
10.   Mamdani, E.H. and Assilian, S., An Experiment in Linguistic Synthesis with a Fuzzy Logic Controller, *Int. J. of Man-Machine Stud.*, vol.7, no.1, pp.1-12, 1975.
11.   Takagi, T. and Sugeno, M., Fuzzy Identification of Systems and Its Applications to Modeling and Control, *IEEE Trans. on Syst., Man, and Cybern.*, vol. 15, pp.116-132, 1985.
12.   Pontryagin, L.S., Boltyanskii, V.G., Gamkrelidze, R.V., and

Mishchenko, E.F., *The Mathematical Theory of Optimal Processes*, Interscience Publishers, Inc., New York, 1962.

13.   Humphreys, D.E., Vehicle Hydrodynamics & Maneuvering Model for the FAU Ocean Explorer Vehicle (OEX), V.C.T. Technical Memorandum 96-05, Vehicle Control Technologies, Inc., 1996.

14.   Humphreys, D.E., Correlation of VCT Maneuver Model with FAU Ocean Explorer In-water Test Data, V.C.T. Technical Memorandum 97-05, Vehicle Control Technologies, Inc., 1997.

15.   Hung, J. Y., Gao, W., and Hung, J. C., Variable Structure Control: A Survey, *IEEE Trans. on Ind. Electron.*, vol.40, no.1, pp.2-21, 1993.

16.   Slotine, J.J., Sliding Controller Design for Nonlinear Systems, *Int. J. of Control*, vol.40, no.2, pp.421-434, 1984.

17.   Slotine, J.J. and Sastry, S.S., Tracking Control of Nonlinear Systems Using Sliding Surfaces with Application to Robot Manipulators, *Int. J. of Control*, vol.38, no.2, pp.465-492, 1983.

18.   Utkin, V.I., Sliding Modes in Control and Optimization, Springer-Verlag, NY, 1992.

19.   Glower, J.S. and Munighan, J., Designing Fuzzy Controllers from a Variable Structures Standpoint, *IEEE Trans. on Fuzzy Syst.*, vol.5, no.1, pp.138-144, 1997.

20.   Song, F. and Smith, S.M., Cell State Space Based Incremental Best Estimate Directed Search Algorithm for Takagi-Sugeno Type Fuzzy Logic Controller Automatic Optimization, *IEEE Int. Conf. on Fuzzy Syst.*, pp.19-24, San Antonio, TX, 2000.

21.   Song, F. and Smith, S.M., How Blind Can a Blind Fuzzy Logic Controller Design Be? Analysis of Cell State Space Based Incremental Best Estimate Directed Search Algorithm, *IEEE Int. Conf. on Fuzzy Syst.*, pp.134-139, San Antonio, TX, 2000.

22.   Song, F. and Smith, S.M., Cell State Space Based Incremental Best Estimate Directed Search Algorithm for Robust Fuzzy Logic Controller Optimization with Multi-model Concept, *IEEE Int. Conf. on Fuzzy Syst.*, pp.1001-1004, San Antonio, TX, 2000.

23.   Song, F. and Smith, S.M., Design of Sliding Mode Fuzzy Controllers for an Autonomous Underwater Vehicle without System Model, *OCEANS'2000 MTS/IEEE*, pp.835-840, Providence, RI, 2000.

24.   Song, F. and Smith, S.M., Automatic Design and Optimization of Fuzzy Logic Controllers for an Autonomous Underwater Vehicle, *OCEANS'2000 MTS/IEEE*, pp.829-834, Providence, RI, 2000.

25.   Wu, J.C. and Liu, T.S., A Sliding-Mode Approach to Fuzzy Control Design, *IEEE Trans. on Control Syst. Techn.*, vol.4, no.2, pp.141-150, 1996.

26.   Xu, M., High Performance and Robust Control, Ph.D. Dissertation, Florida Atlantic University, 1996.

27.   Miller, G. A., The Magical Number Seven, Plus or Minus Two: Some Limits on Our Capacity for Processing Information, *The Psychological Review,* vol.63, no.2, pp.81-97, 1956.

28.   Rondeau, L., Levrat, E., and Bremont, J., Analytical Formulation of the Influence of Membership Functions Shape, *IEEE Int. Conf. on Fuzzy Syst.*, pp.1314-1319, New Orleans, LA, 1996.

29.   Lee, C.C., Fuzzy Logic in Control Systems: Fuzzy Logic Controller-Part I, II, *IEEE Trans. on Syst., Man and Cybern.*, vol.20, no.2, pp.404-435, 1990.

30.   Cordon, O., Herrera, F., and Peregrin, A., Applicability of the Fuzzy Operators in the Design of Fuzzy Logic Controllers, *Fuzzy Sets and Syst.*, vol.86, no.1, pp.15-41, 1997.

31.   Kiendl, H., Non-translation-invariant Defuzzification, *IEEE Int. Conf. on Fuzzy Syst.*, pp.737-742, Barcelona, Spain, 1997.

32.   Runkler, T.A., Extended Defuzzification Methods and Their Properties, *IEEE Int. Conf. on Fuzzy Syst.*, pp.694-700, New Orleans, LA, 1996.

13 APPLICATION OF FUZZY LOGIC FOR CONTROL OF HEATING, CHILLING, AND AIR CONDITIONING SYSTEMS

Reza Talebi -Daryani

13.1   INTRODUCTION

The building energy management (BEMS) concept was introduced in the early 1970s during the world's first big energy crisis. The oil crisis was the driving force of the intelligent building. It was the first sign of the rising awareness that energy resources are exhaustible. The second driving force of intelligent building was the raising awareness of environmental pollution by inefficient consumption of energy in production lines as well as in buildings in the beginning of the 1980s.

The expansion of computer technology in the early 1980s introduced a "smart" or "intelligent" building which was one step towards a new digital computer era. It provided energy efficiency as well as optimum environmental conditions. Managing the high tech buildings in an energy efficient manner and to the occupants' satisfaction would have become an impossible task without intelligent control systems. On the other hand, an intelligent building is one that creates an environment that maximizes the efficiency of the occupants of the building while at the same time allowing effective management of energy resources with minimum costs. The intelligence of a building depends on the elements that go to make up its intelligence. There are at least three attributes that an intelligent building should possess:
1.   The building should know what is happening inside and immediately outside.
2.   The building should decide the most efficient way of providing a convenient comfortable and productive environment for the occupants.
3.   The building should „ response" quickly to occupants' requests.

These attributes may be translated into a need for various technology and management systems. The successful integration of these systems will produce the intelligent building containing a building automation system in order to enable the building to respond to external climate factors and conditions. Simultaneous sensing, control, and monitoring of the internal environment and storage of the data generated as knowledge of the building performance in a central computer system, is an important feature of an intelligent building .[1]

This chapter is organized as follows. Section 13.2 describes general features of the building energy management systems. Section 13.3 is devoted to the Fuzzy control vs distributed digital control (DDC) for an air condition system. Section 13.4 discuss the fuzzy control for the operation of a complex chilling system. The description of various fuzzy control blocks are presented in this section. Fuzzy control for energy management of a heating center is introduced in section 13.5. Finally section 13.6 provides the conclusion of this chapter.

13.2   BUILDING ENERGY MANAGEMENT SYSTEM (BEMS)

13.2.1   System Requirements

The thermodynamic processes involved either intend to ensure the well being of the occupants of the building or consist of ancillary production conditions of a physical nature. This should be controlled by means of a technical future-oriented automation system which is physically ideal, economical, cost effective and efficient in terms of energy consumption. An integrated building automation system should also include all other technical and administrative processes that may be automated for reasons of security and rationalization, in order to increase the productivity of the building. The assignment of the BEMS is to run the building in such a way that following requirements as the state of the art should be fulfilled: [2]

- reducing the energy consumption and environmental pollution
- security for man, machine, production and environment
- improving the efficiency of the process and reducing processing time
- improving transparency of the process features by useful instrumentation
- operation-oriented maintenance management of technical installations in order to   increase machine running time and reduce maintenance costs

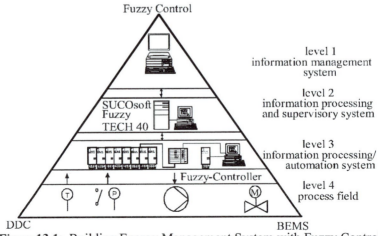

Figure13.1: Building Energy Management System with Fuzzy Control.

- historical and dynamic data processing, presentation, and analysis
- ensuring the well being of the occupants in order to improve productivity
- ensuring ancillary production and research-oriented climate conditions
- reducing energy consumption by optimal operation of the system.

To realize all of these functions, it is evident that a powerful control and automation system with different levels of information processing, as shown in Figure 13.1, must be installed in the building.

A perfect integrated building automation system allows both physical and functional access to all the data in the building. Integration cannot be said to exist unless data communication among the various systems is possible in accordance with requirements. For this purpose, an analysis of information requirements is essential i.e., safe access to the right information in a structured fashion at the right place.

Functions suited to inclusion in an integrated building energy management system could be, for example as shown in Figure 13.2:

- heating, ventilation, air conditioning, cooling (HVAC) - automation systems in buildings
- supervisory systems for energy management and operational tasks
- automation and networking of production facilities
- video monitoring and personal surveillance
- clocking-in systems, fire alarms
- optical and acoustic information processing
- lighting control, elevator control
- further administrative, communication and data processing functions
- maintenance and facility management.

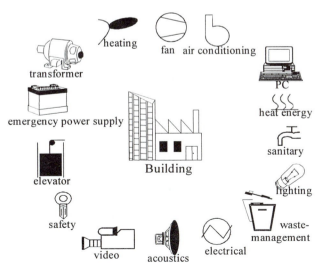

Figure13.2: Application Field for Integrated Building Automation System.

13.2.2   System Configuration

At the heart of building energy management is the building supervisory control system, which consists of a hierarchically organized, function oriented control system having separate intelligent automation units. The following aspects have to be taken into account:

- each level must be able to operate independently
- data interchange must be reduced to a minimum
- the operational readiness of machinery may not be impaired by a breakdown in communication interchange.

Regarding the control aspects, a powerful supervisory system is based on distributed intelligence. The distributed intelligence concept is a concept where intelligent outstations (controllers) are connected with each other by a communication bus (network). The building supervisory control systems with it's distributed intelligent is configured into four hierarchical Information processing levels, as shown in Figure 13.1.

13.2. 3   Automation Levels

1. Information and Management Level
At this level, physical and technical data relating to the building and emanating from the lower control and automation levels may be accessed in a condensed form and processed. In the main, this is done on workstations using user-friendly software interfaced to the various automation levels. The initial function of this level is to analyze the operating status of the systems.

2. Supervisory Control Level
The main function of this level is to control, monitor and log the processes within the building as a whole, but it serves also for configuring automation units and measurements and control units at a level three and setting their parameters. Further functions at level two are:

- data processing, data recording
- maintenance management of technical installations
- energy management

3. Automation Level
The automation level houses the distributed intelligence for mathematical - and physical-based operation functions as outstation multicontrollers. The purpose of the distributed digital control (DDC) is to monitor and control the most important statuses and processes within the building.

The control system, which also provides programmable controller (PLC) functions, allows a logical link to be set up in the form of time or status elements, in   order to guarantee optimum performance and security of installations. This level consists of a number of DDC controller outstations. Each panel is fully programmable and autonomous in operation. They coordinate communication upwards to the central computer, horizontally to other outstations.

4. Field Level

This level is the most basic level and houses the sensors and actuators which are, to a large extent, directly linked to the automation systems at level three. Most of these sensors and actuators are only available as analogue units so communication with levels two and one is only possible via level three. Within the field of application of the intelligent building, more and more bus-compatible-systems will be available on the market in future.

13. 3 AIR CONDITIONING SYSTEM: FLC vs DDC

The aim of the realized project, introduced here, was the application of a temperature cascade control system based on fuzzy logic for a common non linear air conditioning. The inner controller of the cascade control system has fuzzy PID characteristics.

In order to reduce the number of rules, the integral part of the controller is realized in the output of the fuzzy controller. The realization of the integral part through a second output is a unique solution with three advantages : reduction of rules, easy adjustment of the integral part, and use of the additional information of the input.

The additional inputs besides temperature are also to be fuzzified in order to give the controller a better characteristic for fine tuning.

The controller is adjustable over the whole output range independently from the nonlinear working characteristics of the process. The inner controller consists of 99 rules. The fuzzy controller can be adjusted independent of a working point so that the adjustment for the whole output range is optimal.

Fuzzy controller is superior to the digital controller for open loops with intensive nonlinear characteristics like the air supply system. The resulting improvement of the fuzzy control loop behavior is proved by comparing the system result with the loop response of a digital control system.

13.3.1 Process Description

The air conditioning system (Figure.13.3), with its control loops for a temperature cascade control, is used for education and research on building energy management systems. In its structure it is equivalent to common industrial applications. The room temperature ϑ is controlled by conditioned supply air. The supply air system consists of motorized dampers (Damp), a preheat exchanger (Ph), a chiller (Ch) and a second heat exchanger (Sh) to condition the air temperature. The steam humidifier (Hu) and filter (Fi) are not taken into account for the temperature control. [3]

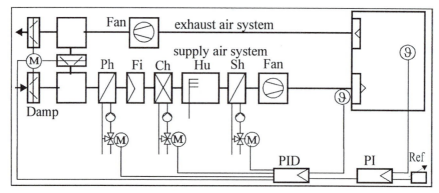

Figure 13.3: Cascade Control Schematic of an Air Conditioning System.

13.3.2 Process Control

The open loop of the supply air temperature of an air conditioner is known within the HVAC- system as one of the most difficult controlled open loops on account of its large degree of time delay and the nonlinear characteristics of the heat exchangers, as shown in Figure 13.4.

Figure 13.5 illustrates the open loop gain of the system as a function of the control valve position. In order to compensate the nonlinear behavior of the system, a cascade controller is used in order to compensate for the nonlinear characteristic of the heat exchanger and improve the control loop behavior.

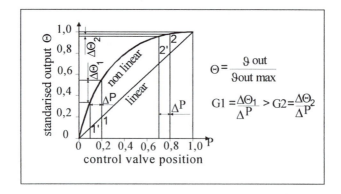

Figure 13.4: Heat Exchanger Operation Characteristic [4].

The structure of the cascade controller shown in Figure 13.3 permits a subdivision of the open loop and the solving of the control problem in several steps with simpler control circuits. The nonlinearity of the open loop is partially compensated outwards by the cascading of the room air temperature loop, but the nonlinear characteristics of the heat exchangers still remain.

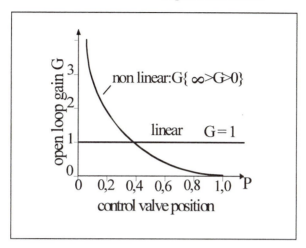

Figure 13.5: Nonlinear Open Loop Gain of the Heat Exchanger.

13.3.3 Digital PID Controller

Equation 13.1.a describes a digital PID controller for A/C system:[3]

$$y_K = y_{K-1} + q_0 * e_K + q_1 * e_{K-1} + q_2 * e_{K-2} \qquad (13.1.a)$$

We can generalize the Equation 13.1a in the following way:

$$y_K = P_{1*}y_{K-1} + q_0 * e_K + q_1 * e_{K-1} + q_2 * e_{K-2} \qquad (13.1.b)$$

The structure of Equation 13.1b has some generalized features: there are only three free determinable parameters, q_0, q_1, q_2, for the optimal working behavior of the control loop. If there is an integral part in the control algorithms, we have to introduce the parameter P_1 with the value of 1.

A digital controller for the air conditioning system is optimized by means of practical adjustment rules of Takahashi. The response behavior of the processes were received and evaluated [3]. The disadvantage of this method is that the control loop is only optimized for a fixed working point.

Now we can recognize that optimal control loop behavior is only guaranteed when we extend the parameters of the PID controller in order to cover the whole range of the working point of the controlled process.

The following method ensures an enhanced DDC/PID algorithms The current set point error ek in combination with q_0 is presenting the current state of the process, where set point errors e_{k-1} and e_{k-2} are presenting the passed states of the process, and therefore are presenting the dynamic behavior of the system. In order to optimize the control loop behavior, it is important to extend Equation 13.1.b with further set point errors of the control loop. Additional information about the control loop behavior can be obtained if we also consider all the control output values (y $_{k-1}$...y $_{k-n}$) to any state of the system which occur the set point errors. Now we have to extend the PID algorithms as follows.

$$y_k = p_1 y_{k-1} + ... + p_n y_{k-n} + q_0 e_k + q_1 e_{k-1} + ... + q_n e_{k-1} \qquad (13.2)$$

Equation 13.2 consists of n optimization parameters. Using Equation 13.2 for real AC control problems is a time consuming process, because there are many optimization parameters which have to be evaluated and calculated. Now we can realize that the experimentally oriented optimization methods and all other empirical methods used by the control industry will fail, in order to fulfil the requirements for the PID algorithm of n degree.

The consequence is that almost all nonlinear control loops have in fact a stable dynamically behavior for the whole working range of the process, but the quality of the control loop is very poor because of the very weak gain factor of the control loop.

The target is to enable an operator to adjust an air conditioning controller optimally for the whole working range. It will be described in the following section how to implement an HVAC technician's knowledge and experience onto a controller by using fuzzy logic. The linguistic variables and rules of a Fuzzy controller are similar to the technician's memory power and therefore easier to formulate than any abstract mathematical formula. Local changes in the sets of rules generate local changes in the characteristics of the fuzzy controller.

13.3.4   Fuzzy Cascade Controller

For a fuzzy PID supply air temperature controller the following four input variables as shown in Figure 13.6 are utilized:[3]

- set point error e;
- differential of process variable (dx/dt);
- difference of set point error Δe (increase or decrease) ; and
- reference output u_o.

The set point error (e) is defined as the difference between the set point ref. and the process value x (ϑ_s) according to Equation 13.3 for a maximum range of $e_{max} = \pm 5K$ with seven sets. If the actual set point error (e) is on a larger scale, the set point error (e) will be determined on $e = e_{max}$, so that, apart from these limits, the controller generates a maximal output.

$$e = \text{ref} - x \qquad (13.3)$$

The fuzzy system supports a maximal of seven sets with Λ-, Π-, Z- and S - functions. For fine control, thin sets are utilized around the set point, e = 0 (ns, zr, ps). The set width increases with distance from the set point for rough control. In case of a large set point error the control system should first bring the process variable quickly with a rough control near to the set point. Second the controller has to zoom in carefully with a fine control. This fine tuning method for the sets helps to avoid the overshoot and undershoot of the process variable. Because of the extreme nonlinear characteristic of the preheat exchanger, the definition range of the set point error is ± 5 K. If the process variable is moving with a maximal speed towards the set point for a big set point response, the controller has to start very soon with the control mode, i.e. at ± 5 K set point error before reaching the set point.

For surpressing noise changes of the output signal caused by noise signals in the range e = ±0.08, the Π-function is chosen for the set zero (zr). The valve is spared by such a defined dead band, otherwise any slight set point error will cause a change of the controller output. Subsequent to this dead band the value of the set point error e will be smaller than ± 0.15 K.

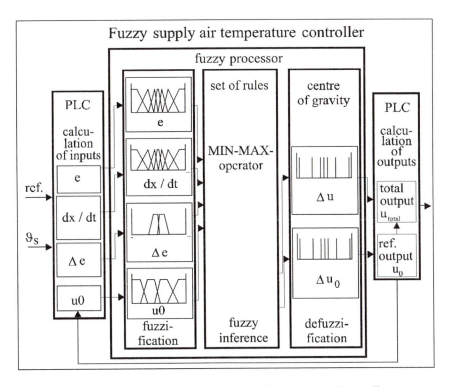

Figure 13.6: Fuzzy PID Supply Air Temperature Controller.

The second input differential of process value (dx/dt) is calculated by Equation 13.4 and is determined in the range ± (1 K/10s). This is congruent with the real maximal speed of the process variable. For the second input variable there are also seven sets defined:

$$(dx / dt) = \frac{x(k) - x(k-1)}{Tc} \tag{13.4}$$

with $x(k)\equiv$ process variable in cycle, $x(k-1)$ $\equiv$ process variable in cycle k-1, $Tc\equiv$ scan time.

The third input is difference of set point error (Δe), calculated according to Equation 13.5 (Δe_{max} is defined in the range $\pm(1K/10s)$, according to the differential of process variable). Only two sets are defined for this variable. It is possible to recognize with this input variable whether the process variable is moving towards the set point (set n) or whether the set point error is increasing (set). This additional information is necessary for set e = zero because it accepts negative and positive measured values.

The input difference of set point error Δe is associated with the patches e = zr $\cap$ (dx/dt) = ns or ps only. For example the set point error is positive but still within the set e = zero. That means the process value is a bit too deep (set Δe = negative). The set (dx/dt) = positive small shows an increasing process value. In this case the output signal should not change as the process value zooms into the set point. If the set point error is negative within the set e = zero, the process value is slightly too high. On an increasing process value the set dx/dt is still positive small. But in this case the set point error is increasing (set Δe = positive). The controller output must be reduced, in order to avoid on overshooting of the process value:

$$\Delta e(k) = \frac{|e(k)| - |e(k-1)|}{Tc} \tag{13.5}$$

The fourth input is the reference output (u_0) which is not utilized in common PID controllers. The reference output shows the controller which unit is the active one (= heat exchanger, chiller or damper). By combining the specific rules for each unit the controller behavior can be adapted locally in its control range by changing single rules. The Fuzzy controller utilizes the following two outputs:

- change of controller output Δu and
- change of reference output Δu_0.

The defuzzification results from the method of the center of gravity. The output change of controller output Δu is defined in the range ±50% and contains 7 singletons.

The fuzzy PD characteristics are included in this variable. The addition of the Fuzzy PD output (Δu) to the reference output u_0 gives the complete output of the inner controller:

$$u_{total} = u_0 + \Delta u \tag{13.6}$$

The inner controller output (u_{total}) is a positional signal in contrast to a speed signal. Note that the end position 4095 bit of the total controller output is possible immediately only if the reference output (u_0) is 2047 or larger. In this application a sequence wiring with four units is utilized. Therefore a change of the fuzzy controller output of percent causes a bounce over two units only. The second output change of the reference output (Δuo) is defined in the range $\pm 2\%$ and contains 7 singletons. This variable generates the fuzzy integral characteristics. The integral characteristics are implemented in the calculation of the reference point u_0. The reference output is calculated as:

$$u_0(k) = u_0(k-1) + \Delta u_0 \tag{13.7}$$

Within the rule base the sets of set point error and the sets of differential of process variable (dx/dt) are associated with the sets of change of reference output (Δu_0). In comparison the common realization of the integral part is in need of an additional variable at the input which is calculated through the rectangular integration Equation 13.8. These statements correspond because Δu_0 of Equation 13.7 and $(Tc/Tn).e(k)$ of Equation 13.8 are nearly the same.

$$u(k) = u(k-1) + \frac{T_c}{T_n}.e(k) \tag{13.8}$$

Δu_0 is calculated once per scan time cycle which corresponds to the scan time Tc and the integral acting time Tn of is a digital integral control algorithm, and presented in the order of the membership functions of Δu_0. The realization of the integral part through a second output shows three major advantages:
- reduction of rules;
- easy adjustment of the integral part; and
- use of the additional information of the D-input.

The reduction of rules is a result of an removed integral input to the output of the fuzzy controller. It is self evident that the handling of a controller with fewer rules is easier, but there is another aspect for the optimization which gives a rather easy adjustment of the integral part.

After a set point bounce or any disturbances, the controller has to bring the reference output u_0 near to the next available steady condition quickly. The sets of change of reference output Δu_0 are associated with the sets of the variable e and additionally with the sets of the variable (dx/dt). The control behavior is clearly improved by using the additional information of the D-input.

The Fuzzy Lead Controller with PI characteristics

The lead controller for room temperature control loop has considerably fewer input variables and therefore fewer rules than the inner controller for the supply temperature control loop. Instead of the fuzzy PID characteristics, a fuzzy PI characteristics is implemented only (Figure.13.7). The fuzzy lead controller utilizes one input and two outputs:

- set point error of the room temperature e_{tr};
- change of reference set point Δref_0; and
- set point change Δref.

The input e_{tr}, is defined for a range of ± 4 K with 7 sets. The output set point change Δref is defined in the range ± 100 percent and contains 7 singletons. Because of the large range of the set point change, both end positions, 15°C and 35°C, are immediately available. Within this output there is a fuzzy PI behavior only implemented. In order to realize the integral characteristics the same method as for the inner controller is used (Equation. 13.9).

$$ref_0(k) = r_0(k-1) + \Delta ref_0 \tag{13.9}$$

The second output change of reference set point Δr_0 generates the fuzzy integral behavior, with a scan time of 30 s. This variable is defined for a range of $\pm 0{,}6K \equiv \pm 3\%$ and contains 7 singletons. This fuzzy controller has seven rules only. Each rule associates one input set with two output sets.

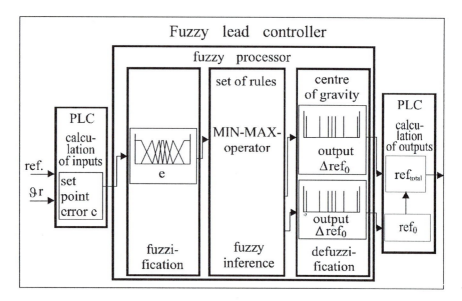

Figure 13.7: The Lead Controller with Fuzzy PI Characteristics.

13.3.5 DDC vs FLC

To compare the control loop behavior, the method of control time (T_C) measurement is chosen here. The inner controller is optimized for set point responses. The control time T_C is 900 s with the digital controller (Figure 13.8) whereas the fuzzy controller needs only 460 s (Figure. 13.9) for the same task. The reason for this large time difference is the big overshoot The fuzzy controller brings the process variable   to the new set point and avoids an overshooting of the process variable as a result of the specific rules.

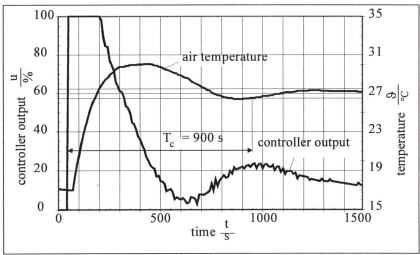

Figure 13.8: Digital Control Loop Behavior for a Set Point Step Response.

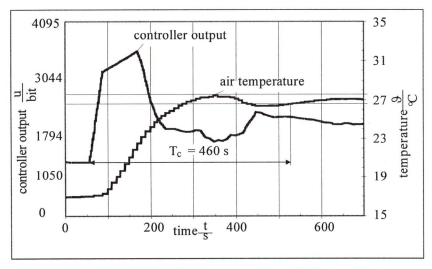

Figure 13.9: Fuzzy Control Loop Behavior for a Set Point Step Response.

**13.4   FUZZY CONTROL FOR THE OPERATION MANAGEMENT
OF A COMPLEX CHILLING SYSTEM**

The optimization   potentials for the operation of a chilling system within
building supervisory control systems are   limited to the   abilities of the
programmable logical controller (PLC) functions with their binary-logic-
oriented operations. Little information about thermal behavior of the building
and the chilling system is considered by the operation of chilling systems with
plc-based control strategies. The main goal of this project was, to replace the plc
strategy by fuzzy control. A concept of knowledge engineering by measuring
and analyzing of system behavior is necessary, since no expert knowledge exists
for formulating the fuzzy rules.

The focus of the optimization strategy by Fuzzy control is to ensure an
optimal operation of a chilling system. Optimal operation means:

• reducing operation time and operation costs of the system; and
• reducing cooling energy generation and consumption costs.

Different optimization strategies have been defined for developing proper
fuzzy controllers. Missing expert knowledge and online measurement of
different physical values and their evaluation are the basis for the fuzzy control
system. Few rules for each controller are necessary in order to have fine tuning
of the fuzzy control system. Three fuzzy controllers are necessary in order to
reach maximum efficiency by operation of different components of the chilling
system. The realized fuzzy control system is able to forecast the maximum
cooling power of the building and also to determine the cooling potential of the
outdoor air. Operation of the systems by fuzzy control enormously reduces the
cost of cooling power. The system has been successfully commissioned and
remarkable improvement of the system behavior has bee reached. This project
opens new application fields for the market of building automation. [5]

13.4.1   Process Description

The chilling system described here supplies chilled water to the air
conditioning systems (AC systems) installed in different research laboratories
and computer rooms at the Max Plank Institute for Radio Astronomy in Bonn.
The amount of cooling power for the building is the sum of internal cooling and
the external cooling load, which depends on outdoor air temperature (T_{out}) and
sun radiation through the windows. The cooling machines installed here use the
compression cooling method. The principle of a compression cooling machine
can be described in two thermodynamic processes. In the first step of the cooling
process, the heat energy will be transferred from the system to an evaporator of
the chilling system and therefore the liquid gas will evaporate by absorbing the
heating energy. After the compression of the heated gas in the second part of the
process, the gas condenses again by cooling the gas through the air cooling

system. The chilling system as shown in Figure 13.10 consists of these components: three compression cooling machines, three air cooling systems and two cooling load storage systems. During the operation of the cooling machines, the air cooling systems will be used in order to transfer the condensation energy of the cooling machine to the outdoor air space. If the outdoor air temperature is much lower than user net return temperature on heat exchanger one, the air cooling system should serve as a free cooling system.

The additional cooling load storage systems are installed in order to load cooling energy during the night, and therefore reduce the cost of electrical power consumption. They also supply cooling energy during the operation time, if a maximum cooling energy is needed and cannot be provided by existing cooling machines. In both cases the cooling storage system does not reduce energy consumption, but rather the cost of energy consumption.[4]

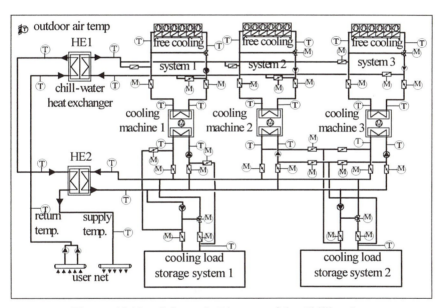

Figure 13.10: Schematic Diagram of the Chilling System.

13.4.2 Process Operation with FLC

Thermal Analysis of the Building and Chilling System

The aim of the thermal analysis of the building is to find measurable information for the current cooling load. Measurement of current cooling power of the building as shown in Figure 13.11 has proved that there is not a significant correlation between outdoor air temperature T_{out} and the current cooling power. The current cooling power will increase if T_{out} gets higher than 23°C. In the summertime, when the T_{out} increases to about 34°C, the current cooling power will be more influenced by the T_{out}. So the T_{out} can be used for forecasting the maximum cooling power. Additional information needed for analyzing the thermal behavior of the building is the return temperature of the user net (Tr_{-un}). Any change of total cooling load will influence the Tr_{-un} and is an important input for the fuzzy controller.

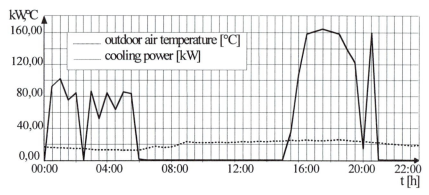

Figure 13.11: Course of Current Cooling Power and Outdoor Air Temperature.

The Design of the Fuzzy Control System

Considering the cooling potential of the outdoor air, the free cooling system should run before the cooling load storage system (CLS) and cooling machines. This has to be considered by the fuzzy controller for the operation of cooling machines. The CLS should run during the daytime before any cooling machine, if the cooling load of the building is expected to be low. Optimization strategy for the discharge of CLS will ensure that there will not be a peak in the electrical power. The cooling machines should run at their lowest level.

Three different FLC have been developed with a total number of just 70 rules. The designed software -based FLC with the SUCOsoft fuzzy tech tool[6], has been translated into a graphical-orientated mathematical and logical programming language.[7] All the operation instructions implemented in the Supervisory level of a BEMS will be transferred to the chilling system through the automation level, as shown in Figure 13.1.[8]

13.4.3 Description of the Different Fuzzy Controllers

Fuzzy Control Block 1

The optimal starting point for the discharge of the cooling load storage system depends on the maximum cooling power demand, which can differ every day. For calculation of the maximum cooling power, T_{out} must be processed by this fuzzy controller, since the maximum cooling power in the summertime will be influenced by T_{out}. A feedback of current cooling power calculated by Fuzzy control block 2 (as shown in Figure 13.12) is also necessary, in order to estimate the maximum cooling power. The input variables of the controller 1 are:

- Out door air temperature $T_{out;}$
- Differential of T_{out}: dT_{out}/dt;
- Current cooling power of the cooling machines.

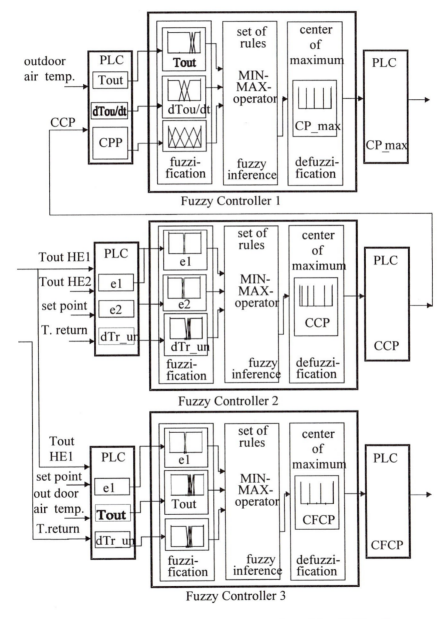

Figure 13.12: Fuzzy Control System for Operation of the Chilling System.

The second fuzzy variable is calculated by Equation 13.10:

$$dT_{out} / dt = \left(T_{out}\left(k\right) * T_{out}\left(k-1\right)\right) \ / TC \qquad (13.10)$$

With T_{out} (K) = Outdoor air temperature by $K^{Th.}$ cycle, Tout (k-1) = Outdoor air temperature by k-1 $^{Th.}$ cycle, TC = Scan time.

Fuzzy Control Block 2

The fuzzy controller 2 is the important part for the optimization of the Control system in order to use the cooling potential of the outdoor air before starting any cooling machine. This controller consists of 21 rules with the 3 input variables:

- Set point error" e1" at heat exchanger 1,
- Set point error "e2" at heat exchanger 2,
- Difference between user net return temperature(Tr_{-un}) , and $T_{set\ point:}$ Δ Tr-un

The third input variable presents the difference between user net return temperature and Set point, which is determined by Equation 13.11:

$$\Delta\ Tr_{-un} = Tr_{-un} - T_{set\ point} \qquad (13.11)$$

Calculation of Δ Tr-$_{un}$ is necessary because $T_{set\ pint}$ is variable and, therefore, ΔT_{run}, contains the real information about the cooling load of the building. If e1 is zero, or negative, then the capacity of the free cooling system is sufficient for the required cooling power. The output signal of FC 2 will be zero. In other cases, FC 2 is responsible for the operation of the cooling machines. In cases, where the capacity of the free cooling system is not enough, e will have values of NS, so other rules will determine the output of the controller. In that case the third input variable Δ Tr-$_{un,}$ is more weighted for the output value of the controller, because ΔTr-un represents the real alternation of the cooling load.

Fuzzy Control Block 3

This control block is necessary in order to use the cooling potential of the outdoor air and run the air cooling systems of the cooling machines as free cooling systems. The cooling potential depends on the difference between user net return temperature $Tr_{-un,}$ and the outdoor air temperature T_{out}. The input variables of the control block 3 are:

- Difference between Tr_{-un} and set point, $\Delta Tr_{-un;}$
- Set point error e1 at heat exchanger 1;
- Different between T_{out}, and Tr-$_{un}$, ΔT_{out}.

An important aspect for the formulation of the rules for this controller is the cooling potential of the system, which is represented by the input variable 3, ΔTout. The higher the value of this variable is, the fewer free cooling system components are necessary in order to supply the demanded cooling power for the building. Producing the cooling power by free cooling system reduces the cost of the cooling energy and the operation time of the cooling machines. This controller consists of 29 rules.

13.4.4 System Performance and Optimization with FLC

Figure 13.13 shows the course of user net supply temperature before the optimization of the system operation by fuzzy control. The alternation of the supply temperature is between 10.5°C and 4.8°C. The reason for such a set point error range is in the discontinuous operation of the system by PLC. Figure 13.14 presents the course of the supply temperature after commissioning the fuzzy control system. The course of the supply temperature indicates a remarkable improvement of the system behavior. This relatively constant supply temperature will ensure research conditions in the building.

Operation of free cooling systems by fuzzy control could reduce the cost of energy production by factor 27 in comparison with the cost of cooling machines.

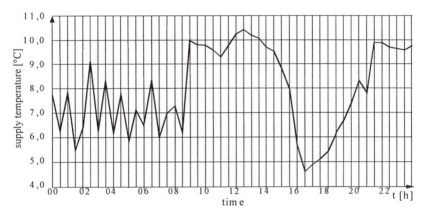

Figure 13.13. Course of Supply Temperature with PLC Operated System.

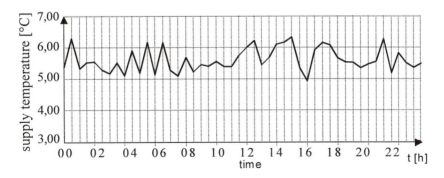

Figure 13.14. Course of Supply Temperature with FLC Operated System.

13.5   APPLICATION OF FUZZY CONTROL FOR ENERGY MANAGEMENT OF A CASCADE HEATING CENTER

Generation and consumption of heat power for domestic demand should consider economical and ecological aspects. Optimal demand - oriented heat power generation by a cascade heat center requires sustainable evaluation of measurement information of the whole system.

Fuzzy logic provides, by evaluation of the thermal behavior of the heating system, a powerful rule base for decision making in order to guarantee optimal operation of the heating system. Analysis of the dynamically thermal behavior of the building and the heating center is necessary, in order to select existing measurement information as input variables for different fuzzy controllers.

The supply temperature control loop of the system is designed and commissioned as a nonlinear fuzzy PID controller for a nonlinear thermal process. This kind of controller can be described as a robust control system.

The control system is optimized through the whole working range of the system and ensures a maximum of control loop quality by a very short response time of any alternation in the process, and at a negligible overshooting of the process value during the control operation. The whole system consists of three different fuzzy controllers with the following functions: a fuzzy PID controller for a hot water supply temperature control loop; a fuzzy controller for optimal evaluation of heat power demand; and a fuzzy controller for the operation of a cascade heat center with high efficiency and lowest contaminated exhaust emission. This control and operation system provides demand -oriented heating energy with minimum fuel consumption and therefore with a minimum of contaminated exhaust gas emission.

13.5.1   The Heating System

Description of the Heating System

The heating system (which has to be optimized as written here) supplies heating power and domestic hot water for a public school. The system is known as a cascade heat center as demonstrated in Figure 13.15. The system consists of two heaters with controllable gas burners and a hot water boiler. The heating system supplies through a hot water distributor, heating energy for autonomous sub control loops and for a hot water boiler.

The control loops in different zones of the building are a digital control system and are already in operation. The zone-oriented control loops operate only during the lecturing and business hours. After business hours, or in summertime, the heating systems should only provide heat energy for the hot water boiler in an efficient way from economical and ecological points of view as described previously.

Operation of the Heating System

Reducing fuel consumption by optimal operation of the system means simultaneously reducing the operating cost of the system and reducing environmental pollution. Investigation by [9,10] as illustrated in Figure 13.16 shows the thermal efficiency of the system by low range operation capacity of the heater. Reducing the environmental pollution is only possible when the system-operation-oriented -emission of the exhaust gas is as little as possible. As we can see from Figure 13.17, the emission of contaminated exhaust gas has its maximum at start and stop phase of the operation. The main goal of the project described here was to reduce the heat capacity of the system by an intelligent control and operation strategy, and also reduce the frequency of the start/stop operation mode of the system.

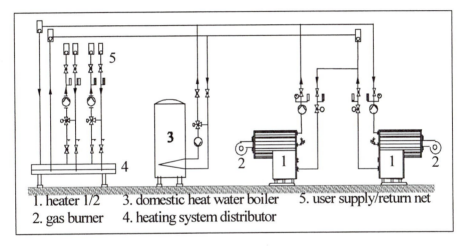

1. heater 1/2   3. domestic heat water boiler   5. user supply/return net
2. gas burner   4. heating system distributor

Figure 13.15: Simplified Flow Diagram of a Cascade Heating Center.

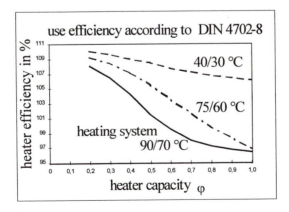

Figure 13.16: Thermal Efficiency of the Heating System.

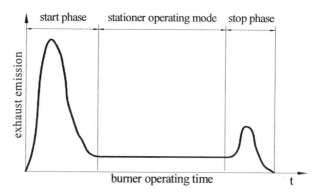

Figure 13.17: Operation Mode Oriented Exhaust Emission.

13.5.2   FLC for System Optimization

State of the Operation Strategy for a Cascade Heating Center

One of the operating strategies is the serially operated heating system. Release of the second heater is guaranteed when the first heater reaches maximum operating capacity. This kind of operating mode reduces the unwanted start/stop phases of the second heater, and therefore guarantees reduced pollution. The disadvantages of this operating strategy lie in the reduced thermal efficiency of the first heater.

The second operating strategy is the parallel operation of the system. As soon as the first heater reaches its basis capacity of the demanded heat energy, the second heater will be released for simultaneous operation. As soon as there is less demand for heating energy, the second heater will stop, then start again when the demand of heating energy increases. The disadvantages of this operation strategy lie in the high frequency for the start/stop phase of the second heater, and the resulting increase in emission of the contaminated exhaust gases.

Optimal Operation Strategy for the Heating System

As soon as the controller output indicates a heat energy, the first heater starts with its lowest operating range (12.5 % heat energy). In the next phase, the first heater is in control mode. As soon as the first heater provides a total heat capacity of 25 % (this is a significant value), that each heater can operate in its basis power range of 12.5 % .

This operating strategy ensures that none of the heaters will reach its maximum capacity, and the start/stop frequency of heater 2 will be very low. To reach this goal of optimal operation, different fuzzy-logic-based evaluation of the thermal behavior of the system is necessary, which will be described in the following section. Figure 13.18 illustrates the different start /stop phase of the system.[11,12]

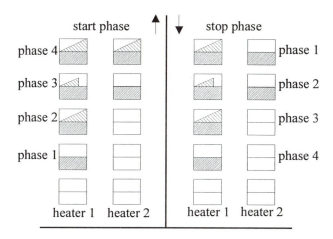

Figure 13.18: Operation Strategy for a Cascade Heating Center.

13.5.3   FLC Description

In order to realize the described optimization strategy by fuzzy control, it is important to determine the precise start/stop operating range of the heating system. The lowest operation range of the first heater should be modified by a hysteresis in order to reduce the frequency of start/stop by varying the low energy demand. Release of the heater from lowest operation range to the variable operation range must also consider the real demand of heating energy.

If the outdoor air temperature is very low, the fuzzy control system must release the control mode of the heater immediately. If the gradient of the heat demand is not as big, the fuzzy control system should release the control mode of the heaters with a time delay.

This strategy ensures that for a long period of time the heater will be in stationary operation mode and the exhaust emission will be reduced.

The release of the second heater at lower outdoor air temperature should be immediate, in order to avoid the operation of the first heater in a high range of heat capacity. If the control mode of the first heater is in operation at mild outdoor air temperature, the release of the second heater should also delay in order to reduce to start/stop frequency of the second heater.

The open loop of the supply temperature of a heating system has a big time delay and nonlinear characteristics. Therefore the same type of fuzzy PID controller will be implemented for control loop as has been used for the supply air temperature control loop as, described in section 13.3.

The fuzzy control system introduced here is a software solution within an existing industrial building energy management system. Figure 13.19 shows the fuzzy control System for optimal operation of the system.

Fuzzy-Control-Block I

This control block controls the system's supply temperature set point with PID characteristics. This controller has four input variables:

- set point error e, calculated by Equation 13.3; and
- differential of process value (dx/dt), calculated by Equation 13.4; and
- difference of set point error Δe , calculated by Equation. 13.5.

The fourth input is the reference output (u_0). The controller can be adjusted independently of a working point. Fuzzy controller I utilizes two outputs:

- change of controller output Δu; and
- change of reference output Δu_0.

The integral characteristics are implemented in the calculation of the reference point u_0. [3]

Fuzzy Control Block II

This fuzzy controller calculates the current power demand capacity and burner delay time. The following input variables are fuzzified by the fuzzy control block II:

- Outdoor air temperature $\vartheta_{Out;}$
- Set point temperature; and
- Current flow capacity $c_{ft.}$

The fuzzy controller utilizes two outputs:

- Heating power demand value Q_t
- Burner control delay time dt.

The definition range of the first input variable is between -12°C and +22°C. The second input variable is the set point, which indicates the highest set point of all the active control loops. The current flow capacity c_{ft}, of the system as the third input variable can be calculated by Equation 13.12:

$$cft = \frac{\sum\limits_{i=1}^{n}(cf(i))}{cf_{total}} * 100 \qquad (13.12)$$

C_f represents the flow capacity of each control value by the current valve position so that the total flow capacity is easily calculated by measuring all control valve positions in the system. The calculated current heating power demand by this fuzzy control block is the input variable for fuzzy control III as a release criterion for the start phase of the heating system.

The second output variable of the fuzzy control block II serves as a time delaying system for release of the variable control mode of the first heater. The control mode of the heater will be released earlier by low outdoor air temperature than by mild outdoor air temperature.

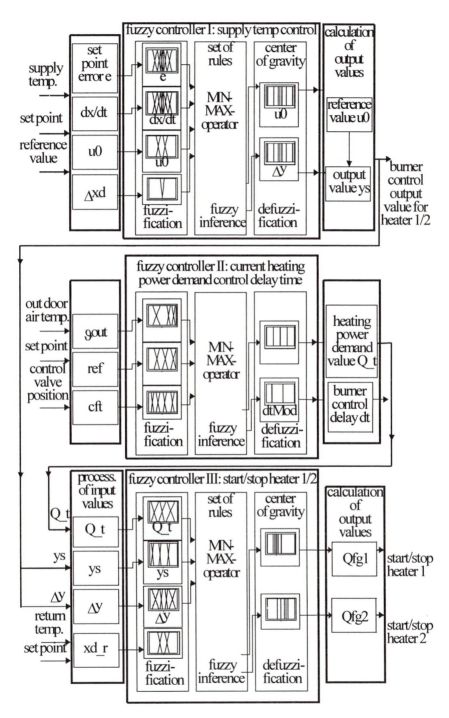

Figure 13.19: Fuzzy Control System for the Cascade Heating Center.

Fuzzy Control Block III

This fuzzy controller determines the start/stop points of the heaters. The third fuzzy control block is the important part of the fuzzy system for the determination of the precise start/stop modus of the two heaters.

The following input variables are used by the fuzzy control block III:

- Current Heating power demand value Q_t;
- Controller output ys from fuzzy control block I;
- Changing of the controller output dy from fuzzy control block 1; and
- Heating power demand gradient e_r.

The fuzzy controller utilizes two outputs:

- threshold value Qfg1 for heater 1 and
- threshold value Qfg2 for heater 2.

The first input value, i.e., current heating power demand value Q_t, is the most important factor for the evaluation of the start/stop - point of the heaters. The   second   input   value   has   been   calculated   by   the   first control block and determines the position of the control range of the heaters. As soon as this value is higher than 40 percent, the start phase of the second heater will be released. The third input presents the set point error of the system as well as its dynamic behavior. The fourth indicates the gradient of the thermal energy, which is necessary in order to keep the set point temperature of the system constant. The crisp value of this input is calculated as:

$$e\_r = ref - x\_r \qquad (13.13)$$

Where, in Equation 13.3, *ref* $\equiv$ reference set point temperature, and x_r: $\equiv$ system return temperature.

For the determination of the start/stop point of the heaters the threshold values of both heaters are evaluated by the Fuzzy controller III, as the output values. The calculation of the crisp number of the threshold values is realized by Equation 13.14 . As soon as the threshold value Qfg1 for heater 1 is over a limit of 12.5 % of the total energy demand, the heater's start phase will be released. Decreasing of the threshold value Qfg1 for heater 1 under 12.5 % of the total energy demand , is a crisp signal for stopping of the heater 1.

$$Qfg_{(k)} = Qfg_{(k-1)} + dQfg \qquad (13.14)$$

with $Qfg_{(k-1)}$: $\equiv$ threshold value Qfg by cycle k-1 and dQfg: $\equiv$ change of the threshold value Qfg .

The release of the second heater is due to the output value of the fuzzy control block I, when its value reaches 40 percent.

To prove the reliability of the fuzzy control block III, Qfg2 has been calculated for different outdoor air temperatures as shown in Figure 13.20. For this experiment, the alternation of the outdoor air temperature had to be simulated. As we can see from Figure 13.20, increasing the threshold value Qfg2 is quite fast by lower outdoor air temperatures (here -5°C). At higher outdoor air temperatures, e.g., 2°C, to reach the Qfg2 is slow, and at a $\vartheta_{out} = 10°C$, very slow.

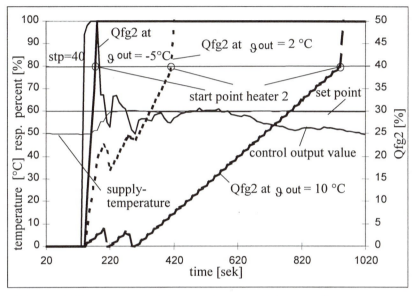

Figure13.20: Course of Qfg2 at Different Outdoor Air Temperatures.

13.5.4 Temperature Control: Fuzzy vs Digital

In order to compare the features of the fuzzy system with the existing DDC system, the heating system was operated during summertime for providing domestic hot water in the building. For providing hot water only the first heater was in operation. Heater 1 had four start/stop phases as shown in Figure 13.21 where this system was operated by DDC system. For the same process, as we can see from Figure 13.22, the fuzzy control-operated heater has only one start/stop phase. Considering the pollution effect of the heaters, this is a remarkable improvement of the systems' features from an ecological point of view.

Operation of the system with fuzzy control ensures that the release of the control mode of the heater is time-delay-oriented (ca. 300 sec.), which avoids shooting the supply temperature over the set point and reduces the number of the start/stop phases of the heater. Also, the working point of the fuzzy PID controller could keep very low (0%) in order to reduce the speed of the controller output for low energy demand.

Figure 13.21 shows the system's behavior for control and operation mode realized by digital control system. Figure 13.22 shows the system's behavior for control and operation mode realized by fuzzy control system.

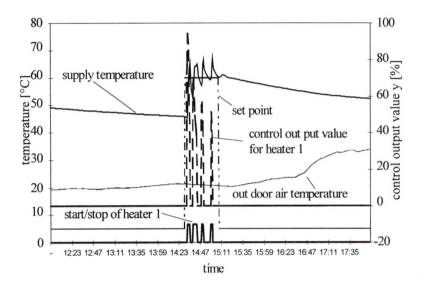

Figure 13.21: Heating System's Behavior Operated by Digital Control.

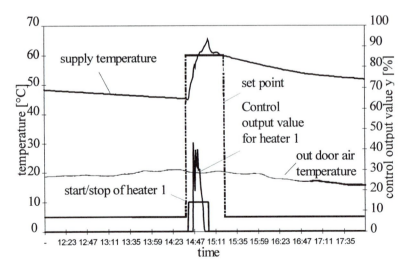

Figure 13.22: Heating System's Behavior Operated by Fuzzy Control.

13.6   CONCLUSIONS

The objective of building energy management system (BEMS) is to achieve more efficient building operation at reduced labor and energy costs while providing a safe and more comfortable working environment for building occupants. In the process of meeting these objectives, the BEMS has evolved from a simple supervisory control system to a totally integrated computerized control and management system.

Today's BEM system requires use of soft computing methodologies in order to cope with the automation and control problems of the intensive nonlinear technological processes in buildings. The first step in reaching this goal was developing a control system for an air conditioning system based on Fuzzy logic. A new fuzzy PID characteristic has been developed. The realization of the integral part of the fuzzy PID through a second output is a unique solution with three advantages: reduction of rules, easy adjustment of the integral part and use of the additional information of the D-input.

The control system is optimized through the whole working range of the process and ensures a maximum of control loop quality by a very short response time of any alternation in the process and at a negligible overshooting of the process value during the control phase. The fuzzy controller showed enormous advantages in processes with intensive nonlinearity and is superior to the digital controller. Proving the applicability of fuzzy logic for energy management tasks, two new operation and optimization strategies for a complex chilling and heating system have been realized and implemented into the existing industrial BEMS. The focus of the optimization strategies for both projects was

- reducing operation time and operation costs of the system;
- reducing cooling energy generation - and consumption costs;
- forecast the maximum cooling power of the building;
- determine the cooling potential of the outdoor air;
- optimal evaluation of heat power demand of the system;
- optimizing the heating system's thermal features from the economical and ecological point of view and reducing fuel consumption; and
- Increasing thermal efficiency of the system by lowest exhaust emission.

Based on the thermal analyses of the building and the chilling and heating systems, different optimization strategies have been defined for developing proper fuzzy controllers. Analyzing and evaluating the thermal behavior of the system was necessary in order to formulate proper input and output variables for different fuzzy controllers.

The developed fuzzy control and operation system could fulfill all the formulated requirements, and has been successfully commissioned ; remarkable improvement of the system behavior has been reached. The projects described in this chapter open new application fields for fuzzy logic and fuzzy control in the market of building automation and building management.

REFERENCES

1.   Talebi-Daryani, R., *Intelligent Building for Integrated Building Automation and Building Energy Management System*, KEIO University, Yokohama, Japan, June 1999.

2.   Talebi-Daryani, R., *Digtitale Gebaeudeautomation und Fuzzy Control*, University of Applied Sciences, Cologne, 1995.

3.   Talebi-Daryani, R. and Plass, H., Application of Fuzzy Control for Intelligent Building Part I: Fuzzy Control for an AC System, *Proc. of the World Auto. Conf.*, 745-750, TSI Press Series, Albuquerque, NM, 1998.

4.   Talebi - Daryani, R., *Control Engineering for Mechanical Engineers*, 115 – 120, University of Applied Sciences, Cologne, 1990.

5.   Talebi - Daryani, R. and Luther, C., Application of Fuzzy Control for Intelligent Building, Part II: Fuzzy Control of a Chilling System, *Proc. World Auto. Conf.*, 751-756, TSI Press Series, Albuquerque, NM, 1998.

6.   *SUCO Soft Fuzzy TECH 4.0, FT4-400-DX2 Application Guide*, Kloeckner Moeller, Bonn, Germany, 1994.

7.   *Application Guide and Handbook for GPL, METASYS System*, Johnson Controls International, Essen ,Germany, 1994.

8.   Talebi-Daryani, R. and Luther, C., Anwendung der Fuzzy Technologie in der Gebaeudeautomation Teil II, *Sondernummer MSR-Technik, TAB(Technik am Bau)*, 41-48, Bertelsmann Fachzeitschriften, Guetersloh, Germany, 1997.

9.   *Plannungsanleitung für Gas-Brennwertkessel Veromat, Product Information*, Viessmann, Allendorf, Germany, 1997.

10.   *Energiewirtschaftliche Beurteilung für Heizungsanlagen*,VDI-Richtline 3808, VDI- Düsseldorf , 1993.

11.   Talebi-Daryani, R. and Olbring, M., *Kesselfolgeschaltung mit Fuzzy Control*, ISH- Jahrbuch 1999, 142-155, Bertelsmann Fachzeitschriften, Guetersloh, Germany, 1999.

12.   Talebi - Daryani, R., and Olbring. M., Application of Fuzzy Control for Energy Management of a Cascade Heating System, 618-625, in: *Soft Computing, Multimedia, and Image Processing*, TSI Press Series, Albuquerque, NM, 2000.

14 APPLICATION OF ADAPTIVE NEURO-FUZZY INFERENCE SYSTEMS TO ROBOTICS

Ali Zilouchian and David Howard

14.1   INTRODUCTION

During the past three decades, fuzzy logic has been an area of heated debate and much controversy. Zadeh, who is considered the founding father of the field, wrote the first paper in fuzzy set theory [1], which is now considered to be the seminal paper of the subject. In that work, Zadeh was implicitly advancing the concept of human approximate reasoning to make effective decisions on the basis of available imprecise linguistic information [1]-[3]. The first implementation of Zadeh's idea was accomplished in 1975 by Mamdani, which demonstrated the viability of fuzzy logic control (FLC) for a small model steam engine [4]. After this pioneer work, many consumer products as well as other high tech. applications have been developed using fuzzy technology. A list of industrial applications and home appliances based on FLC can be found in several recent references [5]-[13].

However, the design of an FLC relies on two important factors: the appropriate selection of knowledge acquisition techniques, and the availability of human experts. These two factors subsequently restrict the application domains of FLC. In this chapter, the application of adaptive neuro fuzzy inference systems (ANFIS) [14]-[16] to robot manipulators is presented to overcome such restrictions. Both kinematics and control of robot manipulators are addressed to demonstrate the applicability of ANFIS in the design, implementation, and control of industrial processes.

Within the past fifteen years, the utilization of NN and FL to aid the controls as well as kinematics mapping of robotic manipulators has been investigated by many researchers [17]-[35]. Ngyen, Patel, and Khorasani describe the solution of forward kinematics equations using NN [17]. The authors therein have used four different neural networks including back propagation and counter propagation to check their hypothesis. Wang and Zilouchian [18] presented the solution of the forward and inverse kinematics using Kohonen self organization neural network. Further investigations have been conducted in the area of NN to solve the kinematics equations as well as the control of robot [19]-[24].

On the other hand, the FL has been utilized for the solution of inverse kinematics as well as control of robot manipulators by several investigators [24]-[36]. Nedungadi [24]-[25] presented the inverse kinematics calculations of a four-degree of freedom (DOF) planner robot using Fuzzy Logic. Kim and Lee [26] investigated the inverse kinematics of redundant robot using fuzzy logic. Further, in 1993, Xu and Nechyba [27] proposed a general method for the calculation of inverse kinematics equations of an arbitrary n-DOF manipulator through FL approach. Lim, and Hiyama [28] presented the initial work related to control of robot manipulators using FL. In addition, Martinez, Bowles, and Mills[29] propose a fuzzy logic position system for a thee-DOF articulated robot arm. The mapping between a robot's end effector coordinate and joint angle was also successfully implemented using fuzzy logic [30]. Lea, Hoblit, and Yashvant [31] have implemented fuzzy logic controller for a remote manipulator system of the space shuttle. Kumbla, and Jamshidi [32] evaluated the hierarchical control of a robotic manipulator using fuzzy logic, which also included a fuzzy solution to the inverse kinematics equations. Several other researchers including Nianzui, Ruhui, and Maoji [33], Moudgal, et al., [34] and Lee [35] have also investigated the control of robot manipulators using fuzzy logic.

In this chapter, an alternate and attractive approach for solution of inverse kinematics as well as the control of a robotic manipulator is presented. The control of a robotic manipulator is hampered by complex kinematics and non-linear motion. The proposed solution will solve these problems by using: (i) the simple forward kinematics equations to train a fuzzy associative memory (FAM) to map the inverse kinematics solution, and (ii) test data from the DC motor to train the fuzzy controller. The individual ANFIS controller for each joint generates the required control signals to a DC servomotor to move the associated link to the new position. The proposed hierarchical controller is compared to a conventional proportional-derivative (PD) controller. The simulation experiments indeed demonstrate the effectiveness of the proposed method. The detailed work can be found in the Howard's thesis [36]. The chapter is organized as follows. In section 14. 2, the concept of ANFIS is introduced. In section 14.3, the solution of inverse kinematics using ANFIS is presented. Section 14.4 pertain to the controller design of a microbot with ANFIS. Finally, section 14.5 includes the conclusions and remarks related to the proposed method.

14.2 ADAPTIVE NEURO-FUZZY INFERENCE SYSTEMS

An adaptive neuro-Fuzzy Inference System (ANFIS)[14]-[16] is a cross between an artificial neural network and a fuzzy inference system (FIS). An artificial neural network is designed to mimic the characteristics of the human brain and consists of a collection of artificial neurons. An adaptive network is a multi-layer feed-forward network in which each node (neuron) performs a particular function on incoming signals. The form of the node functions may vary from node to node. In an adaptive network, there are two types of nodes: adaptive and fixed. The function and the grouping of the neurons are dependent on the overall function of the network. Based on the ability of an ANFIS to learn from training data, it is possible to create an ANFIS structure from an extremely limited mathematical representation of the system. In sequel, the ANFIS architecture can identify the near-optimal membership functions of FLC for achieving desired input-output mappings. The network applies a combination of the least squares method and the back propagation gradient descent method for training FIS membership function parameters to emulate a given training data set. The system converges when the training and checking errors are within an acceptable bound.

The ANFIS system generated by the fuzzy toolbox available in MATLAB allows for the generation of a standard Sugeno style fuzzy inference system or a fuzzy inference system based on sub-clustering of the data [37]. Figure 14.1 shows a simple two-input ANFIS architecture.

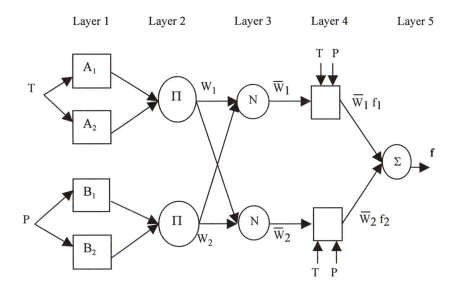

Figure14.1: ANFIS Architecture for a Two-Input System.

The above ANFIS architecture is based on a Sugeno fuzzy inference system. The sugeno FIS is similar to Mamadani format except the output memberships are singleton spikes rather than a distributed fuzzy set. Using singleton output simplifies the defuzzification step.

The ANFIS network shown in Figure 14.1 is composed of five layers. Each node in the first layer is a square (adaptive) node with a node function computed as follows:

$$O_i^1 = \mu_{Ai}(T) \tag{14.1}$$

Where T is the first input vector and μ is the membership function for that particular input. Layer two only consists of circle (fixed) nodes. The output of each node is the product of the two membership functions:

$$O_i^2 = W_1 = \mu_{Ai}(T)\mu_{Bi}(P) \tag{14.2}$$

Layer three only contains circle (fixed) nodes with their normalized firing strengths in the following form:

$$O_3^i = \overline{W}_i = \frac{W_1}{W_1 + W_2} \quad i = 1,2 \tag{14.3}$$

The fourth layer (square nodes) is computed from the product of consequent parameter set and the output of the third layer as:

$$O_4^i = \overline{W}_i(p_iT + q_iP + r_i) \tag{14.4}$$

Finally, layer five, consisting of circle nodes is the sum of all incoming signals.

$$O_{5,1} = \sum_i \overline{W}_i f_i \tag{14.5}$$

The above adaptive architecture is functionally equivalent to Sugeno fuzzy model. This ANFIS structure can update its parameters according to the gradient descent procedure. Other ANFIS structures corresponding to different types of FIS and defuzzification mechanism h are also proposed by the researcher [16]. However, throughout this chapter, we shall utilize the above first order Sugeno fuzzy model for the microbot application due to its transparency and efficiency.

14.3   INVERSE KINEMATICS

In general, most of controllers in robots require the inverse kinematics solution to determine the required joint angles. The inverse kinematics solution provides each joint angle based on the position of the end effector [38]. Due to the design of most robots, there are multiple joint angles that provide the same end effector position.

There are various methods to generate the inverse kinematics equations such as inverse transform, iterative and geometric approach [38]-[40]. The method used in this chapter is based on the using the position vector "**p**" described below as outlined in wolovich [38]:

$$\begin{bmatrix} p_x \\ p_y \\ p_z \end{bmatrix} = \begin{bmatrix} ec_1c_2 + fc_1c_{23} \\ es_1c_2 + fs_1c_{23} \\ h + es_2 + fs_{23} \end{bmatrix} \tag{14.6}$$

Solving these equations gives the following solutions to the inverse kinematics problem:

$$\theta_1 = a\tan 2\left[\frac{P_y}{P_x}\right] \quad \text{Or} \quad \theta_1 = a\tan 2\left[\frac{-P_y}{-P_x}\right] \tag{14.7}$$

$$\theta_2 = a\tan 2\left[\frac{(p_z - h)(e + fc_3) \mp fs_3\sqrt{p_x^2 + p_y^2}}{(p_z - h)fs_3 \pm (e + fc_3)\sqrt{p_x^2 + p_y^2}}\right] \tag{14.8}$$

$$\theta_3 = a\tan 2\left[\frac{\pm\sqrt{4e^2 f^2 - \left[(p_z - h)^2 + p_x^2 + p_y^2 - e^2 - f^2\right]^2}}{(p_z - h)^2 + p_x^2 + p_y^2 - e^2 - f^2}\right] \tag{14.9}$$

The above inverse kinematics equations show the multiple solutions for the microbot joint angles. In this section two different inverse kinematics solutions using FL and ANFIS are discussed. As was already mentioned, the main advantage of such methods is the establishment of the inverse kinematics mapping without any access to the robot kinematics equations.

14.3.1 Solution of Inverse Kinematics Using Fuzzy Logic

In this section, the inverse kinematics of the micro-robot using fuzzy rules will be developed. Equations 14.6 from the appendix describe the forward kinematics model of the microbot. For some nominal robot configuration θ_1, θ_2, θ_3, the first variations of Equation 14.6 is given as:

$$\delta P_x = \frac{\partial P_x}{\partial \theta_1}\delta\theta_1 + \frac{\partial P_x}{\partial \theta_2}\delta\theta_2 + \frac{\partial P_x}{\partial \theta_3}\delta\theta_3 \tag{14.10a}$$

$$\delta P_y = \frac{\partial P_y}{\partial \theta_1}\delta\theta_1 + \frac{\partial P_y}{\partial \theta_2}\delta\theta_2 + \frac{\partial P_y}{\partial \theta_3}\delta\theta_3 \tag{14.10b}$$

$$\delta P_z = \frac{\partial P_z}{\partial \theta_1}\delta\theta_1 + \frac{\partial P_z}{\partial \theta_2}\delta\theta_2 + \frac{\partial P_z}{\partial \theta_3}\delta\theta_3 \tag{14.10c}$$

Or in the matrix form:

$$\begin{bmatrix} \delta P_x \\ \delta P_y \\ \delta P_z \end{bmatrix} = \begin{bmatrix} A_1 & A_2 & A_3 \\ B_1 & B_2 & B_3 \\ C_1 & C_2 & C_3 \end{bmatrix}\begin{bmatrix} \delta\theta_1 \\ \delta\theta_2 \\ \delta\theta_3 \end{bmatrix} \tag{14.11}$$

$$A_1 = -[es_1 c_2 + fs_1 c_{23}] \quad B_1 = -[ec_1 c_2 + fc_1 c_{23}] \quad C_1 = 0 \tag{14.12a}$$

$$A_2 = -[es_2 c_1 + fc_1 s_{23}] \quad B_2 = -[es_1 s_2 + fs_1 s_{23}] \quad C_2 = ec_2 + fsc_{23} \tag{14.12b}$$

$$A_3 = -[fc_1 s_{23}] \quad B_3 = -fs_1 s_{23} \quad C_3 = fc_{23} \tag{14.12c}$$

In Equation 14.10, $\delta\theta_1, \delta\theta_2$ and $\delta\theta_3$ are small variations of the joint angles from their corresponding nominal values ($\delta\theta_i \ll 1$). Accordingly, 14.11 is the linearized version of 14.6 for a given operation point; similar to the method proposed in Nedungadi [24]. Therefore, Equations 14.11 and 14.12 provide useful information for the development of heuristic fuzzy rules in order to achieve the desired end effector position of the robot. In order to develop the rules, the following seven fuzzy sets are defined:

Positive Big   (PB)
Positive Medium   (PM)
Positive Small   (PS)
Zero   (Z)
Negative Small   (NS)
Negative Medium   (NM)
Negative Big   (NB)

Due to the fact that equation 14.11 represents the linearized kinematics of the microbot in the nominal operation point, the principle of superposition is utilized when these equations are applied to the individual angle variations. This superposition concept will be utilized to develop the FAM for the microbot. The following fuzzy rules are proposed in order to determine $\delta\theta_1$ for a given $\delta P_x, \delta P_y$ and δP_z.

Table 14.1: Fuzzy Rules.

$\delta\theta_1$		NB	NM	NS	Z	PS	PM	PB
	NB	PB	PM	PS	Z	NS	NM	NB
	NM	PB	PM	PS	Z	NS	NM	NB
	NS	PB	PM	PS	Z	NS	NM	NB
A_i	**Z**	Z	Z	Z	Z	Z	Z	Z
	PS	NB	NM	NS	Z	PS	PM	PB
	PM	NB	NM	NS	Z	PS	PM	PB
	PB	NB	NM	NS	Z	PS	PM	PB

Notice the entries of the FAM are obtained through graphical inspection of a three-dimensional graph. Similar tables should be developed for B_1 and C_1 in order to determine $\delta\theta_1$. Each entry in the FAM represents a fuzzy associative memory rule of the following form:

If (A_1 is NM) and (δP_x is PS) then $\delta\theta_1$ is NS

Thus each of 3 banks (i.e., look up tables) A, B, C comprises 49 rules, which results in 147 rules in order to determine the inverse kinematics solution. However, the rules can be reduced in order to reduce the computation effort. However, the above method does not provide a systematic approach for the solution of inverse kinematics. In order to build the banks A, B, C, we need to obtain a 3-D graphical representation of FAM. It can be constructed and implemented similar to Kim and Lee [26]. In the next section, an alternate methodology (ANFIS) is proposed.

14.3.2   Solution of Inverse Kinematics Using ANFIS

In this method, a set of training samples is collected from measurements. This is done by moving the robot to different desired end effector positions (P_x, P_y, P_z), and measuring the corresponding angles. By utilization of such data an ANFIS is trained with a sufficient number of data points. The block diagram of the proposed method is shown is Figure 14.2 as follows:

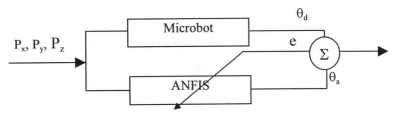

Figure 14.2: ANFIS Inverse Kinematics Block Diagram.

In the training phase, the membership functions and the weights will be adjusted such that the required minimum error is satisfied. In sequel the trained ANFIS can be utilized in order to provide fast and acceptable solutions of the inverse kinematics of the Microbot for various applications, such as on-line control of the robot.

14.3.3   Simulation Experiments

The ability to determine the requirement joint angles has been discussed previously, which also showed one way of generating the solutions to the inverse kinematics equations. The following method is an alternate approach to mapping the inverse kinematics solutions. The forward kinematics equations are relatively straightforward and easy to generate. The method chosen uses the forward kinematics equations to generate a collection of data relating the joint angles to the resulting Cartesian coordinates of the end effector. These data are then used to generate and train an ANFIS.

The method used to generate the training data is similar to using the microbot manipulator to create its own inverse kinematics solution. Instead of solving the inverse kinematics equation, various angles are applied to the robot and the resultant Cartesian coordinates (P_x, P_y & P_z) are determined. This output is directly related to the angles input, and is one solution to inverse kinematics equation. This allows the ability to select the desired angles at the design stage and creates a unique mapping between the angles and the resultant Cartesian coordinates. This method allows the creation of the inverse mapping based on actual information from the manipulator.

In the microbot manipulator the angle θ_1 is dependent on the X and Y. This is confirmed by the physical configuration of the robot. In the simulation, the test data were created from the possible ranges of X and Y. Figure 14.3 shows the desired mapping between X, Y and θ_1.

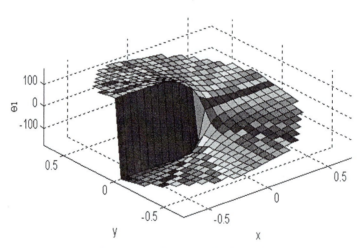

Figure 14.3: Desired θ_1 Mapping.

The following method was used to obtain the fuzzy mapping for θ_1. The first step was to allow the ANFIS algorithm to generate the membership functions and number of rules. Next, the ANFIS was trained with the test data and the resultant accuracy was determined. Figure 14.4 shows the initial membership functions and the resultant mapping after training for 50 epochs. The initial fuzzy inference system resulted in an overall root mean squared accuracy of 0.4374, which was not considered acceptable.

The ANFIS was then regenerated using a greater number of membership functions. The membership functions presently used are

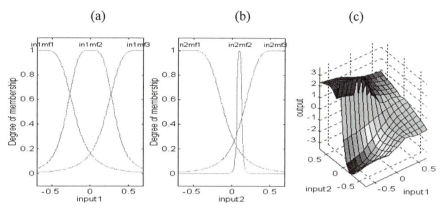

Figure 14.4: Initial θ_1 Membership Functions and Mapping.

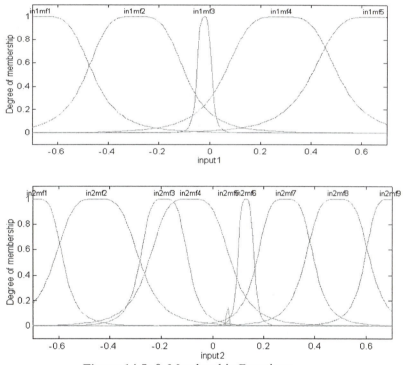

Figure 14.5: θ_1 Membership Functions.

These membership functions produce an output surface, which is shown below. Also included is the surface showing the error between the fuzzy angle and the desired angle. The maximum error was 6.93° and occurred with X, Y equal to [-0.7, 0.058]. The RMS error was 0.0134. The error associated with the base joint (θ_1) is very critical since, with the design of the microbot manipulator, there is no other joint that can directly compensate for this error.

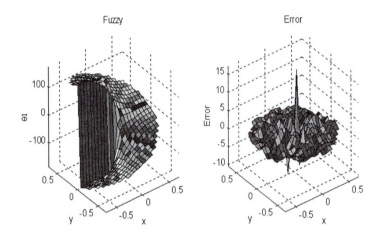

Figure 14.6: θ_1 Surfaces.

The same method was used to generate the fuzzy mapping for θ_2. Both of the angles θ_2 & θ_3 are dependent on *X, Y* and *Z*. Based on the training data, the mapping between *X, Y, Z,* and θ_2 are presented as follow:

(a)   (b)   (c)

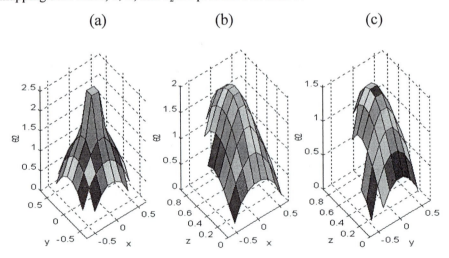

Figure 14.7. Desired θ_2 Mapping.

To generate the inverse kinematics mapping for θ_2 & θ_3, the ANFIS algorithm generated the initial membership functions and number of rules. The ANFIS was trained with the test data containing X, Y and Z and the corresponding angle θ_2 or θ_3. Figures 14.8 and 14.9 shows the present membership functions and surfaces for θ_2 respectively:

(a)   (b)   (c)

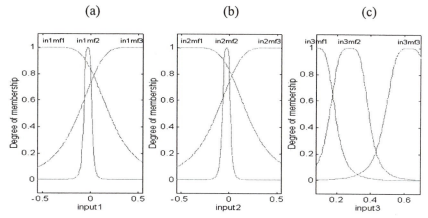

Figure 14.8: θ_2 Membership Functions.

(a)   (b)   (c)

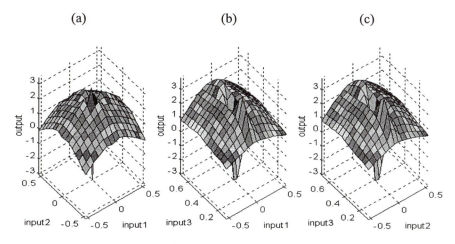

Figure14.9: θ_2 Fuzzy Inverse Surfaces.

The desired inverse kinematics mappings for θ_3 are presented in figures 14.10 (a, b, c). Figure 14.11 shows the present membership functions. The resultant inverse kinematics mapping surfaces for θ_3 are also shown (Figure 14.12). Figures 14.13 and 14.14 show the errors between the desired angles and the fuzzy generated angles for θ_2 and θ_3, respectively.

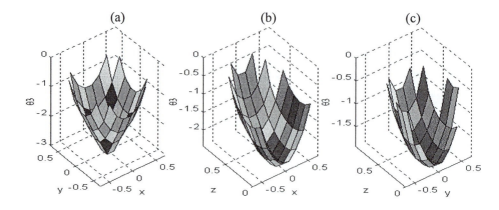

Figure 14.10: Desired θ_3 Mapping.

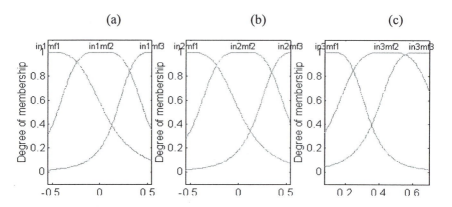

Figure 14.11: θ_3 Membership Functions.

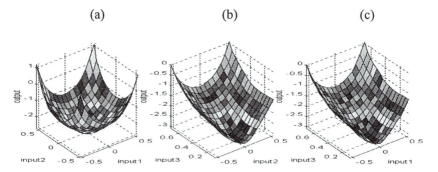

Figure 14.12: θ_3 Fuzzy Inverse Surfaces.

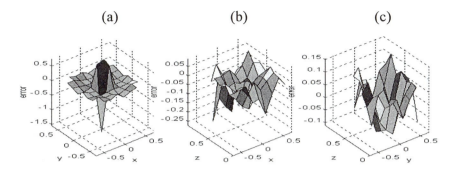

Figure 14.13: θ_2 Error Plots.

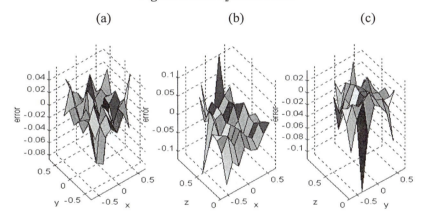

Figure 14.14: θ_3 Error Plots.

Since the inverse kinematics solutions for angles θ_2 or θ_3 are interdependent, the impact of the error is more difficult to describe. Figure 14.15 shows the impact of the overall error using all three fuzzy angles to determine the final end effector position. Each figure represents multiple end effector positions with the links of the manipulator shown. The desired link position and the resultant position due to the fuzzy angles are also shown in Figure 14.15.

14.4 CONTROLLER DESIGN OF MICROBOT

Robotic manipulator control is predominately motion control using classical servomechanism control theory [38]. Due to the non-linearity of the manipulator motion a wide variety of controls schemes have been devised. Classical schemes include computed torque, resolved motion, PID decoupled, model reference adaptive and resolved motion adaptive, to name a few [38]-

[41]. These schemes can be very complicated and require intensive computer resources. For instance, the computed torque technique uses the Legrange-Euler or Newton-Euler equations of motion of the manipulator to determine the required torque to servo each joint in real time to track the desired trajectory as closely as possible [40]. Model reference adaptive control requires an accurate reference model of the manipulator to compare against and an adaptive mechanism to store the adjustable feedback gains. For comparison purposes, the simple PD decoupled control scheme will be used, which is described in the following sections. The corresponding end effector position's error surfaces are shown in Figure 14.16.

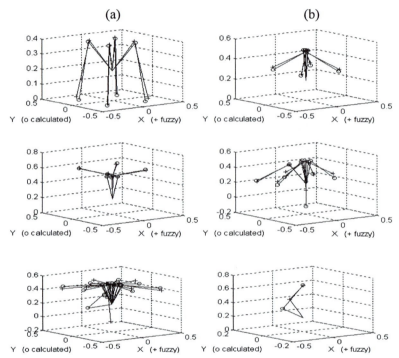

Figure 14.15: Overall Error Impact.

14.4.1 Design of a Conventional Controller

In this subsection, the design of a conventional controller for the microbot is presented. The controller is designed considering each joint as uncoupled, and usually neglects the effects of motion. To provide a comparison for the fuzzy controller, the microbot manipulator will be controlled with a conventional proportional-derivative type controller. Figure 14.16 shows the schematic of the conventional proportional-derivative type controller used in MATLAB to generate the comparisons. The proportional-derivative values for the individual joint were determined by considering each joint as uncoupled from the other, and calculating the approximate corresponding values. The proportional-derivative values were then fine tuned on-line and shown in Table 14.2.

In a conventional controller, the designer would normally generate the required trajectory of the individual links and end effector. These trajectories are based on the function of the manipulator, object avoidance, velocities and desired accelerations. For these simulations, the desired trajectory is not required since overall performance is being evaluated. Also, in the proposed conventional controller the required joint angles are calculated using the complex inverse kinematics equations. The inverse kinematics Equations 14.7-14.9 are utilized in the conventional controller to convert the desired Cartesian coordinates of the end effector into joint angles. There is no way to determine the comparative computing resources used to solve the inverse kinematics equations during the simulations, so no quantitative comparison will be made.

Table 14.2: Conventional Control Action: P-D Values.

	Initial Values		Final Values	
	Proportional	Derivative	Proportional	Derivative
Joint/Motor 1	100	5	100	8.5
Joint/Motor 2	139	8.5	150	15
Joint/Motor 2	153	10.4	153	10.4

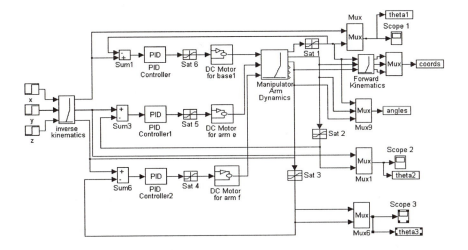

Figure 14.16: Conventional P-D Controller.

Based on the physical limitations of the microbot robot the joints have a defined acceptable range. The base, θ_1, can rotate from -170° to 170°. The elbow joint θ_2 and the forearm joint θ_3 can also be adjusted from -170° to 170°. However, due to the existence of the floor, the links and end effector have the additional constraint of not going below an elevation of zero. This is accounted for in the conventional controller by selecting the inverse kinematics solution of θ_2 to be the positive angle and using MATLAB saturation modules.

Based on the allowable ranges of the joints and the possible uses for a microbot robot, various position changes have been selected to test the fuzzy controller. These desired changes were simulated in Matlab and provide a wide range of change including multiple joint movements at one time. The plot of the simulation results for the conventional controller is shown in section 14.4.3, with the fuzzy logic controller results.

14.4.2 Hierarchical Control

One of the advantages of fuzzy logic control is that the controller can be designed in the same way a person would think of doing the control. In the case of a robotic manipulator point-to-point or trajectory position control, one could think of a supervisor telling the end effector where to go by specifying the required joint angles. In sequel, each joint could individually move to the desired angle. Therefore, the proposed approach is a hierarchical control scheme. A fuzzy mapping of the inverse kinematics solution will supervise the individual joints. A fuzzy controller will direct each individual joint to the desired position as shown in Figure 14.17.

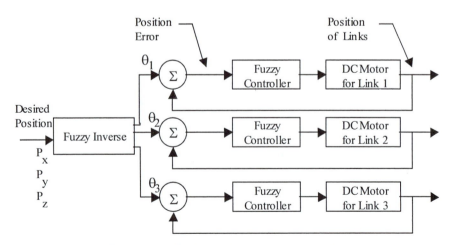

Figure 14. 17: Hierarchical Fuzzy Controller.

14.4.3. ANFIS Controller for Microbot

In this subsection, the ANFIS controller is proposed in conjunction with the switching curve for each joint of the robot in order to provide the desired end effector position of the microbot. Due to the interaction and coupling between the links of the manipulator, applying a constant torque to each joint will simplify the control of the other joints. In addition, applying the maximum torque should produce the fastest response. To produce a fast response, the premise is to apply a maximum torque in the required direction, then at the last possible instance apply a braking torque to stop the motion. One method of determining when to apply the maximum torque is called "switching surfaces."

The concept of sliding surface ("switching curve" or "variable structure") for nonlinear systems is well known and was originated by several Russian mathematicians (e.g., Aizerman and Gantmacher, Filippov). However, the application of the switching surface creates chattering [39]. In order to smooth out such effects, the ANFIS algorithm has been utilized. For each joint of Microbot, the switching curve is a relation between the position error and the rate of change of position. The switching curve can be generated analytically if an accurate model of the motor and joint is known. However, in most cases such a model is unknown. The practical method is to determine the switching curve by using the actual manipulator. In this method, the maximum torque is applied to the manipulator and the resultant position error vs the rate of change of position is plotted. This plot is then inverted to obtain the switching curve as shown in Figure 14.18. For a DC motor application, for all conditions on one side of the curve, a maximum positive voltage is applied. The other side of the switching curve receives a maximum negative voltage. The switching curve is therefore a very steep transition surface that could induce instability or chattering in the controller. To reduce the possibility and effects of this chattering, a transition region is created. The transition region is essentially twice the distance that the manipulator can travel during one sampling period. The size of the transition region can be determined from the data that created the switching curve. The next step in the algorithm is to generate test data and produce the fuzzy equivalent of the switching curve for each joint. The resultant surfaces are shown in Figure 14.19. The switching curve is the basis of the proposed fuzzy controller, but to enable the FLC to handle a dynamic system with improved response, gains are applied to the inputs of the FLC. These gains essentially allow the transition region (not curve) to rotate and account for a fast or slowly changing process. The value of the gains will be determined using the functioning system and tuned on-line.

The same algorithm was used to create the fuzzy inference systems for the remaining joints. The simulation results for point-to-point control of microbot (cases 1 – 4) compare the conventional PD controller to the fuzzy logic controller of the robot utilizing the switching curve surfaces.

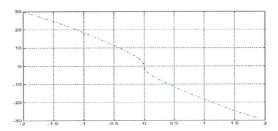

Figure 14.18: Switching Curve for Base Joint and Motor.

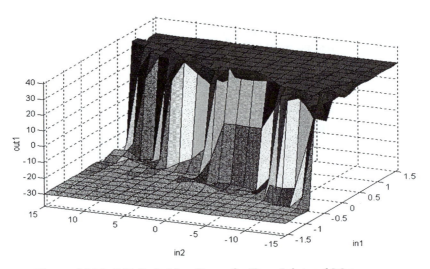

Figure 14.19: 3-D Switching Curve for Base Joint and Motor.

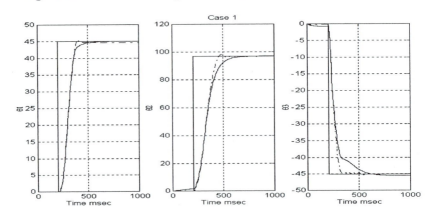

Figure 14.20: Simulation Results: Case 1.

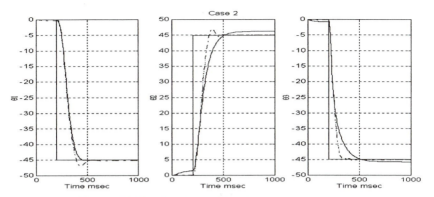

Figure 14.21: Simulation Results: Case 2.

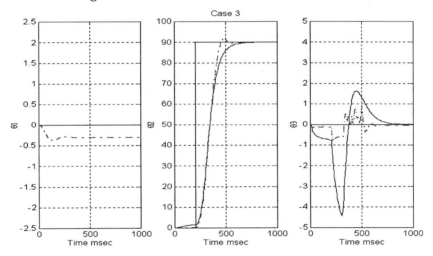

Figure14.22: Simulation Results: Case 3.

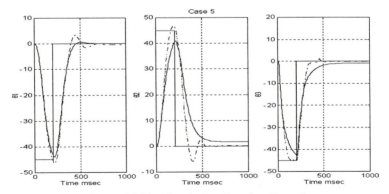

Figure 14.23: Simulation Results: Case 4.

The results of comparing the P-D controller with the fuzzy logic controller (shown in Figures 14.20 through 14.23) show that the FLC performs well. However, the simulation also displays that coupling exists between the elbow and the shoulder joints of the microbot manipulator. One method of accounting for this coupling is to use a feed-forward controller for each of the joints as described in Howard[36].

Table 14.3: Point-to-Point Control of the Microbot

	Desired final position			Actual final position with PID controller			Actual final position with ANFIS controller		
	X	Y	Z	X	Y	Z	X	Y	Z
1	0.0	0.15	0.812	.1504	.0154	.816	.1517	.1521	.8095
2	0.433	-.433	0.412	.4296	-.4296	.4199	-4334	.4315	.4076
3	0	0	0.9	0.	0	.899	.0003	0	.9
4	.7	0	0.2	0.699	0	.214	0.7	.0015	.1962
5	0.15	0.15	0.812	.1499	.1499	.8111	.1525	.1509	.8101
6	0.0	0.5	0.5	0.0	.4972	.2056	.0016	.4943	.1933

To evaluate the FLC in terms of robustness and sensitivity, additional case studies were performed. The original case 1 was performed again with a 100% increase in mass of link 3. Additionally, case 1 was performed again with a change in the time constant (25%) of the elbow joint motor. The simulation results show the effectiveness of the ANFIS for these cases also. Due to space limitation these cases are not reported here and can be found in reference [36].

Cases 1 through 4 demonstrate the controller performance for the end effector point-to-point control of the microbot. In order to determine the overall performance of the FLC, case 5 was created using a planned trajectory. The desired trajectory in this case is to move the end effector from the home position (0.7, 0, and 0.2) to an intermediate position (0.5, 0, and 0.2), then to the final position (0, 0.5, and 0.2) via a circular trajectory.

Case 5 simulated a simple trajectory following scenario. Figure 14.24 displays the PID controller against the FLC with the fuzzy inverse kinematics. In Figure 14.25, the Cartesian positions (X,Y,X) of the end effector are shown. Figure 14.26 compares the PD and FLC using fuzzy inverse kinematics solutions. In addition, case 5 has been simulated for PID and FLC utilizing the calculated inverse kinematics, and Figure 14.27 displays the results. Figure 14.27 shows that while the end point error is slightly higher during the transit, the FLC is actually superior to the P-D controller. Analysis of both figures

(14.26 and 14.27) indicates the significant difference is due to fuzzy inverse kinematics portion. Therefore, the inverse kinematics of the robot should be solved analytically in order to increase the accuracy of the trajectory performance. It should be point it out here, that an analytical solution of inverse kinematics is indeed much more accurate that the NN solution if the inverse kinematics equations are available. Otherwise, the NN inverse kinematics could be utilized. Of course the proper selection of data for the mapping of the inverse kinematics using NN is an important consideration.

Table 14.4: Position Error (L_2 Norm)

	Desired final position			Error	
Case	X	Y	Z	PD	ANFIS
1	0.15	0.15	0.812	.0048	.0044
2	0.433	-.433	0.412	.0125	.0063
3	0	0	0.9	0.009	.0003
4	.7	0	0.2	.0191	.0056
5	0.15	0.15	0.812	.0011	.0039
6	0.0	0. 5	0.5	.0116	.0178

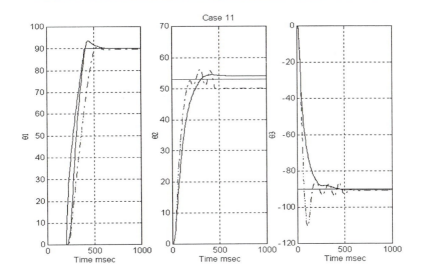

Figure 14.24: Case 5: Simulation Results (Angle Plots).

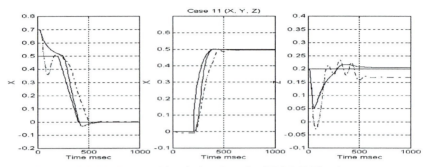

Figure14.25: Case 5: Simulation Results (X, Y, Z Plots).

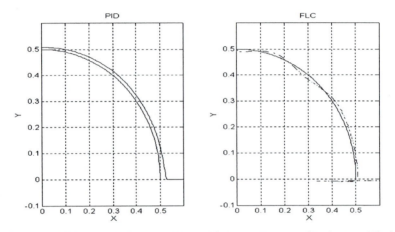

Figure 14.26: Case 5: P-D vs. FLC with Fuzzy Inverse (Trajectory Plot).

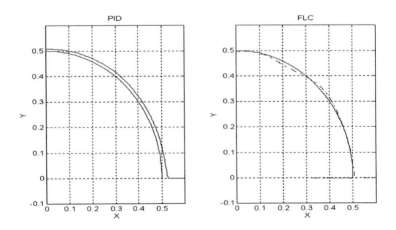

Figure 14.27: Case 5: P-D vs. FLC with Calculated Inverse (Trajectory Plot).

14.5 CONCLUSIONS

In this chapter, ANFIS has been utilized to generate the solutions to inverse kinematics equations of a microbot as well as providing a basic controller. The overall goal was to create a simple algorithm that could be used to control robotic manipulators with a minimum of theoretical modeling. The simulation experiments indicated that the mappings created by neuo-fuzzy algorithm adequately produced the solutions to the inverse kinematics equations. The fuzzy associative memories (FAMs) were utilized for the selection of a minimum number of rules and membership functions, and could be increased to improve the accuracy. By the use of the ANFIS algorithm, the training of the FAMs was relatively easy and straightforward.

The individual ANFIS controller for each joint generates the required control signals to a DC servomotor to move the associated link to the new position. The proposed hierarchical controller is compared to a conventional proportional-derivative (PD) controller. The simulation experiments indeed demonstrate the effectiveness of the proposed method. The analysis performed showed that the FLC had superior performance in regard to robustness of parameter changes. The proposed FLC performed essentially the same, in regard to changes in weight of the end effector and changes in the motor time constants, as the conventional P-D controller. Finally, the intention was to provide a simple algorithm to control a robotic manipulator with minimal or no modeling of the system. Hierarchical control was chosen since this method is very similar to the human thought process. ANFIS was chosen due to the adaptive nature and use of training data to create a fuzzy inference system.

Overall, the algorithm is (1) build the desired manipulator; (2) use the manipulator to generate training data for the mapping between the Cartesian space and the joint angle space; (3) use the manipulator to create the switching curves for the implementation of ANFIS controller of the individual joints and (4) fine tune the controller to achieve the desired performance.

REFERENCES

1.   Zadeh, L.A., Fuzzy Sets, *Information and Control*, Vol. 8, 338–353, (1965).
2.   Yager, R. and Zadeh, L.A. (eds.), *An Introduction to Fuzzy Logic Applications in Intelligent Systems*, Kluwer Academic Publishers, Boston, (1992).
3.   Zadeh, L.A., Making the Computers Think Like People, *IEEE Spectrum*, (1994).
4.   Mamdani, E. H., Application of Fuzzy Algorithms for Control of Simple Dynamic Plant, *Proc. of IEE*, Vol. 121, No.12, (1974).

5.   Surgeno, M. (ed.), *Industrial Applications of Fuzzy Control*, North-Holland, Amsterdam, (1985).

6.   Marks II,R. (ed.), *Fuzzy Logic Technology and Applications*, IEEE Press, Piscataway, NJ, (1994).

7.   Gupta, M. and Sinha, N.(ed.), *Intelligent Control Systems.: Theory and Applications*, IEEE Press, Piscataway, NJ, (1996).

8.   Kosko, B. *Fuzzy Engineering*, Prentice Hall, Upper Saddle River, NJ, (1997).

9.   Relics, A.(ed.), *Applied Research in Fuzzy Technology*, Kluwer Academic Publishers, Boston, (1994).

10.   Kaufmann, A. and Gupta, M. (eds.), *Introduction to Fuzzy Arithmetic Theory and Applications*, Van Nostrand Reinhold, NY, (1985).

11.   Nguyen, H., Sugeno, M., Tong, R., and Yager, R., *Theoretical Aspects of Fuzzy Control*, John Wiley & Sons, NY, (1995).

12.   Zilouchian, A., Hamono, F., and Jordnidis, T., Recent Trend and Industrial Applications of Intelligent Control System Using Artificial Neural Networks and Fuzzy Logic,.in: Tzafestas, S. (ed), *Method and Application of Intelligent Control*, Kluwer Academic Publishers, Boston, (1997).

13.   Diaz-Robainas, R., Zilouchian, A., and Huang, M., Application of Fuzzy Pattern Recognition to Functional Mapping and Controller Design, *Int. J. of Intelligent Automation and Soft Computing*, Vol. 5, No. 2., 95–109, (1999) .

14.   Jang, J.S., Self-Learning Fuzzy Controllers Based on Temporal Back Propagation, *IEEE Trans. on Neural Networks*, Vol. 3, No. 5, (1992).

15.   Jang, J.S. and Sun,C., Neuro-Fuzzy Modeling and Control, *Proc. of IEEE*, Vol. 83, No. 3, 378–406, (1995).

16.   Jang, J.S., Sun, C., and Mizutani, E., *Neuro Fuzzy and Soft Computing*, Prentice Hall, Upper Saddle River, NJ, (1997).

17.   Nguyen, L., Patel, R., and Khorasani, K., Neural Network Architectures for the Forward Kinematics Problem in Robotics, *Proc. of IEEE Int. Conf. on Neural Networks*, 393–399, (1990).

18.   Wang, D. and Zilouchian, A., Solution of Kinematics of Robot Manipulators Using a Kohonen Self Organization Neural Network, *Proc. of IEEE Int. Symp. on Intell. Control*, (1997).

19.   Liu, H., Iberall, T., and Bekey, G., Neural Network Architecture for Robot Hand Control, *IEEE Control Syst.*, Vol. 9, No. 3, 38–43, (1989).

20.   Eckmiller, R., Neural Nets for Sensory and Motor Trajectories, *IEEE Control Syst.*, Vol. 9, No. 3, 53–59, (1989).

21.   Nagata, S., Sekiguchi, M., and Asakewa, K., Mobil Robot Control by a Structured Hierarchical Neural Network, *IEEE Control Syst.Mag.*, Vol. 10, No. 3, 69 –76, (1990).

22. Handelman, D., Lane, S., and Gelfand, J., Integrating Neural Networks and Knowledge-Based Systems for Intelligent Robotic Control, *IEEE Control Syst. Mag.*, Vol. 10, No. 3, 77–86, (1990).

23. Rabelo, L.C. and Avula, X., Hierarchical Neuo-controller Architecture for Robotic Manipulation, *IEEE Control Syst. Mag.*, Vol. 12, No. 2, 37–41, (1992).

24. Nedungadi, A., Application of Fuzzy Logic to Solve the Robot Inverse Kinematics Problem, *Proc. of Fourth World Conf. on Robotics Research*, 1–14, (1991).

25. Nedungadi, A., A Fuzzy Robot Controller-Hardware Implementation. *IEEE Int. Conf. on Fuzzy Syst.*, 1325–1331, (1992).

26. Kim, S.W. and Lee,J.J., Inverse Kinematics Solution based on Fuzzy Logic for Redundant Manipulators, *IEEE Int. Conf. on Intell. Robotics and Syst.*, 904–910, (1993).

27. Xu,Y., and Nechyba, M., Fuzzy Inverse Kinematics mapping: Rule Generation, Efficiency and Implementation, *IEEE Int. Conf. on Intell. Robotics and Syst.*, 911–918, (1993).

28. Lim, C. M. and Hiyama, T., Application of Fuzzy Logic Control to a Manipulator, *IEEE Trans. on Robotics and Automation*, Vol. 7, No. 5, (1991).

29. Martinez, J., Bowles, J., and Mills, P., A Fuzzy Logic Positioning System for an Articulated Robot Arm, *IEEE Int. Conf. on Fuzzy Syst.*, Vol. 1, 251–257, (1996).

30. Kim, S.W., and Lee, J.J., Resolved Motion Rate Control of Redundant Robots Using Fuzzy Logic, *IEEE Int. Conf. on Fuzzy Syst.*, 333–338, (1993).

31. Lea, R. N., Hoblit, J., and Yashvant, J., Fuzzy Logic Based Robotic Arm Control in *Fuzzy Logic Technology & Application*, Mark, R. (ed.), IEEE Press, Piscataway, NJ, (1994).

32. Kumbla, K. K. and Jamshidi, M., Control of Robotic Manipulator Using Fuzzy Logic, *Proc. of IEEE Int. Conf. on Fuzzy Logic,* (1994).

33. Nianzui, Z., Ruhui, Z., and Maoji, F., Fuzzy Control Used in Robotic Arm Position Control, *IEEE Int. Conf. on Fuzzy Syst.*, (1994).

34. Moudgal, V. G., Kwong, W. A., Passino, K. M.,and Yurkovich, S., Fuzzy Learning Control for a Flexible-Link Robot, *IEEE Trans. on Fuzzy Syst.*, Vol. 3, No. 2, (1995).

35. Lee, S., Industrial Robotic Systems with Fuzzy Logic Controller and Neural Network, *IEEE Int. Conf. on Knowledge-based Intell. Electr. Syst.*, Vol. 2, 599–602, (1997).

36. Howard, D., Application of Fuzzy Logic for the Solution of Inverse Kinematics and Hierarchical Control of Robotics Manipulators, MS Thesis, Florida Atlantic University, Boca Raton, FL (1997).

37. *Fuzzy Logic Toolbox User's Guide,* MathWorks, Inc., (1995).

38.   Wolovich, W. A., *Robotics: Basic Analysis and Design*, Holt, Rinehart and Winston, NY, (1987).
39.   Slotine, J.J. and Li,W., *Applied Nonlinear Control*, Prentice-Hall, Englewood Cliffs, NJ, (1991).
40.   Lewis, F.L., Abdallah, C.T., and Dawson, D.M., *Control of Robot Manipulator*, Macmillan, NY, (1993).
41.   Baily, E. and Arapostathis, M., Simple Sliding Mode Control Scheme Applied to Robotics Manipulators, *Int. J. Control*, Vol. 45, No. 4, 1197–1209, (1987).

15

APPLICATION OF SOFT COMPUTING FOR DESALINATION TECHNOLOGY

Mutaz Jafar and Ali Zilouchian

15.1 INTRODUCTION

This chapter will discuss the application of artificial neural networks (ANNs) and fuzzy logic control (FLC) in the desalination industries, with particular emphasis on implementation of soft computing to a real-time direct seawater intake reverse osmosis (RO) plant. The chapter discusses the following as well:

- The use of back propagation learning techniques as well as radial basis function networks (RBFN) to predict critical water parameters for three different types of water intakes;
- Techniques for learning strategy for RBFN that involve a combination of supervised and unsupervised learning for redistribution of centers of receptive fields;
- The design of an intelligent control software environment for the development of a hybrid combination of NN, and Fuzzy Logic Controller (FLC) for real time RO plants;
- The implementation of the designed soft computing methodology for a prototype direct seawater intake RO plant; and
- The use of adaptive neuro-fuzzy inference system (ANFIS) for optimization of membership functions of the variables

At a time of intensive demand for producing fresh water at a reasonable cost, addressing automation, process control and cost optimization of desalination plants have become increasingly evident. large scale desalination processes must perform at high standards due to the increasing cost of high water quality production, equipment utilization, and rising government regulations on labor protection and the environment.

In this chapter, the recent innovation and technological advances in the design, implementation and application of soft computing methodologies to several desalination processes are addressed. Such advances are mainly due to the recent developments of intelligent control design approaches for the integration of sensory information, computation, human reasoning and decisionmaking. The principal partners in such an intelligent system include fuzzy logic (FL), and neural network (NN). In particular, the application of these approaches to RO desalination plants is presented. Various issues related to the design and implementation of soft computing methodologies including the trade off among tolerance, precision and uncertainty are also addressed.

The application of NN for quality control of RO plants is one of the main subjects of this chapter. Two NN predictive models are proposed based on back propagation and RBFN algorithms. These models are applied to three different types of RO feed intakes plants in order to verify the applicability of the NN models. The predictive models are studied using actual operating data for all three RO processes in order to predict various parameters of the plants including system recovery, total dissolved solids and ion concentration in brine stream. A proposed NN predictive model is presented based on redistributed RBFN centers using integration of supervised learning of centers and unsupervised learning of output layer weights. Extensive simulations are presented to demonstrate the effectiveness of the proposed method.

As a case study, the design and implementation of an intelligent control methodology for a direct Atlantic Ocean RO system located in Boca Raton, Florida, is also presented in this chapter. The operation of the prototype plant indeed demonstrated the effective and optimum performance of the proposed design for two types of membrane modules, spiral wound (SW) and hollow fine fiber (HFF), under forced diverse operating conditions. The system achieved a constant recovery of 30% and salt passage of 1.026% while salt concentration of six major salts as kept below their solubility limits at all times. The implementation of the proposed intelligent control methodology achieved a 4% increase in availability and reduction in manpower requirements as well as reduction in overall chemical consumption of the plant. Therefore, it is believed that implementing the developed control strategy can decrease the cost of producing fresh water.

This chapter is organized as follows: in section 15.2, the general background on desalination technology is provided with emphasis on the RO process. Section 15.3 presents the use of NN for prediction applications. Section 15.4 presents three case studies of prediction of critical parameters for RO plants. Section 15.5 will discuss implementation of a novel soft computing methodology to a seawater plant using a hybrid combination of FLC and NN. The final section of this chapter will discuss implementation of ANFIS to optimize different membership functions.

15.2   GENERAL   BACKGROUND   ON   DESALINATION   AND REVERSE OSMOSIS

Desalination methods are classified into two major processes: thermal and non-thermal. Thermal distillation involves phase changes and it includes multi-stage flash (MSF), vapor-compression (VC), and multi-effect (ME) [1]. Nonthermal processes do not involve phase change and include reverse osmosis (RO), electrodialysis (ED) and ion exchange (IE). Other less significant and cost intensive processes include vacuum freezing and refrigerant freezing [1], [2].

RO is defined as the separation of one component of a solution from another component by means of pressure exerted on a semipermeable membrane. RO achieves the finest level of filtration available by acting as a barrier to all dissolved salts, organic as well as most inorganic molecules with a molecular weight greater than 100. On the other hand, water molecules can pass freely through the membrane creating a purified product stream. Rejection of dissolved salts is typically 95 to 99% achieved at transmembrane pressure that ranges from 200 pounds per square inch (PSI) for brackish water to 1000 PSI for seawater. RO is applied to various applications such as brackish and seawater desalination, wastewater purification, biomedical separations, and food and beverage processing [1–3]. For an ideal aqueous electrolyte solution, Vant Hoff's law [1] theoretically defines the osmotic pressure (π) by a relation of the form:

$$\pi = nRTx_s / v_w \tag{15.1}$$

where n is the number of ions per molecule of solute, R is the universal gas constant, T is the absolute temperature, x_s is the salt mole-fraction, and v_w is the molar volume of the water. Equation 15.1 gives reasonable approximations of osmotic pressures for many solutions. Figure 15.1 shows the osmotic pressure as a function of total dissolved solids (TDS) of sodium nitrate, chloride, sulfate and seawater at 25°C.

15.2.1 Critical Control Parameters

Permeate flux, system recovery, and TDS are three of the most important factors that indicate the performance of the RO system. As an example, a low permeate flow combined with high salt passage could indicate colloidal fouling, metal oxide fouling or membrane scaling. Low permeate flow combined with normal salt passage may indicate biological fouling of the membrane. It is therefore essential for early detection of potential problems and proper adjustment of operating variables in such a way that fouling or scaling does not occur [6].

Control parameters may vary from the desired values causing harmful, and possibly severe, effects on membrane elements and the materials and components of the RO system [7]. Therefore, it is essential to identify any operating conditions that might lead to system failure. The following parameters have a direct effect on an RO system performance.

15.2.1.1 Temperature

Temperature variations have a determining factor on the osmotic pressure, membrane compaction rate, and hydrolysis rate. In general higher temperatures will increase the internal osmotic pressure and therefore lead to lower recovery ratio and permeate concentration. To estimate the effect of temperature on permeate flow rate of an element provided that the pressure remains constant, the temperature correction factor (TCF) can be found from using [4]:

$$TCF = \frac{Q_{25}}{Q_T} = EXP[U(\frac{1}{298} - \frac{1}{(273+T)}] \tag{15.2}$$

where Q_{25} is the permeate flow rate at 25°C, Q_T is the permeate flow rate at actual temperature T and U is a membrane factor.

15.2.1.2 Pressure

The effective net pressure is one of the most critical parameters to the RO membrane since the membrane element comes directly after the high pressure pump. System pressure provides brine pressure for the designed mass transfer of water and salts and affects compaction rate of the module. In general, the pressure is adjusted to achieve the desired recovery and salt rejection. A more general definition to Vann Hoff's law [5] for osmotic pressure is defined as:

$$\pi = 1.205\Phi(T + 273.15)\Sigma \overline{m}_i \tag{15.3}$$

where $\overline{m}_i$ is the molal concentration of ionic and nonionic constituents in the feed and Φ is the osmotic coefficient, which can be calculated based on the operating conditions [5]. The mean osmotic pressure ($\Delta\pi$) can be determined from feed osmotic pressure (π_f) and brine (π_b) as:

$$\Delta\pi = \frac{\pi_f + \pi_b}{2} \tag{15.4}$$

The effective driving pressure (P_{eff}) is defined as the difference between the applied pressure (P_f) reduced by pressure losses (P_l) due to piping and mean osmotic pressure reduced by product osmotic pressure (π_p):

$$P_{eff} = (P_f - P_l) - \Delta\pi - \pi_p \tag{15.5}$$

The effects of temperature and pressure variations are shown in Figure 15.2 for FILMTEC FT30 membrane [4] using a synthetic seawater solution.

15.2.1.3 Recovery

Recovery is defined as:

$$Y \equiv \frac{\text{Permeate Flow}}{\text{Feed Flow}} = \frac{TDS_{b\,rine} - TDS_{feed}}{TDS_{b\,rine} - TDS_{product}} \tag{15.6}$$

where TDS is the total dissolved solids of the ions considered.

15.2.1.4 Feed pH

Feed pH affects the permeate flux, salt rejection, hydrolysis and alkaline scale control. Cellulose acetate (CA) reacts slowly with water that forms alcohol and an acid. The rate of this reaction depends on feed temperature and pH and is defined as hydrolysis [8,9]. A pH control can maximize the lifetime of cellulose acetate membranes by operating at pH 6 or less, and therefore minimizing the hydrolysis rate. For Aramid membranes pH control can minimize the carbonate scale formation.

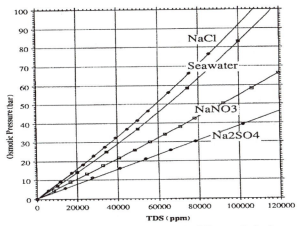

Figure 15.1: Osmotic Pressures of Different Solutions at 25°C
(Courtesy of The DOW Chemical Company, 1993).

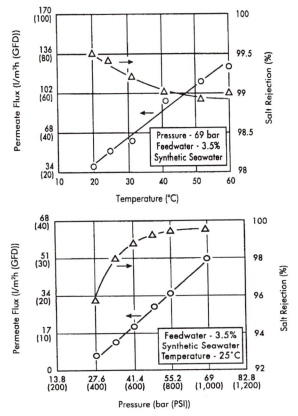

Figure 15.2: Effects of Feed Temperature and Pressure on Permeate Flux
and Salt Rejection (Courtesy of The Dow Chemical Company, 1993).

15.2.1.5 Salt Rejection

Cationic and anionic ions are usually repelled from the surface of the membrane to a distance proportional to its valence. This is due to the phenomena of di-pole that is formed between the charged ion and a surface with an equal and like charge. This electro-chemical interaction between the membrane and the salts causes the rejection of the salts. The net effect of the repulsion of the salts is the formation of a thin layer of pure water at the surface, which in turn is pushed through the pores by the applied pressure [9]. Membranes usually achieve high overall salt rejections (97%); however, small variations exist for each cation and anion. Mathematically, salt passage can be defined as:

$$\% \text{ Salt Passage} = \frac{\text{Permeate Salt Concentration}}{\text{Feed Salt Concentration}} \times 100 \qquad (15.7)$$

$$\% \text{ Salt Rejection} = 100 - \text{Salt Passage} \qquad (15.8)$$

The permeate concentration can be defined as [4]:

$$C_p = C_{fc} . \overline{p}_f . TCF . \frac{S_E}{Q_p} \qquad (15.9)$$

where Q_p is the permeate flow, S_E is membrane surface area, $\overline{p}_f$ is concentration polarization, TCF is temperature correction factor, and C_{fc} is brine side concentration. Effects of pH and feed water salinity on permeate flux and salt rejection for FT30 membrane is shown in Figure 15.3 for a synthetic seawater solution.

15.2.1.6 Scaling

Scaling and colloidal and biological fouling of the RO membranes can seriously impair system performance by lowering salt rejection and product recovery [4], [5]. Certain waters frequently contain troublesome constituents such as barium, hydrogen sulphide, or strontium. These constituents have to be controlled and kept at minimum levels in order to maintain the RO plant at a proper operation level. Basically, there are two parameters, which should be measured, monitored, and controlled: scaling and silt density index (SDI). Calcium sulphate and calcium carbonate scaling prevention involves calculation and prediction techniques in order to keep salts below their solubility index. Molal concentrations of all ions of the feed must be measured and/or predicted in order to determine the stability index.

The ion product (IP) of a salt $A_m B_n$, where A and B are the molal concentrations, is defined as [4]:

$$IP = A^m B^n \qquad (15.10)$$

And the ionic strength of feed water I_f and that of the concentrate stream I_c are defined as [5]:

$$I_f = \frac{1}{2} \sum_{i=1}^{N} \left[m_i . z_i^2 \right] \tag{15.11}$$

$$I_c = I_f . \frac{1}{1-Y} \tag{15.12}$$

where m_i is the ith ion concentration in mol/kg and z_i is the charge associated with it. Based on the index calculations and the type of ion being considered in addition to other limiting factors determined by the system, operating variables of the plant have to be adjusted in such a way that scaling will not occur. The precipitation of the dissolved salts should be kept below the solubility limit by adjusting parameters such as pH of the feed and system recovery.

The following are the most common salts that have a scaling potential in RO desalination:

- Calcium Carbonate ($CaCO_3$)
- Calcium Sulfate ($CaSO_4$)
- Barium Sulfate ($BaSO_4$)
- Strontium Sulfate ($SrSO_4$)
- Calcium Fluoride (CaF_2)
- Silica (SiO_2)

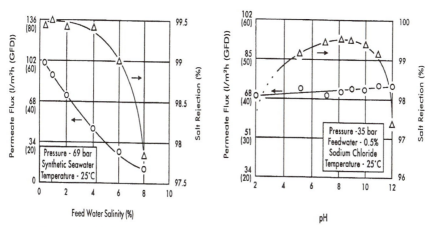

Figure 15.3: Effects of Feed Water Salinity and pH on Permeate Flux and Salt Rejection (Courtesy of The Dow Chemical Company, 1993).

15.3   PREDICTIVE MODELING USING NEURAL NETWORKS

In the development of predictive model for desalination plants, application of NN is essential due to non-linearity and complexity of interactions between operating variable [7]. In particular, prediction of product quality of RO process variables is a key factor to decreasing membrane degradation and the overall efficiency of an RO system. We next consider three case studies for different types of RO feed intakes.

Predictive modeling can be more efficient if scaling and cross validation are carried out prior to training the network. In scaling, the input and output training vectors are normalized in such a way that they fall within a specific range. In the work presented, training vectors were normalized to the range of [-1,1], and having a zero-mean. The input normalization variable P_{norm} is given by:

$$P_{norm} = \frac{2*(P_i - P_{min})}{(P_{max} - P_{min}) - 1}$$ (15.13)

Any future input-target vectors are normalized by the same method. The output network is then converted back to the original values.

On the other hand, cross validation provides a guided criterion for selection of network parameters and validation of the training data set chosen [17]. When a training set is picked from the available data, there is a need to validate the model on a data set that is different from the training set. Overfitting can occur when too many parameters are selected, while underfitting can occur when few parameters are used. The particular model that gives the best performance, is then trained on the full training data set and generalization is then tested on the verification set. The available data is partitioned to two sets, training and test set. The training set is further partitioned into two sets: training and validation. The network is then simulated with the selected training data set.

The first step in back propagation learning is to initialize synaptic weight and threshold levels for the different nodes. Wrong choice of initial weights and threshold levels can lead to a premature saturation [10]. This effect causes delay in convergence and is different from local minima. On the other hand, proper choice of initial values can lead to a more generalized approximating network. Throughout the back propagation simulation presented, the network was initialized based on the technique of Nguyen and Widrow [11].

15.3.1   Redistributed Receptive Fields of RBFN

Redistributions of centers to locations where input training data possess significant effects can lead to more efficient RBFN. The proposed method herein is based on clustering of input space vectors and computing weights of Euclidian distances. Histogram equalization within each cluster will determine the center and width of each receptive field. The supervised part of the algorithm includes redistribution of the centers and widths of receptive fields over the input space and computation of weight and bias matrix. The unsupervised part of the algorithm includes computation of the output weight and bias matrix. Figure 15.4 shows a general structure of RBFN network. Summary of the various steps involved in the adaptive receptive field training is shown in Figure 15.5.

15.3.1.1 Data Clustering
Clustering partitions a data set into subgroups that have similar input-output pairs. K-means and fuzzy C-means clustering are two of the most common methods that are frequently used with radial basis. Specht [13] used an effective clustering method based on determination of radius of influence. Training data

are first normalized to a value between [-1,1]. A radius of influence (r) is then specified, and the first point establishes a new cluster center at x_i. Each vector x in the input space is considered one at a time and if a sample with an absolute distance x-x_i to the nearest cluster is $>r$, then the vector center becomes the center of the new cluster. If the absolute distance of x-x_i of the sample is less than the distance to any other cluster center and is $\leq r$, the vector is assigned to that cluster. This procedure performs clustering in a noniterative way and requires only one pass through the training set.

15.3.1.2 Histogram Equalization

Histogram equalization is a method used in digital image enhancement techniques. Basically, histogram equalization stretches the contrast of an image by uniformly redistributing the gray values. For the selected k^{th} output running sums of the ith input vector are evaluated by:

$$y_k = \sum_{i=1}^{k} y_i \qquad (15.14)$$

A plot of running sums is evaluated and the number of radial basis functions evenly divides the Y-axis. The values of the resultant input vector will be used to determine the new Euclidian distances over the input space.

Once the Euclidian distances are determined, the first layer weights are then computed. This involves solving N nonlinear equation of the form,

$$\left\| x - x_\alpha \right\|_i = \sqrt{\sum_{i=1}^{N} \left[x_i - x_{\alpha i} \right]^2} \qquad (15.15)$$

where $\left\| x - x_\alpha \right\|$ is the desired Euclidian distance of the ith cluster, N is the number of clustering centers, x_i is the average mean of the input cluster, and $x_{\alpha i}$ are the weights to be calculated. The next step involves solving N nonlinear equations to solve for the weight matrix. The method of generalized reduced gradient nonLinear optimization (GRG2) code developed by Ladson and Waren [14] is used for solving the set of nonlinear equations. The solution to the minimization of the function y_k in Equation 28 determines the set of weights $x_{\alpha i}$.

$$y_k = \sum_{i=1}^{N} \left\| x - x_\alpha \right\|^2 \qquad (15.16)$$

15.3.1.3 Widths of Receptive Fields

In all of the simulations carried out on different data sets, it was found that the choice of a single global width value (σ^2) gave better results than separate widths for each cluster [15]. The value of σ^2 in our study was set at the average standard deviation of the input vector histogram.

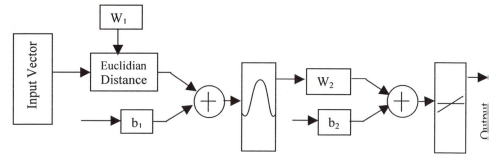

Figure 15.4: RBFN Network Architecture.

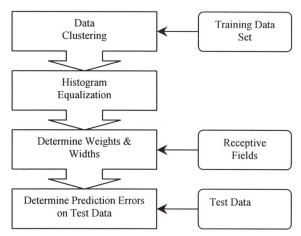

Figure 15.5: Redistributed Receptive Fields Learning Steps.

15.4   CASE STUDIES

15.4.1   Example 15.1: Beach Well Seawater Intake

The system is two train membrane seawater type that is operated by the Water Desalination Department of the Kuwait Institute for Scientific Research in Doha, Kuwait. Each line is designed to operate independently to produce 300 cubic meter per day (m^3/D) fresh water fed from a common seashore well. Average feed temperature of the beach well was 24°C. Feed flow was at $35m^3/h$ passing through a pretreatment stage of chemical dosing which included anti-scalant, sulfuric acid and sodium meta-bi-sulfate followed by a bag filtration. Average feed pressure to the membrane was 58 and 70 bar for train one and two, respectively. Feed conductivity from the well was 54 micro Siemens per centimeter (mS/cm) with an SDI <2. In this example we shall predict the permeate TDS using a nine-element input vector consisting of temperature, feed TDS, feed flow, pH, SDI, permeate TDS, permeate conductivity, feed pressure, and brine pressure.

15.4.1.1 Simulation Results

To determine the number of layers and the corresponding number of hidden neurons for the back propagation network, the criterion used is smallest number of neurons that yield a minimum RMS error with least number of iterations. Baugham and Liu [12] found that adding a second layer to the network could significantly improve the performance of the network, while adding a third layer required longer training. The values of the learning rate η and momentum coefficient α were arbitrarily set at small values. The network was simulated on 500 training data set for various numbers of hidden neurons in each layer. The number of layers chosen was two and the number of neurons in each layer was selected based on the lowest RMS error.

Proper choice of the learning η and momentum coefficient α could yield convergence that has good generalization characteristics with least number of iterations. An experimental procedure is used here for the determination of the two coefficients. Using a two hidden layer with 28:13 configurations, Figure 15.6 shows the results for the RMS error for $\alpha \in \{0.05, 0.1, 0.5\}$ for two values of learning rate parameter $\eta \in \{0.1, 0.6\}$.

The learning curves indicate that the smaller values of learning rate resulted in slower convergence to the set error minimum. Increase of momentum coefficient gave a faster convergence while too high of a momentum rate caused instability ($\alpha = 0.85$ not shown here). Learning rate of 0.1 and momentum coefficient of 0.5 resulted in fast convergence and lowest overall RMS error minimum. Results of the cross validation curves are shown in Figure 15.7.

The results, in general, are satisfactory since the test and validation errors have similar characteristics. No over or underfitting is visible and, therefore, we conclude that the training set and parameters chosen are adequate.

The simulation was carried out using MATLAB R11 with neural network toolbox v3.0 [16]. L_2 and $L\infty$ norms error criteria were used for comparison and error analysis [18]. Figures 15.8 and 15.9 show the predicted TDS of the product water using back propagation and RBFN. Table 15.1 summarizes the error norms for the two methods.

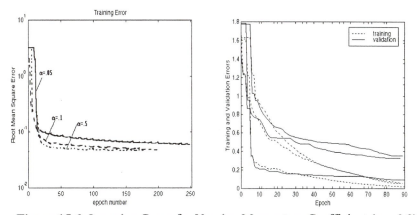

Figure 15.6: Learning Curve for Varying Momentum Coefficient ($\eta = 0.3$).

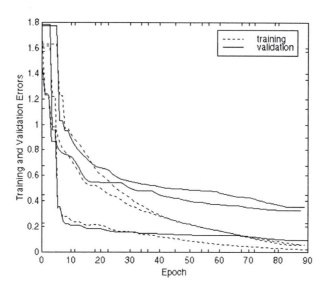

Figure 15.7: LMS and Corresponding Validation Curves for Selected Values of Learning Rate (0.05 Top Curves, 0.1 Middle Curves, 0.5 Bottom Curves).

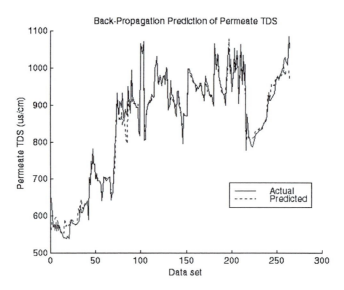

Figure 15.8: Back Propagation Prediction of TDS.

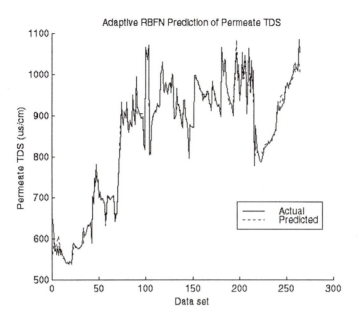

Figure 15.9: Redistributed RBFN Prediction of Permeate TDS.

15.4.2   Example 15.2: A Ground Water Intake

The system is a two-stage RO system that is operated by the City of Boca Raton Public Utilities in Boca Raton, Florida. The number of parallel pressure vessels was four in stage one, and two in stage two, and each vessel had seven elements. The membrane element type used was Hydronautics ESNA. Membrane material was Polyamide, assembled as a thin film composite with a negative charge. The size of each element was 4 by 40 inches and the total active membrane area was 400 ft^2. Maximum element recovery was 21% while the salt rejection was 80%. Maximum operating pressure for the plant was 400 PSI while the maximum SDI allowed was 4. Plant capacity was 37 gallon per minute (GPM) fed from the well at 43.75 GPM to meet the designed system recovery, which was set at 85% at 25°C.

The feed was ground water set at 44 GPM fed to a pretreatment station of dual media filter, composed of silica and garnet, and 5-μm cartridge filter. Chemical addition composed of scale inhibitor and sulfuric acid for pH adjustment to range between 6.0 and 6.4. The plant was operated and tested over a two-year period for the first study. Input variables used in the input vector included feed temperature, feed flow, feed conductivity, feed pressure, and permeate conductivity. We attempted to predict the permeate flow rate as our output variable. The model for both back propagation and RBFN was developed in similar manner as in Example 15.1. Results of back propagation and RBFN predictions of permeate flow are shown in Figures 15.10 and 15.11. Summary of comparison between the two algorithms is shown in Table 15.2.

15.4.3   Example 15.3: A Direct Seawater Intake

The system is a surface seawater intake located at the Florida Atlantic University research facility at Gumbo Limbo Research Park in Boca Raton, Florida. The schematic diagram of the RO plant built at FAU research laboratory is shown in Figure 15.12. The intake line is 4 inches wide and is 700 yards deep into the Atlantic Ocean. Two membrane elements were used for the research study: spiral wound (SW) and hollow fine fiber (HFF) membranes.

The membrane type of the HFF configuration was a B10 Aramid made by Du Pont chemical. The membrane can achieve salt rejection higher than 98.5% under a nominal operating pressure of 800–1000 psi. Temperature range allowed is 32–95°F and the pH allowable range is 4–9. Permeate Productivity is 250 GPD with maximum and minimum brine rates of 2400 and 800 GPD, respectively. Pressure vessel material is filament-wound fiberglass epoxy and weighs 10 Kg when filled with water.

The membrane type of the SW configuration was a cellulose triacetate (CT) made by Toyobo. The membrane can achieve salt rejection of 99.4% under a nominal operating pressure of 60 bar. Temperature range allowed is 5–40°C; pH allowable range is 3–8. Permeate productivity is 792 GPD with maximum and minimum brine rates of 2 and 10 m^3/day, respectively. Pressure vessel material is fiberglass epoxy and weighs 23 Kg when filled with water. Feed conductivity from the ocean ranged between 57–59 mS/cm with an SDI <4. The pre-treatment stage included sulfuric acid for pH adjustment to range of 5.5–7, followed by a 5-micrometer cartridge filter. Average feed temperature from the ocean inlet was 27.6°C and average feed pressure to the membrane was 60 bar.

We shall attempt to predict the recovery of the RO system, which will be used as part of the fuzzy control methodology described in the next section. A ten-variable input vector was used for this simulation and it consisted of feed temperature, feed flow, pH, feed conductivity, pressure inlet, transmembrane pressure, salt passage, permeate flow, permeate conductivity, and permeate pressure. RBFN prediction of system recovery is shown in Figure 15.12.

15.4.3.1 Scaling Simulation

The ionic strength of feed water was determined from feed water analysis of major constituents as:

$$I_f = \frac{1}{2}\left\{([Ca^{++}]+[Mg^{++}]+[SO_4^{--}])\times 4 + ([Na^+]+[K^+]+[HCO_3^-]+[Cl^-])\right\} \quad (15.17)$$

The ionic strength of the concentrate I_c can be calculated from the predicted system recovery using equation 15.12.

Solubility product (K_{sp}) of all salts can be determined as a function of the ionic strength I_c. The predicted I_c is shown in Figure 15.13 for the HFF membrane system as a function of I_f.

Calcium Sulfate

The ionic product (IP_c) for calcium sulfate ($CaSO_4$) in the concentrate stream can be calculated from the predicted recovery Y as:

$$IP_c = [(^m Ca^{++})_f \times \frac{1}{1-Y}] \times [(^m SO_4^{--})_f \times \frac{1}{1-Y}] \qquad (15.18)$$

Results of the ion product IP_c are compared with the solubility product Ksp for $CaSO_4$. For the example above at 28% recovery, Ksp is equal to $1.8e^{-3}$. Therefore, Ipc=0.45 Ksp and no scaling will occur at this point. If $IP_c > 0.8$ K$_{sp}$ scaling will form and the system must go into a lower pH set point or lower recovery.

Barium Sulfate
This salt is the most insoluble alkaline sulfate, and may lead to precipitations when present in feed water. The ionic product (IP_c) for barium sulfate ($BaSO_4$) in the concentrate stream can be calculated from the predicted recovery Y as:

$$IP_c = [(^m Ba^{++})_f \times \frac{1}{1-Y}] \times [(^m SO_4^{--})_f \times \frac{1}{1-Y}] \qquad (15.19)$$

Figure 15.14 shows the RBFN-predicted IP_c value for calcium sulfate and barium sulfate of the Aramid membrane (HFF).

Calcium Carbonate
$CaCO_3$ can be determined using the Stiff and Davis Stability Index (S&DSI) by finding the difference between the value of the pH of the concentrate and the pH at which the stream is saturated with calcium carbonate:

$$S \& DSIc = pHc - pHs \qquad (15.20)$$

A negative value of S&DSI$_c$ indicates that $CaCO_3$ tends to dissolve; however, if the calculations turns out to be positive, adjustments must be made to the system. The addition of sulphuric acid to the feed solutions converts the carbonate ion bicarbonate and converts bicarbonate to CO_2 and therefore decreases the S&DSI$_c$ by the following reaction [5]:

$$H_2SO_4 + 2HCO_3^- \rightarrow 2H_2O + 2CO_2 + SO_4^= \qquad (15.21)$$

Strontium Sulfate
Predictions of $SrSO_4$ are performed in a similar manner to $CaSO_4$ using the following equations:

$$IP_c = [(^m Sr^{++})_f \times \frac{1}{1-Y}] \times [(^m SO_4^{--})_f \times \frac{1}{1-Y}] \qquad (15.22)$$

Silica
Dissolved silica SiO_2 ions are present in most feed waters and can polymerize and cause scaling to the membrane when supersaturated. The SiO_2 concentration in the brine is calculated from:

$$SiO_{2c} = SiO_{2f} \times \frac{1}{1-Y} \qquad (15.23)$$

Table 15.2: Summary of Error Norms for Example 15.2 (permeate prediction)

Method	$\varepsilon \infty$	ε_2	Total Neurons	Remarks
Back propagation	2.084	0.876	45	$\alpha=0.45; \eta=0.2$
Redistributed RBFN	2.3	0.582	5	RI=0.3; $\sigma^2=0.92$

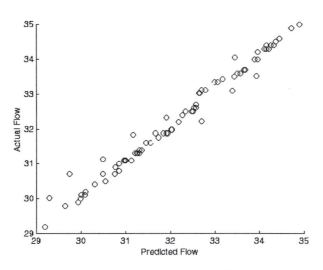

Figure 15.10: Actual and Predicted Permeate Flow for Back Propagation network

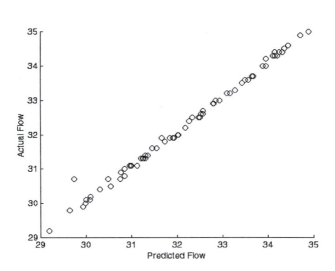

Figure 15.11: Actual and Predicted Permeate Flow

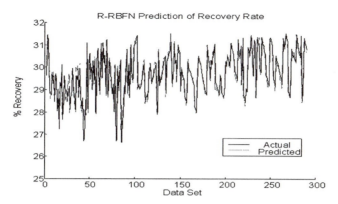

Figure 15.12: Redistributed RBFN Prediction of Recovery (SW)

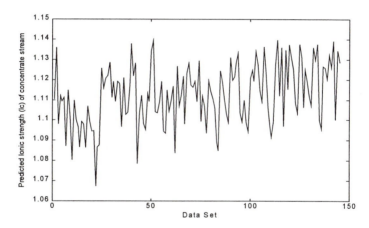

Figure 15.13: Predicted ionic Strength Ic of Concentrate Stream

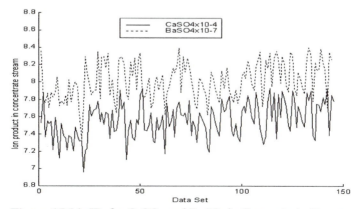

Figure 15.14: IP_c for $CaSO_4$ and $BaSO_4$ in Concentrate Stream

15.5   FUZZY LOGIC CONTROL

RO desalination is a nonlinear process, which has to operate under specific conditions that are of utmost importance for plant operation optimization. The nonlinearity relates to changing process characteristics such as feed total dissolved solids and pH, which in turn cause changes to product quality and quantity parameters such as salt rejection and recovery. An RO desalination system is usually designed based on a defined set of data analysis such as flow, temperature, and feed water composition. However, in reality, plant operation has to be flexible in order to respond to changing variables. Integration of key process information into the control strategy's decisionmaking and prediction can yield an increase in the lifetime of the membranes, availability, and efficiency and optimize plant performance.

The nature of the membrane separation process and the characteristics of the membrane system impose a number of constraints on the system [19], [20]. These constraints require continuous monitoring and control if the system is to perform economically over a long period. The main operational constraints for RO desalination are as follow:

1. Pretreatment control for suspended solids to obtain biological and chemical stability;
2. Operation between a minimum temperature to provide required flux and a maximum temperature allowed by the membrane specifications;
3. Operation between minimum brine flow to avoid concentration polarization and maximum flow with respect to desired recovery;
4. Operation at a pressure to obtain desired mass transfer and equalization of pressure drop; and
5. Chemical characteristics of feed water and dynamics of mass transfer.

Loss in salt rejection and loss of permeate flow are the main problems encountered in RO plant operation. It is of utmost importance that corrective measures are taken as early as possible [21]. Some of the parameters, such as temperature and feed water salinity change naturally. Other parameters may change as a consequence of other changes that are present in the system. Once the problem has been identified, causes must be identified and corrective measures must be taken by the system. The RBFN predictions of recovery and salt rejection described in previous sections were used as part of the input to the FIS. Measured values from sensor information make up the remaining input values to the FIS. These inputs include the following:

1. Temperature
2. Feed TDS
3. Feed pH
4. Feed flow
5. Feed pressure
6. Brine pressure
7. Permeate conductivity
8. Salt rejection

9. Recovery
10. Predicted Scale Index

The fuzzy system was implemented using Mamdani architecture [22] with the following FIS properties:

And method: min
Or method: max
Implication: min
Aggregation: max
Defuzzification: centroid

The hardware system configuration is shown in Figure 15.15 and consisted of a fuzzy controller and a Siemens PLC interface to the controlled plant. A programmable logic controller (PLC) controlled input-output interface signals, and provided for data transfer to the fuzzy controller. The PLC included two 16-channel analog input and output cards and eight 32-channel digital input and output cards. The CPU used 32-bit architecture for all arithmetic and comparison operations and an expanded register set. Figure 15.16 shows the schematic diagram of the developed fuzzy controller. Figure 15.17 shows the process layout of the prototype RO plant built at FAU research laboratories with feed intake from the Atlantic Ocean.

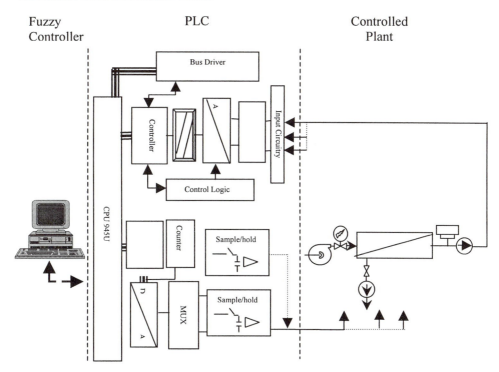

Figure 15.15: Hardware Setup for the RO Plant.

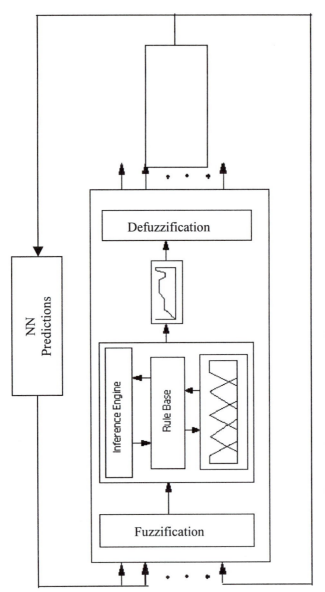

Figure 15.16: Scheme of the Hybrid Control System.

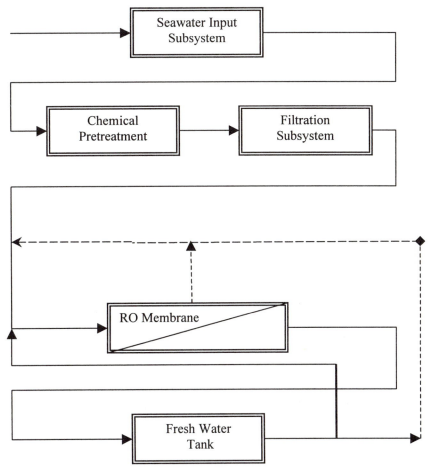

Figure 15.17: Process Layout of the Implemented RO System.

15.5.1   Chemical Dosing Control

There are two factors that can influence the acid dosing required: desired pH value and the actual and/or predicted scaling. The pH set point was set to a value determined by S&DSI to be equal to –0.1 for minimum carbonate scaling of the HFF membrane, and by the lowest hydrolysis rate for the CTA membrane. Two methods were used for scaling calculations: predicted and actual ion concentration in the brine stream. In the predicted method, RBFN was used to predict the ion product and ion strength of the brine stream and compared with the solubility product of the particular salt in question as discussed previously. The fuzzy controller provided an external analog input to the acid-dosing pump based on the defuzzified output conclusion.

15.5.1.1 Fuzzy Rule Base

The rule base consisted of fuzzy rules based on two inputs and one output. The set point is determined by the pH value desired, the output of the system consisted of external dosing rate signal to the acid pump, whereas the inputs to the fuzzy system were as follows:

Error = the pH set point minus the process pH.

Error change = difference between present error of process output (e1) and error of previous output (e2).

15.5.1.2 Membership Functions

Having formulated the control variables, we next define the membership functions of the linguistic set. Figures 15.18 and 15.19 show the membership of the linguistic properties for each attribute. Input membership functions were either triangular or trapezoidal, whereas singleton membership functions were used for output attributes throughout. The block diagram of the fuzzy controller for the chemical dosing pump is shown in Figure 15.20.

15.5.1.3 Decision Matrix

The rules are formulated and combined to form a decision table for the fuzzy controller to accommodate different situations transpired by the system. Table 4 shows the rule base for the two inputs and output of the dosing rate. The defuzzified output is converted to a crisp value using the centroid method. The decision surface for the fuzzy control of chemical dosing for the two inputs is shown in Figure 15.21.

15.5.1.4 Results and Discussion

Changes in pH set point as a result of ion index calculations and disturbances, such as feed pH and feed flow rate and overall system recovery, were tested and recorded. The results of these changes contributed vital information for controller tuning and improvement of system operation. Figure 15.23 shows the pH value, control activity, and error signal for chemical dosing pump.

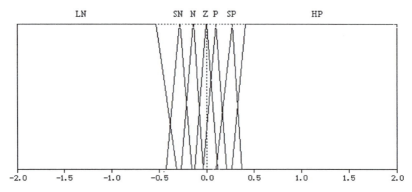

Figure 15.18: Membership Functions of the Seven Linguistic Properties for the Attribute "Error".

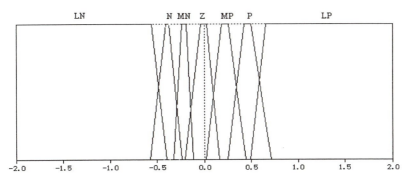

Figure 15.19: Membership Functions of the Seven Linguistic Properties for the Attribute "Error Change".

Table 15.3: Look-Up Table for Chemical Dosing Pump

Error	Δ Error						
	HN	SN	N	Z	P	SP	LP
LN	VH	H	H	H	SH	SH	SL
N	VH	SH	H	SH	M	M	LP
MN	L	SH	SH	SH	M	SL	LP
Z	L	M	SH	M	SL	SL	LP
MP	L	M	M	SL	SL	L	VL
P	SL	M	M	L	L	L	VL
LP	SL	SL	SL	L	L	VL	VL

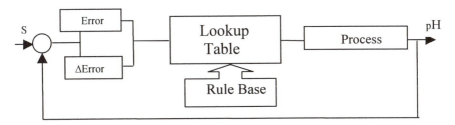

Figure 15.20: Block diagram of the Fuzzy Controller for the Chemical.

Changes in feed TDS and temperature may cause changes to system recovery and scaling potential and could result in carbonate saturation. Changes to feed TDS were induced by feeding the brine stream back into the feed line in forcing variables to change from the normal range tested by the fuzzy controller's response to changes in critical variables in order to raise the feed TDS to simulate changes in the operating conditions. Heating the feed pipe induced change in feed temperature and raised the temperature to approximately 30°C. Higher temperatures increased the S&DSI from –0.1 to 0. This change in the stability index indicated possible scaling potential of CaCO$_3$ and, therefore, needed to be adjusted. The pH set point was lowered from 7.0 to 6.4; as a result the CaCO$_3$ ion concentration remained constant over the operational period as shown in Figure 15.22.

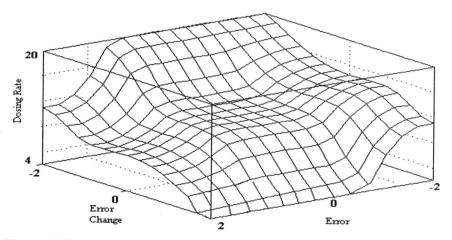

Figure 15.21: Decision Surface for the Fuzzy Control of Chemical Dosing

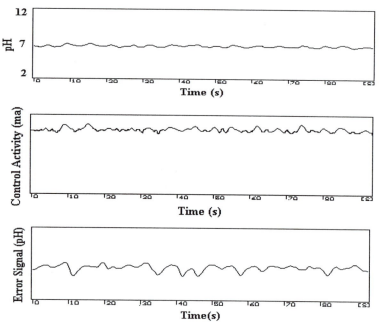

Figure 15.22: pH Value (Top), Control Activity (Middle) and Error (Bottom) Signal for Chemical Dosing Pump.

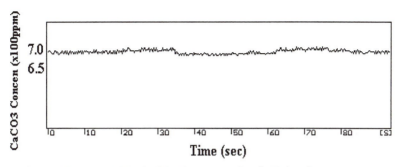

Figure 15.23: CaCO₃ Concentration in Brine Stream.

15.5.2 High-Pressure Control

Control of effective high pressure applied depends on water and salt fluxes, total dissolved solids, brine discharge, permeate flow, permeate conductivity, and feed temperature. The task of the fuzzy controller is to adapt to changes in these variables by recalculating new high pressure applied, taking into account the maximum allowable operating pressure of the membrane. The defuzzified output signal of the fuzzy controller was externally fed into a high pressure valve in the brine stream. This in turn determined the percentage of the opening or closing of the valve based on the six-input variables.

15.5.2.1 Fuzzy Rule Base

The rule base consisted of rules for the six inputs and one output. The set point for normal pressure was determined based on initial values of conductivity, temperature and the effective applied pressure as described previously. Each parameter causes changes to other variables and, therefore, the pressure needs to be adjusted accordingly in order to compensate for these changes and provide a constant water quality – in this case salt rejection.

15.5.2.2 Decision Matrix

The rules are formulated and combined to form a decision table for the fuzzy controller. Examples of such rules are as follows:

1) If temperature is warm and feed TDS is medium and permeate TDS is normal and salt passage is high and feed pressure is medium and salt passage is slightly high, then output is low.
2) If feed TDS is slightly low and permeate flow is high and temperature is normal or temperature is warm and permeate TDS is medium low and salt passage is slightly low and feed pressure is normal and permeate flow is very high, then output is slightly low.

The rules are combined further to form the decision making for the fuzzy controller that accommodates different situations of the six inputs experienced by the system. Table 15.4 lists the look up matrix for two-variable input vectors. The corresponding decision surface is shown in Figure 15.24.

15.5.2.3 Results and Discussion

Changes in temperature, feed TDS, permeate flow, permeate TDS, and overall system recovery were tested and recorded. Figure 15.25 shows the conductivity, temperature, and control activity for the high pressure pump. The system response was then tested by induced variations in TDS and temperature. Higher feed TDS was induced by feeding the brine concentrate back into the feed line, resulting in increasing the feed TDS from 37500 ppm to approximately 40000 ppm. Effects of these changes on permeate flow and control activity are shown in Figure 15.26. The feed pressure raised from 60 bar to 63 bar, which resulted in higher permeate flow and compensated for the changes. Lower TDS was induced by feeding permeate water into the feed stream. This resulted in reduction in the concentration of the feed from 37500 ppm to approximately 33000 ppm. Effects of lower feed water TDS on permeate flow and control activity for high pressure were also tested. Lower feed temperature was induced by cooling the feed water pipe. This reduced the temperature by approximately 5°C. The water was allowed to go back to a normal range of 27°C. Effects of these changes and the high-pressure control activity are shown in Figure 15.27. Changes in feed TDS and feed temperature had diverse effects on the quality of the product. The salt rejection changed drastically and caused variations in feed parameters. The results of control activity of the high pressure due to these changes kept the salt rejection constant over this period. Figure 15.28 shows the salt rejection at constant level of 98.97%. The permeate flow changed due to changes in TDS and feed temperature; however, the flow control signal kept the recovery constant. This will be discussed in detail in the next section.

15.5.3 Flow Rate Control

Control of effective flow rate depends on permeate flow, transmembrane pressure, scaling index, and recovery. The task of the fuzzy controller is to adapt to changes in these variables and control the flow rate required for the feed to the RO membrane. The system continuously monitores and adjusts the flow rate by adjusting the speed (rpm) of the motor drive. This is done in such a way to ensure that the recovery remains constant over the normal operational period. Low permeate flow combined with high salt passage indicates that scaling might occurs; therefore, the recovery in this case must be lowered. A high differential pressure across the membrane is caused by the salts deposits or fouling of the membrane on the feed side. The signal from the FLC drove a variable speed frequency converter, which in turn determined the flow rate required.

15.5.3.1 Fuzzy Rule Base for Flow Control

The rule base consisted of rules for the inputs such as differential pressure (ΔP), feed TDS, salt passage and permeate flow. The set point for normal feed flow was determined based on initial values of pressure, temperature, and recovery ratio. Each parameter caused changes to other variables and therefore, the flow was adjusted accordingly in order to compensate for these changes and provide a constant water quantity. The output of the system consisted of an external signal to the frequency controller, which in turn provided a signal (rpm) to the motor based on concluded output.

Table 15.4: Look-Up Table of Valve Position for Temperature and Permeate Flow

	Permeate Flow						
Temperature	V_Low	Low	M_Low	S-Low	Design	High	V_High
V_Cold	VH	H	H	MH	MH	ML	ML
Cold	VH	H	H	MH	ML	ML	ML
Medium	MH	MH	MH	M	M	L	L
Warm	ML	ML	M	M	ML	L	L
Hot	L	L	ML	ML	L	L	VL
V_Hot	L	L	ML	ML	L	VL	VL

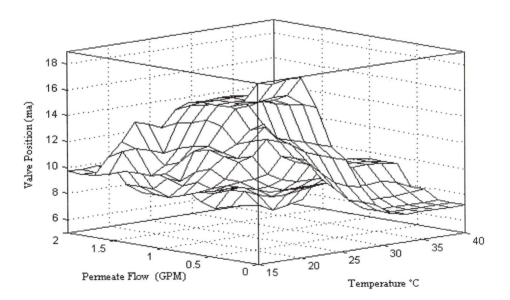

Figure 15.24: Decision Surface for Control of Valve Position for Temperature and Permeate Flow.

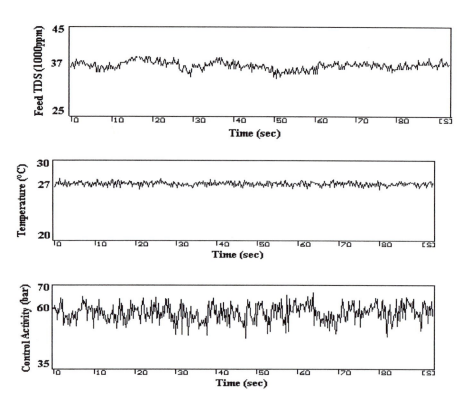

Figure 15.25: Feed Conductivity (Top), Temperature (Middle), and
Control Activity of High Pressure Pump (Bottom).

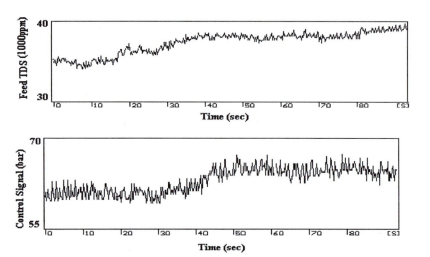

Figure 15.26: Effects of Higher Feed Water TDS and Control Activity
for Applied Pressure.

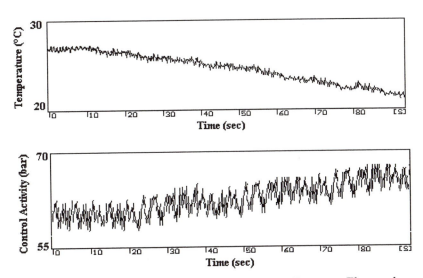

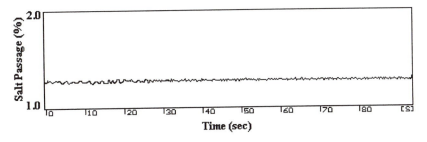

Figure 15.27: Effect of Low Temperature on Permeate Flow and High Pressure Control Signal.

Figure 15.28: Salt Rejection of the RO System Under Changed Conditions.

15.5.3.2 Decision Matrix

The rules are formulated and combined to form a decision table for the fuzzy controller. Examples of such rules are as follows:

1) If recovery is low and permeate flow is low and ΔP is medium low or low and TDS is high and scale is medium and salt passage is normal then speed is fast.

2) If permeate flow is low and feed TDS is medium and ΔP is high and temperature is normal and permeate TDS is low and salt passage is high and feed pressure is high then speed is medium low.

3) If ΔP is high and permeate flow is slightly low and recovery is medium-low and TDS is normal and salt passage is normal and feed pressure is medium high and scale is medium then speed is slow.

The rules are combined further to form the decisionmaking for the fuzzy controller that accommodates different situations of the six inputs experienced by the system.

15.5.3.3 Results and Discussion

The speed control response was tested by induced variations in feed TDS, temperature, flow, and recovery. Effects of these changes and the motor speed control activity are shown in Figure 15.29. The resulting control activity of the motor speed kept the recovery and the scale index of the sulfates constant over the operational period. The system was tested further on high feed TDS for longer period to induce scale formation. The control activity of both the pH and the motor speed kept the solubility index at minimum. Figure 15.30 shows the ion product for $CaSO_4$ and overall system recovery.

15.6   APPLICATION OF ANFIS TO RO PARAMETERS

FLC and NN can be integrated to get the strength of each system and provide for the adaptability and learning aspect to FLC [23]. One of the successful methods of such integration is done through adaptive neuro-fuzzy inference system (ANFIS), which can identify the near-optimal membership function for achieving desired input-output mapping [24]. The network applies a combination of the least squares method and the back propagation gradient descent method for training FIS membership function parameters to emulate a given training data set. The system converges when the RMSE training and checking error are within limits.

15.6.1   ANFIS Simulation Results

The training data were divided into two sets: training and checking. Each set contained the desired input-output pairs of the form (Input vector, desired output). The 200-point input vector included differential pressure and permeate flow as input vector set and recovery as the desired output. Other input-output vector pairs included temperature, feed pressure, feed conductivity, and permeate conductivity as input vectors and salt rejection as the desired output vector. Figure 15.31 shows the training and checking root-mean-square error. The initial and resultant final membership functions for two inputs are shown in Figures 15.32 and 15.33. The results of ANFIS were used to refine the membership functions of different attributes and resulted in better system responses of the FLC.

15.7   CONCLUSION

The main feature of the developed intelligent system is its ability to diagnose and respond to critical variations of key operating parameters to avoid permanent damage to equipment, materials, or modules. In addition, the system utilized output of NN predictions of ion product concentration in the brine stream, as well as online calculated ion concentrations, to control the scale formation of different salts. The control system adapted for changes in the TDS, temperature, permeate flow, feed flow, and pH and was able to keep the recovery ratio at 30% and salt rejection at 98.97% throughout the operational period. This also kept the solubility concentration well below the saturation limits for all salts considered. Furthermore, the control of the pH value kept the $CaCO_3$ ion concentration constant over the operational period.

The use of adaptive neuro-fuzzy inference provided valuable information for optimal membership functions for key variables. The final membership functions were utilized in the actual running of the FLC and provided better and smoother operation, due to overlapping in the regions considered.

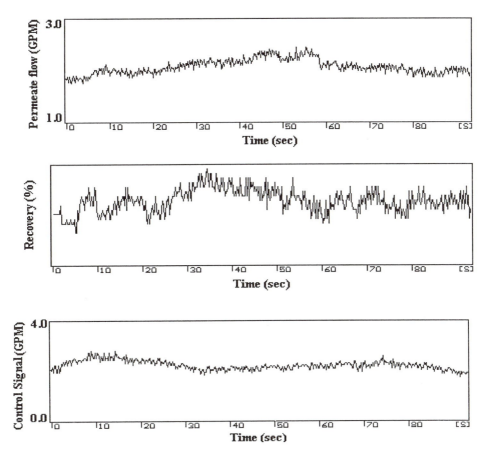

Figure 15.29: Effect of High Permeate Flow on Recovery and Motor Speed Control Signal.

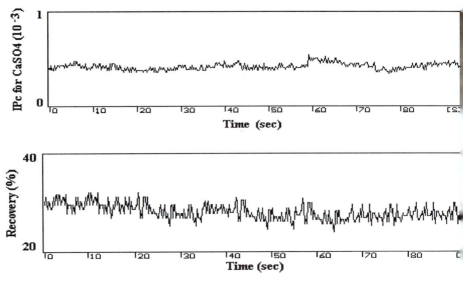

Figure 15.30: Ion Product for CaSO4 (Top), and Recovery (Bottom),
Under Changed Conditions.

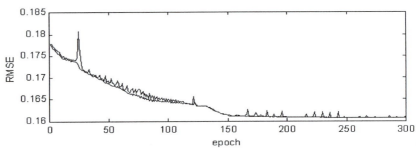

Figure 15.31: ANFIS Training (Top), and Checking
(Bottom) Error Curves for Salt Rejection.

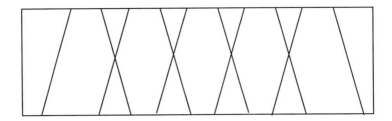

Figure 15.32: Initial Membership Functions for Temperature
and Feed Pressure.

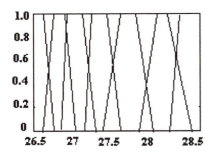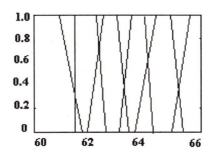

Figure 15.33: Final Membership Functions Using ANFIS for
Temperature (Left), and Feed Pressure (right).

REFERENCES

1.   Hanbury, W.T., Hodgkeiss, T., and Morris R., *Desalination Technology*, Porthan Limited, Glasgow, UK, 1993.
2.   Ben Hamida A., Seawater Pretreatment for Reverse Osmosis Plants, *The Int. Symp. on Pretreatmenrt of Feedwater for Reverse Osmosis Desalination Plants*, March 31–April 2, Kuwait, 1997.
3.   Spiegler, K.S. and El-Sayed, Y.M., *A Desalination Primer*, Balaban Desalination Publications, Santa Maria Imbaro, Italy, 1994.
4.   FILMTECH FT30 Membrane Elements Technical Manual, The DOW Chemical Company, December 1993.
5.   DuPont PERMASEP Products Engineering Manual, Wilmington, DE, 1994.
6.   Jafar, M. and Abdel-Jawad, M., Design, Implementation, and Evaluation of a Fully Automated Reverse Osmosis Plant, *Desalination and Water Reuse*, 8(3): 18–19, 1998.
7.   Jafar M., Zilouchian, A., Ebrahim, S., and Safar, M., Design and Evaluation of Intelligent Control Methodology for Reverse Osmosis Plants, *Proc. of the ADA Biannual Conf.*, Williamsburg, VA, 1998.
8.   Lonsdale, H.K., Merten, U., and R.L. Riley, *Applied Polymer Science*, Vol. 9 pp. 1341–1362, 1965.
9.   Lorch, W., ed., *Handbook of Water Purification*, John Wiley & Sons, 2nd Edition, 1986.
10.   Lee, Y.C. and Kim, M., The Effects of Initial Weights on Premature Saturation in Back Propagation Learning, *Int. Joint Conf. on Neural Networks*, Vol. 1, pp. 765-770, Seattle, 1991.
11.   Nguyen, D. and Widrow, B., Improving the Learning Speed of 2-Layer Neural Networks by Choosing Initial Values of the Adaptive Weights, *Proc. of the Int. Joint Conf. on Neural Networks,* Vol 3, pp. 21–26, 1990.
12.   Baughman, D. R. and Liu, Y. A., *Neural Network in Bioprocessing and Chemical Engineering*, Academic Press, San Diego, CA, 1995.

13.   Specht D.F., A General Regression Neural Network, *IEEE Trans. on Neural Networks*, 2(6): 568–576, 1991.

14.   Lasdon L., Plummer, J., and A. D. Waren , Non-Linear Programming: Mathematical Programming for Industrial Engineers, *Indus. Eng.*, 20, pp. 385–485, 1996.

15.   Wettschereck D. and Dietterich, T., Improving the Performance of Radial Basis Function Networks by Learning Center Locations, in Bruce Spatz, ed., Morgan Kaufmann Publishers, Inc., San Mateo, CA, 1133–1140, 1992.

16.   MATLAB Neural Networks Toolbox, Version 3, The Mathworks Inc.

17.   Stone, M., Cross-Validatory Choice and Assessment of Statistical Predictions *J. Royal Stat. Soc.*, B36, 111–133, 1974.

18.   Ljung, L., *System Identification: Theory for the User*, Prentice-Hall, Inc., Englewood Cliffs, NJ, 1987.

19.   Mindler, A. B. and Epstein, A., Measurement and Control in Reverse Osmosis Desalination, *Desalination,* Elsevier Science Publishers B.V., Amesterdam, The Netherlands, 343–379, 1986.

20.   McIhenny, W. F., Measurement and Control of Feed Water and Product Water Composition, *Desalination,* Elsevier Science Publishers B.V., Amesterdam, The Netherlands, 445–460, 1986.

21.   Jafar, M. and Zilouchian, A., Design and Implementation of a Real-Time Fuzzy Controller for a Prototype Reverse Osmosis Plant, *Proc. of the WAC2000 congress on Automation*, Maui, June 2000.

22.   Mamdani, E.H., Application of fuzzy algorithms for control of simple dynamic plant, *IEEE Proc.*, Vol. 121, No. 12, 1974.

23.   Lin C.-T. and Lee, C. S. G., Neural-Network-Based Fuzzy Logic Control and Decision System, *IEEE Trans. on Computers*, 40(12):1320–1336, December 1991.

24.   Wang, L.-X. and Mendel, J. M., Back Propagation Fuzzy Systems as Non-Linear Dynamic System Identifiers*, Proc. of the IEEE Int. Conf. on Fuzzy Sys.*, San Diego, March 1992.

16

COMPUTATIONAL INTELLIGENCE APPROACH TO OBJECT RECOGNITION

K. C. Tan, T. H. Lee, and M. L. Wang

16.1   INTRODUCTION

Object recognition is an important function required in many intelligent applications, such as autonomous vehicles, medical diagnosis, security, military target detection, and etc., [1-3]. These systems are often equipped with multiple sensors, which can generate data regarding different properties of the scene of interest. The object recognition problem can be described as identifying the corresponding object based upon the image data acquired from these multiple sensors. In many cases, visual information is the most powerful single source of sensory information available to a system for measurement and object recognition. The data received from these different sensors, however, often incomplete, distorted, noisy, or vague. Moreover, the objects to be recognized can have very different appearances in different conditions, such as viewing positions, photometric effects (lighting condition), background environment of objects, or changes in the shapes of objects, which further add to the difficulty of the problem in object recognition. Therefore, the need of a highly intelligent, reliable and efficient processing system to recognize objects from these imperfect data is obvious.

Current research on the topic of object recognition can be roughly divided into three categories, based upon the concept that regularity across different views of one object must be exploited in order to recognize an object by matching the images of the object to the stored internal representation [1]. These approaches differ in their assumptions and ways of obtaining these regularities. The first is the *invariant properties* method [4], which assumes that certain basic properties remain invariant under the transformations or changes that one object is allowed to make. The second is called *alignment* method [5], which recognizes an object by aligning the images of the object with the corresponding stored model. The third approach is the *object decomposition* method [6], which is based on the principle that an object is constituted by generic components and the recognition of the object relies on the recognition of these components on their own.

Object recognition by decomposition of objects into constituent parts assumes that each object can be broken into a smaller set of generic components. All the objects in the recognition space can then be described by the different combination of these basic generic components. Generally, recognition by decomposition method is achieved in two steps. First, the image data are transformed to the generic parts that describe the objects, which is often achieved by edge detectors or segmentation algorithms [5], [7]–[10]. The second step is to combine these parts into a complete object. For this, the approach of *hierarchical features* [11], [12] may be employed to repeat the decomposition process by breaking certain parts into simpler parts. Some low-level parts can be identified first, and then groups of simple parts are identified together to form higher order parts. For example, straight line segments are detected as the most basic parts and then parts such as corners and vertices are obtained according to the already-detected line segments. Combining these corners, vertices can identify higher-level structures, such as triangles. Another approach is *structural descriptions* [7], [13], which employs relations defined among different parts to transit the parts to objects. This method assumes that relations among different parts of an object are easier to capture and thus can be used to recognize an object. For example, the total number of parts of a given type may be invariant of the object: a triangle always has three lines, three vertices, and no free line termination; a human face always has two eyes, one nose, and one mouth at respective positions.

It is often difficult to recognize objects using parts extracted from multiple data sources with high confidence value, since data acquired from the imaging devices are usually imperfect, e.g., the same object can have different images if the imaging devices are put in different positions. Besides the noise influence that exists in almost every stage of image acquisition, objects for recognition in the real world are also often surrounded by environmental objects, which could result in significant degradation on the image quality. Object recognition using decomposition method has received significant attention over the years [7],[8], [13]-[16] and has found physiological supports from research work on the visual cortex of animals [14], [17]-[18]. However, most of these works only focused on low level feature extraction (part decomposition) methods via advanced image processing algorithm. Other procedures such as combining features resulting from the feature extraction and adaptive adjustment of these feature extractors are often ignored.

This chapter presents the state-of-art computational intelligent technique for modulation of feature extractors and high level intelligent combination of features, based upon the use of fuzzy logic and artificial neural networks. The CI based approach is appropriate since fuzzy logic provides a means of dealing with uncertainty, inexactness, and imprecision, which are often encountered during the data acquisition process in object recognition. On the other hand, the artificial neural network is a distributed computing model largely inspired from studies on mechanisms of human neural systems [19]. It is naturally suited for the task of feature extraction and has been found to offer better results than traditional symbolic computing approaches for many problems.

16.2   OBJECT RECOGNITION BY NEURAL FEATURE EXTRACTION AND FUZZY COMBINATION

Based on studies in biology areas, a biological vision system performs substantial pre-processing of image data to focus attention and exclude any irrelevant information [17]. Preference is often given to elements in which the observer is paying attention. Given the same stimulus, the response of certain neurons might increase dramatically due to the focus of attention. It is believed that this preferential treatment of stimulus having interest is caused by state dependent signals, which are originated from visual areas other than the retina, and are said to modulate the response of neurons to any object on which the attention is focused. The signals may come from areas in the visual cortex, or from the higher processing areas in the parietal and temporal lobes. This phenomenon is called *state dependent modulation* [17] and is applied in the area of processing to superimpose its findings or expectations over other areas. This modulation results in the elevation of areas of interest and suppression of any unneeded information from the visual data.

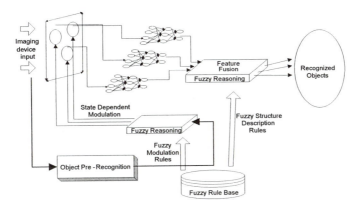

Figure 16.1: A Prototype of Object Recognition by Neural Feature Extractions and Fuzzy Structural Description.

An efficient approach by means of artificial neural network and fuzzy reasoning is proposed in this article to generate the state dependent modulation signals and to improve the feature extraction process. As shown in Figure 16.1, objects are pre-recognized to obtain the category knowledge of image under observation and to yield the information for the creation of state dependent modulation signals. This involves the normalization of image size, filtering of the image, and categorization. A feature may be defined as a rectangular two-dimensional window with a width of M_i and height of N_i. Different features may have different window sizes and the image to be processed by the feature extractor is preprocessed to a normalized size and scale of brightness. The brightest pixel in an image has a value of 255 and the darkest a value of 0. The feature extraction process is achieved using a multi-layer neural network to process the information acquired from the imaging devices. After the features have been extracted from the image data, the results are further processed with feature fusion processing, which uses fuzzy reasoning to combine the results based on predefined fuzzy structural description rules to yield the object recognition results. These fuzzy rules are stored in a rule-base for fuzzy reasoning as well as for the construction of the required modulation signals and structural descriptions. Note that different objects have their respective fuzzy rules and, hence, the recognition process can be executed in parallel for faster execution time if desired.

16.2.1   Feature Extraction by Neural Network

An Artificial neural network (ANN) is a computing model that uses statistical properties instead of logical rules to transform information. It is inspired by the research on neuron physiology and relies on parallel processing of sub-symbols. The basic computation units, neurons, are modeled as a unit that can output a stimulus pulse when the input pulses to this neuron reach a threshold value. The stimulus may come from other neurons or directly from the environment. The stimulus from other neurons is transferred in weighted connections between the neurons. In practical applications, the network topology of these interconnected neurons is often modeled as layered structure.

The proposed neural network for feature extraction has three layers, i.e., one input layer, one hidden layer and one output layer [19]. The neurons in the input layer are directly connected to the inputs with appropriate weights without any activation functions. The neurons in the input layer are fully connected to the neurons in the hidden layer in the forward direction. The activation functions in the hidden layer are chosen as "tangent sigmoid" functions. The neural network has one neuron per input image pixel in the input layer. The image data acquired to be fed to the neural network can be viewed as a two dimensional matrix with each cell giving the gray scale of the corresponding pixel. Since inputs to the neurons in input layers are a one-dimensional array, the two dimensional image data is thus transformed such that every column of the cells is connected sequentially so as to form a single new long column. The feature extraction network is presented with a set of training image samples, which are selected so that some of them give different views of the features which vary in contrast, brightness, perspective of view, etc. Also, other training samples are used to provide feature extraction neural network for learning to distinguish images that are not views of a feature of interest. Here, the back propagation algorithm [20] is used to train the weights and biases of the neural network based on comparison of output values of the network and the target output values, so as to give a confidence value of one to views of a feature, and a zero otherwise.

To apply the trained neural networks for feature recognition, the image to be recognized is first normalized to a fixed size. One specific rectangle area of the image is then clipped out from the image and the data are passed on to the input layer of the neural network. The neuron network works out the confidence value in this specific patch of image data for occurrence of the feature that it had previously learned. Note that the position of the rectangle is controlled by the mechanism of state dependent modulation, and the final output of the feature is a value between [0, 1] which stands for the confidence of the desired feature detected in the input data.

16.2.2 Fuzzy State Dependent Modulation

The idea of state dependent modulation is to focus the attention, or to concentrate the computational power, on stimulus of interests while ignoring or give lower priority to other irrelevant stimulus. The focus of processing depends on the objects on which the attention is focused and the intention of the observer. In the case of object recognition, a system is required to obtain information regarding the specifics of objects that can help to recognize the objects. In the method of recognition by decomposition, especially, the process before feature combination is to extract features that can be further used in making decisions. In applications such as autonomous robot, computational power available for object recognition may be very limited. In order to reduce the computational effort needed, a good approach is to schedule the feature extraction processes according to different priorities. Features that are likely to be useful should be discovered with higher priority, and a larger portion of computational power should be allocated to these feature extractors. One way of generating these priority allocation signals is to use the state dependent modulation signals.

As mentioned in a previous section, the technique of artificial neural network can be applied to recognize certain features of an object from an image for feature extraction. Since features are defined in rectanglular windows and the positions are unknown to the feature extractor, a tedious and time consuming trial-and-error procedure is therefore needed to determine the position of features in an image. This, however, could be largely overcome with the use of fuzzy reasoning [21] as adapted in this work, where state modulation signals are generated via the approach of fuzzy reasoning. As shown in Figure 16.2, the priorities and positions of focus centers of feature extractors can be defined by fuzzy sets. For example, if a face image is presented and used for the recognition, the priority of EYE feature extractor can be regulated by a fuzzy set of *HIGH_PRIORITY*; the focus position of left eye should be at the top left in a human face or, more specifically, a fuzzy set named *LEFT_EYE_POSITION*. The featured extractor is then modulated to focus its computation on left eye position to extract the left eye feature from a face image running at a higher priority.

A fuzzy rule that modulates the feature extraction process in face recognition may be written as

If FACE_OBJECT is YES
then EYE_EXTRACTOR_PRIORITY is HIGH_PRIORITY
if FACE_OBJECT is YES
then EYE_EXTRACTOR_POSITION is LEFT_EYE_POSITION

where *LEFT_EYE_POSITION, YES* are two fuzzy sets that give the degree of a face being presented in an image and the degree of an image being at the left eye position. *HIGH_PRIORITY* is the measure of priority assigned to a feature extractor for allocation of the computational power. Here, the degree of an image is the human face given by another processing module (object pre-recognition), which gives a maximum confidence value of one and a minimum confidence value of zero. In the definition of *LEFT_EYE_POSITION*, the degree of membership is given by calculating the percentage divergence of a 2-D position coordinate from the original coordinate.

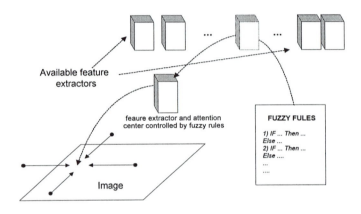

Figure 16.2: State Dependent Modulation by Fuzzy Reasoning.

16.2.3   Combination of Features Extracted from Multiple Sources with Fuzzy Reasoning

Fuzzy reasoning is an inference process that is capable of processing data in a way similar to human decision-making. By employing linguistic variables, fuzzy rules provide a high level and efficient interface for building a system with human knowledge. In a fuzzy inference system, the implication rule is represented in a fuzzy relation, and the inferred conclusion is obtained by applying the compositional rule of inference to the fuzzy implication relation. These two properties allow fuzzy reasoning capable of making reasonable inference even when the conditions of an implication rule are only partially satisfied.

The problem in image data fusion based on the decomposition and feature extraction model can be expressed as follows: Given a number of L images $I_j (j = 1,..., L)$ representing different data on the observed scene, a feature extraction operation is made to extract a set of features out of them. The value that associates i^{th} feature $F_i (i = 1,..., N)$ extracted from the jth image I_j is expressed as M_{ij}. At this processing stage, the image data input spaces are transformed to a feature space. The measures of M_{ij} are then combined to make decisions regarding the images to be recognized based on the fuzzy structural description of objects. Note that each object has a set of rules that describe the object in generic features and different objects have their respective fuzzy rule sets.

This process can be formulated as: For an object k, a decision $C_k = F(M_{11}, M_{12},..., M_{ij},..., M_{LN})$ that yields the degree of an object being detected is given by evaluating its fuzzy description rules. An object is said to be recognized when the feature measures obtained match a set of fuzzy description rules of an object. Employment of information from multiple images is achieved in both the antecedent clauses of fuzzy rules and the defuzzification stage. The Mamdani fuzzy reasoning model [21] may be employed, which consists of the following linguistic rules that describe a mapping from $U_1 \times U_2 \times ... \times U_r$ to W:

$$R_i : IF\ x_1\ is\ A_{i1}\ and...\ and\ x_r\ is\ A_{ir}\ THEN\ y\ is\ C_i \qquad (16.1)$$

where $x_j(j = 1, 2, ..., r)$ are the input variables, y is the output variable, and A_{ij} and C_i are fuzzy sets for x_j and y respectively. Given inputs of the form:

$$x_1\ is\ A'_1,\ x_2\ is\ A'_2,...,\ x_r\ is\ A'_r \qquad (16.2)$$

where $A'_1, A'_2,..., A'_r$ are fuzzy subsets of $U_1 \times U_2 \times ... \times U_r$, the contribution of rule R_i to the output of Mamdani model is a fuzzy set whose membership function is computed by

$$\mu_{c'_i}(y) = (\alpha_{i1} \wedge \alpha_{i2} \wedge ... \wedge \alpha_{ir}) \wedge \mu_{c_i}(y) \qquad (16.3)$$

where α_i is the matching degree (i.e., firing strength) of rule R_i, and α_{ij} is the matching degree between x_j and R_i's condition about x_j:

$$\alpha_{ij} = \sup_{x_j} \left\{ \mu_{A'_j}(x_j) \wedge \mu_{A_{ij}}(x_j) \right\} \qquad (16.4)$$

where $\wedge$ denotes the "min" operator. The final output of the model is the aggregation output from all rules using the max operator:

$$\mu_c(y) = \max \left\{ \mu_{c'_1(y)}, \mu_{c'_2(y)}, ..., \mu_{c'_L(y)} \right\} \qquad (16.5)$$

Note that the output C is a fuzzy set, which can be defuzzified into a crisp output using a defuzzification method, such as the center-of-area (COA) approach [21].

For example, a face may be recognized based on the detection of left eye, right eye, and mouth at the proper positions of an image. A typical rule to combine the feature extraction may be

if (LeftEyeFound is HIGH_CONFI) and (RightEyeFound is LOW_CONFI) then FACEFOUND is FACE_FOUND_MID_CONFI

where *HIGH_CONFI, FACE_FOUND_MID_CONFI,* and *LOW_CONF* in the antecedent clause are linguistic variables that represent the fuzzy sets whose membership functions determine the degree of features for the respective class detected. Here, the fuzzy operator "OR" is chosen to combine all information for the same feature from different sensors, while the fuzzy operator "AND" is used to make the recognition process more selective and reliable.

16.3   A FACE RECOGNITION APPLICATION

In this section, a face recognition problem is studied to validate the proposed neural-fuzzy based object recognition methodology. The block diagram of the proposed methodology is shown in Figure 16.3. Two gray scale images are acquired by capturing a human face before the cameras in different conditions and then normalized to a fixed size. The input images are normalized to a fixed size of 120 × 100, and 256°of gray level: maximum of 255 for the brightest pixel, and minimum of 0 for the darkest pixel. The two images for the same face after the normalization are illustrated in Figure 16.4.

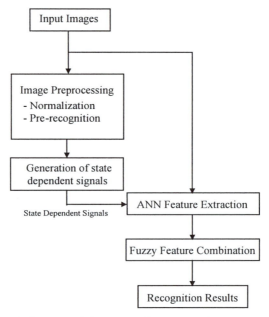

Figure 16.3: Block diagram of the Neuro-fuzzy Based Face Recognition System.

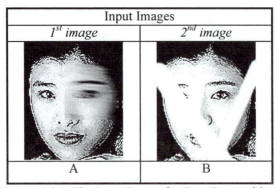

Figure 16.4: The Input Image for Face Recognition.

Results obtained after the pre-recognition process of an image are used to generate the state dependent modulation signals for the feature extractors. The priorities are graded as 16 levels, in which a priority value of 0 gives the highest computational power to a feature extractor while a value of 15 gives the lowest. The fuzzy rules that are used to generate the sate dependent modulation signals are given as follows:

If FACE_OBJECT is YES
then EYE_EXTRACTOR is HIGH_PRIORITY
else EYE_EXTRACTOR is LOW_PRIORITY

If FACE_OBJECT is YES
then MOUTH_EXTRACTOR is HIGH_PRIORITY
else MOUTH_EXTRACTOR is LOW_PRIORITY

The membership functions of *YES*, *HIGH_PRIORITY*, and *LOW_PRIORITY* are showed in Figures 16.5(a) and 16.5(b), respectively.

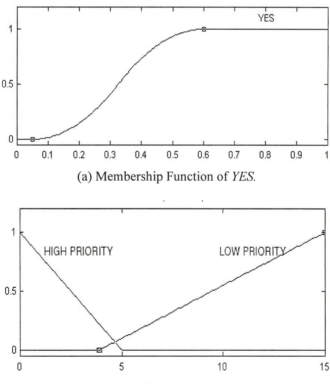

(a) Membership Function of *YES*.

(b) Membership functions of *HIGH_PRIOIRITY* and *LOW_PRIORITY*

Figure 16.5: Membership Functions of the State Dependent Modulation Signals.

The result of the object pre-recognition is given as 0.82 and the fuzzy reasoning for allocating different computational power for the three feature extractors is given in Table 16.1. Here the three extractors are assigned the same highest computation priority, as desired. Three neural networks are constructed and trained to extract the three features from these images. The structure of these neural networks is summarized in Table 16.2. In each neural network, all neurons are arranged in layers and fully connected in the feed-forward direction. The neural network for left eye has a total of $26 \times 14 = 104$ input neurons, and each pixel in the window that defines the feature has a corresponding input neuron in the input layer. The hidden layer has 30 neurons, and the "tangent sigmoid" activation function is used [19]. The output layer has only one neuron, where simple "linear" transfer function is employed.

The neural network for extraction of right eye has exactly the same number of neurons, type of transfer functions, and number of layers. Similarly, the neural network for mouth extraction has $39 \times 20 = 780$ neurons at the input layer and 100 neurons at the hidden layer, respectively. The training of these three neural networks to recognize the features is performed using standard supervised back-propagation algorithm [20], which gives a value of 1 when the feature is presence and a value of 0 otherwise.

Table 16.1: Reasoning Results of State Dependent Modulation.

Extractor	Priority (before rounding)	Priority (0 - 15) (after rounding)
Left eye	1.62	1
Right eye	1.62	1
Mouth	1.62	1

Table 16.2: Structure Summary of the Neural Feature Extraction.

Neural Network	Left Eye Extractor	Right Eye Extractor	Mouth Extractor
Neurons In Input Layer	$26 \times 14 = 104$ None	$26 \times 14 = 104$ None	$39 \times 20 = 780$ None
Neurons In Hidden Layer	30, Tangent sigmoid	30, Tangent sigmoid	100, Tangent sigmoid
Neurons In Output Layer	1, Linear	1, Linear	1, Linear
Training Method	Standard Supervised BP	tandard Supervised BP	Standard Supervised BP

The neural network feature extractors are then executed, which give values between [0, 1] indicating the confidence degree for each feature to be present in the image. The feature extraction results are then combined to obtain an overall recognition output, which indicates the confidence level for recognizing a face from these two images, i.e., a value of one means maximum confidence that a face is recognized and a value of zero stands for the lowest confidence. In this application, the fuzzy description rules used for the combination of feature extraction results are given as follows:

if ((F_{11} is *HIGH_CONFI*) *or* ($F_{21}^{\bullet}$ is *HIGH_CONFI*))
and ((F_{12} is *HIGH_CONFI*) *or* (F_{22} is *HIGH_CONFI*))
and ((F_{13} is *HIGH_CONFI*) *or* (F_{23} is *HIGH_CONFI*))
then FaceFound is *HIGH_CONFI*

if ((F_{11} is *LOW_CONFI*) *or* (F_{21} is *LOW _CONFI*))
and ((F_{12} is *LOW _CONFI*) *or* (F_{22} is *LOW _CONFI*))
and ((F_{13} is *HIGH_CONFI*) *or* (F_{23} is *HIGH_CONFI*))
then FaceFound is *LOW _CONFI*

if ((F_{11} is *LOW_CONFI*) *or* (F_{21} is *LOW _CONFI*))
and ((F_{12} is *HIGH _CONFI*) *or* (F_{22} is *HIGH _CONFI*))
and ((F_{13} is *HIGH_CONFI*) *or* (F_{23} is *HIGH_CONFI*))
then FaceFound is *MID _CONFI*

if ((F_{11} is *HIGH_CONFI*) *or* (F_{21} is *HIGH _CONFI*))
and ((F_{12} is *LOW _CONFI*) *or* (F_{22} is *LOW _CONFI*))
and ((F_{13} is *HIGH_CONFI*) *or* (F_{23} is *HIGH_CONFI*))
then FaceFound is *MID _CONFI*

where $F_{ij}(i = 1, 2; j = 1, 2, 3)$ represents the jth feature extracted from the ith image. The associated fuzzy membership functions are given in Figure 16.6, and the overall results of the fuzzy combinations are shown in Table 16.3. It can be seen that the results give a high confidence value of 0.9 for the first image and a low value of 0.06 for the second image, as expected. The overall confidence value of 0.72 for the face recognition indicates that the recognized image resembles the original image satisfactorily, which is consistent with the quality of the two images as shown in Figure 16.4.

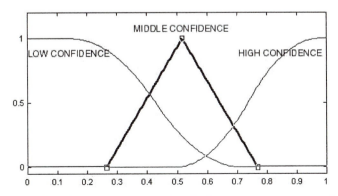

Figure 16.6: Membership Functions for the Fuzzy Combination of Features.

Table 16.3: Results of the Fuzzy Combination.

Feature	1st Image	2nd Image
Left Eye	0.90	0.06
Right Eye	0.48	0.96
Mouth	0.91	0.71
Face Recognition Result	0.72	

16.4   CONCLUSIONS

A CI based object recognition methodology by decomposition, neural feature extraction, and fuzzy structural description from multiple sensory data has been presented. The originality of this article lies in: (1) the feature extraction process is modulated by state dependent modulation signals inspired by biological discovery, and (2) the combination of feature extraction results is realized by fuzzy description rules to address the inherent ambiguity in image data or object description. State dependent modulation has been employed to generate signals for facilitating the feature extraction process by scheduling the computational power between different feature extractors. The validity of the proposed neural-fuzzy based object recognition technique has been illustrated via an application example of face recognition.

It should be noted that object recognition by decomposition requires the objects to have parts clearly distinguishable, which may not be the case for certain applications. For example, the current decomposition method cannot be easily applied to objects that do not decompose into parts naturally, i.e., the decomposition of a loaf of bread. Also, it may be difficult to determine by which criteria the objects should be decomposed. Low-level description based on simple generic parts, such as edges and line segments, often results in structural description that is highly complex, while high-level description often fails to provide enough distinction between different objects.

In most decomposition descriptions, the constituent parts of an object are usually considered to have similar significance to the recognition process. However, this assumption is generally untrue for many applications. For example, in the recognition of human faces, the recognition of eyes is generally more important than other parts of the face. This kind of knowledge should be transferred to the stages of intelligent structural description and recognition. Also, the knowledge of structural description is mostly based upon human perception and cognition, which are built upon the sensory inputs of human beings. It is obvious that the recognition machines have totally different sensory inputs varying many fold from human beings.' How to bridge these gaps and to develop a specific knowledge for machine systems will be an interesting topic for further explorations.

REFERENCES

1. Ullman, S., *High-Level Vision: Object Recognition and Visual Cognition*, Cambridge, MA, MIT Press, 1996.
2. Nagata, T. and Zha, H. B., Recognizing and Locating a Known Object From Multiple Images, *IEEE Trans. On Robotics and Automation*, 7(4), 434, 1991.
3. Hebb, D. O., *The Organization of Behaviour,* John Wiley & Sons, NY, 1949.
4. Mundy, J. L., and Zisserman, A., Towards a New Framework for Vision, *Geometric Invariance in Computer Vision,* MIT Press, 1, 1992.
5. Bennamoun, M. A., Contour-based Part Segmentation Algorithm, *Proc. of the IEEE ICASSP '94*, Adelaide, Australia, 41, 1994.
6. Huttenlocher, D. P. and Ullman, S., Recognizing Solid Objects by Alignment with an Image, *Int. J. Computer Vision*, 5(2), 195, 1990.
7. Bennamou, M., and Boashash, B.A., Structural-Description-Based Vision System For Automatic Object Recognition, *IEEE Trans. On Syst., Man and Cybern. - Part B: Cybern.*, 27(6), 893, 1997.
8. Levine, M.D., *Vision in Man and Machine*, NY, McGraw-Hill, 1985.
9. Canny, J., Computational Approach to Edge Detection, *IEEE Trans. on Pattern Analy. Machine Intell.*, 8, 679, 1986.
10. Paik, J. K, and Katsaggelos, A. K., Edge Detection Using a Neural Network, *Proc. of IEEE Int. Conf. on Acoustics, Speech, Signal*, Tampa, FL, 2145, 1990.
11. Nair, D., and Aggarwal, J. K., Hierarchical, Modular Architectures for Object Recognition by Parts, *Proc. of 13th Int., Conf. on Pattern Recognition,* 1, 601, 1996.
12. Gidas, B., and Zelic, A., Object Recognition via Hierarchical Syntactic Models, *Proc. of The 13th Int. Conf. on Digital Signal Processing*, 1, 315, 1997.
13. Lenaghan, A, Malyan, R., and Jones, G. A., Matching Structural Descriptions Of Handwritten Characters Using Heuristic Graph Search, *IEE Third European Workshop on Handwritting Analy. and Recognition*, October 1998.
14. Siddiqi, K., and Kimia, B. B., Parts Of Visual Form: Computational Aspects, *IEEE Trans. on Pattern Analy. and Machine Intell.*, 17(3), 239, 1995.
15. Rivlin, E., Dickinson, S., and Rosenfeld, A., Recognition by Functional Parts, *Computer Vision, Graphics and Image Processing: Image Understanding*, 62(2), 164, 1995.
16. Horikoshi, T. and Suzuki, S., 3D Parts Decomposition from Sparse Range Data Using Information Criterion, *1993 IEEE Proc. of Computer Vision And Pattern Recognition*, 168, 1993.
17. Maunsell, J. H., The Brains; Visual World: Representation Of Visual Targets In Cerebral Cortex, *Science*, 270, 764, 1995.

18.   Biederman, L., Human Image Understanding: Recent Research and Theory, *Computer Vision, Graphics and Image Processing*, 32, 29, 1985.

19.   Haykin, S., *Neural Networks: a Comprehensive Foundation,* 2nd ed., Prentice Hall, Upper Saddle River, NJ, 1999.

20.   Rumelhart, D.E., Hinton, G.E., and Williams, R.J., Learning Representations by Back Propagation Errors, *Nature*, 323, 533, 1986.

21.   Yen, J. and Langari, R., *Fuzzy Logic:Intelligence, Control, and Information*, Prentice Hall, Upper Saddle River, NJ, 1999.

17 AN INTRODUCTION TO EVOLUTIONARY COMPUTATION

Gerry Dozier, Abdollah Homaifar, Edward Tunstel, and Darryl Battle

17.1 INTRODUCTION

Simulated evolution is quickly becoming the method of choice for complex problem solving especially when more traditional methods cannot be efficiently applied or produce unsatisfactory solutions [1]. Simulated evolution has been shown to be a robust method for developing solutions to a wide variety of complex optimization and machine learning problems [2-7].

Evolutionary computation (EC) is the field of research devoted to the study of problem solving via simulated evolution. Over the past 30 years the field of EC has itself been evolving. Originally, the first generation of EC consisted of three evolution-based paradigms: evolution strategies [8], evolutionary programming [9], and genetic algorithms (GAs) [4]. Each of these evolutionary techniques was developed for solving distinct problems [1, 10, 11].

The second generation of EC techniques consisted of a number of new and equally exciting paradigms. The two most prominent of second generation ECs were methods that evolved populations of data structures rather than string representations [6], and GAs that evolved populations of programs (known as genetic programming) [5, 12]. At present, a third generation of ECs has emerged with the addition of cultural algorithms [13], DNA-based computing [14], particle swarm optimization [15], and ant colony optimization [16]. Each of these methods, like the other ECs of previous generations, has been used to solve a wide variety of problems.

This chapter will focus on two of the more popular types of EC, GAs and genetic programming. In section 17.2 we provide an overview of genetic search and in section 17.3 we present the fundamental theorem that has been used to explain the search behavior of GAs [17]. In Section 17.4, we provide a brief introduction to the field of genetic programming and in section 17.5 we provide a brief summary.

17.2 AN OVERVIEW OF GENETIC SEARCH

GAs [4], as do all EC techniques, differ from more traditional search algorithms in that they work with a number of candidate solutions rather than just one candidate solution or partial solution. Each candidate solution of a problem is represented by a data structure known as an individual. An individual has two parts: a chromosome and a fitness. The chromosome of an individual is made up of genes. The values that can be assigned to a gene of a chromosome are referred to as the alleles of that gene. A group of individuals collectively comprise what is known as a population. For most GAs, the size of the population remains constant for the duration of the search.

Individuals selected from the current population, called parents, are selected based on their fitness and are allowed to create offspring. Usually, individuals with above average fitness have an above average chance of being selected. After selection, reproductive operators such as crossover and mutation are applied to the parents. In crossover, parents contribute copies of their genes to create a chromosome for an offspring. This is analogous to the way offspring of living organisms are created as a genetic mixture of their parents. Mutation requires only one parent. An offspring created by mutation usually resembles its parent with the exception of a few altered genes.

After the children have been created, the candidate solutions that they represent are evaluated and each child receives a fitness. Before the children can be added to the population, some individuals in the current population must die and be removed to make room for the children. Usually, individuals are removed based on their fitness with below average individuals having an above average chance of being selected to die. This process of allowing individuals to procreate or die based on their relative fitness is called natural selection. Individuals that are better fit are allowed to live longer and procreate more often.

An interesting aspect of GAs (and EC in general) is that the initial population of individuals need not be very good. In fact, each individual of an initial population usually represents a randomly generated candidate solution. By repeatedly applying selection and reproduction, GAs evolve satisfactory solutions quickly and efficiently.

GAs can be characterized in terms of eight basic attributes: (1) the genetic representation of candidate solutions, (2) the population size, (3) the evaluation function, (4) the genetic operators, (5) the selection algorithm, (6) the generation gap, (7) the amount of elitism used, and (8) the number of duplicates allowed.

17.2.1   The Genetic Representation of Candidate Solutions

For most GAs, candidate solutions are represented either by binary or real coded chromosomes. In binary coded chromosomes [17], every gene has two alleles. In real coded chromosomes [2, 6, 18], each variable of a chromosome is represented by one gene. These genes may be assigned any value from a k-valued set of alleles. Thus, for real coded chromosomes the set of alleles corresponds to the domain of values that can be assigned to a variable (gene).

It is difficult to compare these two types of representation because, depending on the problem, one representation may be more appropriate than the other. However, one advantage of using a binary coded representation is that a large amount of research has been done on binary coded GAs. Real coded representations have the advantage of being closer to the way candidate solutions are expressed in a problem. Real coded representations typically allow for more accurate solutions as well.

17.2.2 Population Size

The population size [19] is the number of individuals that are allowed in the population maintained by a GA. If the population size is too large, the GA tends to takes longer to converge upon a solution. However, if the population size is too small, the GA is in danger of premature convergence upon a suboptimal solution. This is primarily because there may not be enough diversity in the population to allow the GA to escape local optima.

17.2.3 Evaluation Function

The evaluation function of a GA is used to determine the fitness of an individual. Figures 17.1 and 17.2 show the process that most GAs go through in assigning a fitness value to an individual. The evaluation function used in Figures 17.1 and 17.2 determines the fitness of an individual to be $f(d(x)) = d(x)^2$, where the evaluation function is $f(x) = x^2$ for $2 \geq x \geq 1$. For binary coded representations, $d(ub,lb,l,x) = (ub-lb)\ decode(x)/2^l-1 + lb$, where ub denotes the upper bound of an input value to the evaluation function f, lb denotes the lower bound of an input value to the evaluation function f, $decode$ returns the integer equivalent of the binary representation, and l denotes the user specified length of the chromosome.

In Figure 17.1, the binary coded chromosome, also known as a genotype, of an individual must first be decoded into a candidate solution (phenotype [3]) that the individual represents. Next, the candidate solution is evaluated and the result of the evaluation is assigned as the fitness of the individual. In Figure 17.2, the real coded chromosome of an individual is actually a phenotype. No decoding is necessary.

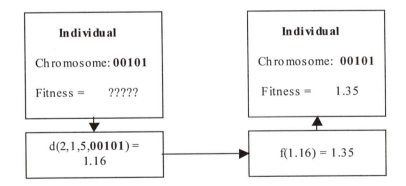

Figure 17.1: The Fitness Assignment Process for Binary Coded Chromosomes ($ub=2$, $lb=1$, $l=5$).

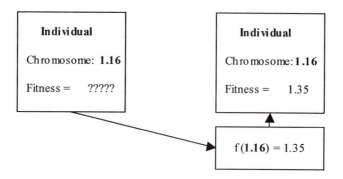

Figure 17.2: The Fitness Assignment Process for Real Coded
Chromosomes.

17.2.4 Genetic Operators

Offspring are created as a result of applying genetic operators to individuals
that are selected to be parents. There are basically two types of operators used in
genetic algorithms: crossover and mutation [2, 4, 17]. Crossover operators create
offspring by recombining the chromosomes of selected parents. Mutation is used
to make small random changes to a chromosome in an effort to add diversity to
the population.

Genetic operators tend to be problem specific; however, the two crossover
operators that will be presented have enjoyed a fair amount of success on a
variety of different problems. These operators were originally developed for
binary coded representations but can be applied to real coded representations as
well.

17.2.4.1 Single Point Crossover

The most common type of crossover operator is called single point crossover
[17]. This operator takes two parents and randomly selects a single point
between two genes to cut both chromosomes into two parts. This point is known
as a cut point. The crossover operator then takes the first part of the first parent
and combines it with the second part of the second parent to create the first
child. Then, in similar fashion, the crossover operator takes the second part of
the first parent and combines it with the first part of the second parent to create a
second child. Figure 17.3 shows an example of how the single point crossover
operator works. The cut point in Figure 17.3 is between the third and fourth
genes. The first three genes of Parent1 are combined with the last four genes of
Parent2 to create Child1. To create Child2, the first three genes of Parent2 are
combined with the last four genes of Parent1. Notice that single point crossover
can only generate a subset of all possible offspring of two parents. This is
because two parents can only be crossed over at one point. For example, Parent1
and Parent2 are unable to produce 1010011 because this would require more
than one cut point. Figure 17.4 shows how single point crossover can be applied

to real coded representations. In Figure 17.4, the alleles for each gene are taken from the set {0, 1, ... , 9}. As in Figure 17.3, the cut point is between the third and fourth genes.

17.2.4.2   Uniform Crossover

```
Parent1    1 0 0 0 0 1 0
Parent2    1 1 1 0 0 0 1

Child1     1 0 0 0 0 0 1
Child2     1 1 1 0 0 1 0
```

Figure 17.3: An Example of Single Point Crossover Between the Third and Fourth Genes Applied to Binary Coded Chromosomes.

```
Parent1    0 1 2 3 4 5 6
Parent2    7 8 9 0 1 2 3

Child1     0 1 2 0 1 2 3
Child2     7 8 9 3 4 5 6
```

Figure 17.4: An Example of Single Point Crossover Between the Third and Fourth Gene Applied to Real Coded Chromosomes.

Another type of crossover used in many GAs is called uniform crossover [11]. In uniform crossover, the value of each gene of an offspring's chromosome is randomly taken from either parent. This process can be repeated to create a second offspring. Uniform crossover is able to produce all possible offspring of two parents. Notice in Figure 17.5 that the value of each gene of Child1 has been taken randomly from one of the corresponding genes of the parents. Also notice that Child1 could not have been created using single point crossover. Figure 17.6 shows how uniform crossover can be applied to real coded chromosomes.

```
Parent1    1 0 0 0 0 1 0
Parent2    1 1 1 0 0 0 1

Child1     1 0 1 0 0 1 1
```

Figure 17.5: An Example of Uniform Crossover Applied to Binary Coded Chromosomes.

```
Parent1    0 1 2 3 4 5 6
Parent2    7 8 9 0 1 2 3

Child1     0 8 9 3 1 5 6
```

Figure 17.6: An Example of Uniform Crossover Applied to Real Coded Chromosomes.

17.2.4.3 Mutation

In mutation, each gene of an offspring is mutated based on, p_μ, the mutation rate [17]. In Figure 17.7, Child1 and Child2 are created via single-point crossover and mutation. The cut point is between the third and fourth gene. Each gene of each offspring is mutated with a mutation rate of 0.01. Notice that Child1 was 1000001 after Parent1 and Parent2 were crossed but had its fourth gene mutated making it 1001001. Similarly, Child2 was 1110010 after Parent1 and Parent2 were crossed but had its first and fifth genes mutated, making it 0110110. Notice that even though the mutation rate is 0.01 it is possible for more than one gene to be mutated because *every gene* is mutated with the same probability. Figure 17.8 is the same as Figure 17.7 except that single point crossover with a mutation rate of 0.01 is performed on real coded chromosomes. Notice that the fourth gene of Child1 and the first and fifth genes of Child2 are mutated by randomly selecting a value from alleles of those genes. A better method of mutating real coded chromosomes is to use Gaussian mutation rather than uniform mutation [3].

```
Parent1    1 0 0 0 0 1 0
Parent2    1 1 1 0 0 0 1

Child1     1 0 0 1 0 0 1
Child2     0 1 1 0 1 1 0
```

Figure 17.7: An Example of Single-point Crossover Between the Third and Fourth Genes with a Mutation Rate of 0.01 Applied to Binary Coded Chromosomes.

17.2.5   The Selection Algorithm

Every GA has a subprocedure, called its selection algorithm [20], which is used to select parents from the current population to be mated with one another to create children that are then evaluated and included in the next population of individuals.

The selection of an individual to become a parent is primarily based on fitness. The better an individual's fitness the greater its chance of being selected to be a parent. The rate at which a selection algorithm selects individuals with above average fitness is commonly referred to as its selective pressure [6]. The rate at which individuals with below average fitness are selected is commonly referred to as the algorithm's diversity of selection. If the selection algorithm does not provide enough selective pressure, the population will fail to converge upon a solution. If there is too much selective pressure, the population may not have enough diversity and converge prematurely.

GA researchers have developed a variety of selection algorithms that provide the type of harmony between selective pressure and diversity needed to enable GAs to search efficiently and robustly. There are basically three types of selection algorithm: (1) proportionate selection, (2) linear rank selection, and (3) tournament selection.

```
Parent1    0 1 2 3 4 5 6
Parent2    7 8 9 0 1 2 3

Child1     0 1 2 5 1 2 3
Child2     4 8 9 3 0 5 6
```

Figure 17.8: An Example of Single point Crossover Between the Third and Fourth Gene with a Mutation Rate of 0.01 Applied to Real Coded Chromosomes.

17.2.5.1 Proportionate Selection

In proportionate selection [20], individuals are selected based on their fitness relative to all other individuals in the population. Proportionate selection works as follows. First, S, the sum of the fitnesses of the individuals in the population, is computed. Then a number, R, is randomly selected within the interval $[0..S]$. Once R has been randomly selected, the fitnesses of individuals chosen at random are added to an accumulator, T, until $T \geq R$. The individual whose fitness, when added to T causes $T \geq R$, is selected to be a parent. To select another parent another R is randomly selected, T is reset to zero, and the process is repeated.

This process of selecting parents is similar to spinning a roulette wheel to determine which individual is chosen to be a parent. The better is an individual's fitness the bigger is the piece of the roulette wheel that is taken up by the individual and the greater is the probability that it will be selected as a parent.

One advantage of using proportionate selection is that its selective pressure varies with the distribution of fitness within a population. A disadvantage is that, as the population converges upon a solution, the selective pressure decreases. This loss of selective pressure may not allow the GA to find better solutions.

17.2.5.2   *Linear Rank Selection*

In linear rank selection [2, 6, 17, 20, 21], the current population of individuals is first sorted from best to worst by order of the fitness they received from the evaluation function. Then each individual in the population is assigned a new fitness, called its subjective fitness (to distinguish it from the candidate solution's raw fitness which is often called its objective fitness), based on applying a linear ranking function to the rank of the individual within the current population. Equation 17.1 is an example of a linear ranking function where *max* and *min* represent the maximum and minimum subjective fitness determined by the user, *r* is the rank of an individual, *P* is the population size and *sf(r)* is the subjective fitness assigned to the individual ranked *r* in a population:

$$sf(r) = (P\text{-}r)(max\text{-}min)/(P\text{-}1) + min. \qquad (17.1)$$

The slope of the above linear ranking function is *(max-min)/(P-1)*. By assigning values to *max*, *min*, and *P* the user is able to determine the slope of the linear ranking function which in turn determines the selective pressure of linear rank selection.

Once subjective fitness values are assigned to the individuals in a population, parents are selected by spinning a roulette wheel similar to the roulette wheel used in proportionate selection. An advantage of using linear rank selection is that the selective pressure, once determined by the user, remains constant. However, a disadvantage is that the population must be sorted. Another disadvantage is that individuals with the same fitness will not have the same probability of being selected.

17.2.5.3   *Tournament Selection*

In tournament selection [6], one parent is selected by randomly comparing *b* individuals in the current population and selecting the individual with the best fitness. A second parent may be selected by repeating the process. The selection pressure of tournament selection increases as *b* increases. Perhaps the most widely used type of tournament selection method is called binary tournament selection. In binary tournament selection, *b* is equal to two. Of the three selection algorithms presented, tournament selection is the most popular method because of its simplicity.

17.2.6   Generation Gap

The generation gap [2, 17] is a real number between 0.0 and 1.0 that represents the fraction of the current population that gets replaced by the offspring. For example, let the population size be 20 and the generation gap be 1.0. This means that, each generation, 20 offspring will be created and that these 20 offspring will replace the 20 individuals of the current population. When the generation gap is somewhere between 0.0 and 1.0 it is necessary to determine which individuals in the current population die. Various approaches have been developed [11] for selecting which individuals will be allowed to be present in the next population and which individuals will be replaced. The most common and probably the easiest strategy is to replace the worst individuals of a population. It is not uncommon to see GAs which only replace one or two individuals per generation. These types of GAs are called steady state GAs.

17.2.7 Elitism

Elitism [2, 17] can also be considered as a real value between 0.0 and 1.0. This value represents the fraction of the best individuals of a population that will not get selected to die. For example, if the population size is 20 and the elitism is 0.1, the best two individuals of the current population do not get replaced.

17.2.8 Duplicates

Individuals that represent the same candidate solution are known as duplicate individuals. It has been shown [2] that eliminating duplicates increases the efficiency of a genetic search and reduces the danger of premature convergence.

17.3   GENETIC SEARCH

The simple genetic algorithm (SGA) [17, 19, 23] is a well known class of GA. SGAs are called simple because they use a binary coded representation, proportionate selection, single-point crossover, mutation, a generation gap of 1.0, elitism of 0.0, and allow an unlimited number of duplicates. Robust parameter settings [22] for population size, crossover, and mutation are within the intervals of [20..50], [0.6..0.95] and [0.001..0.09]. In SGAs, once these parameters are set they remain static.

The Schema Theorem [4] has been used by many GA researchers [2, 17, 20, 23, 24] to explain the quick and efficient search of SGAs and GAs in general. A schema is a similarity template that resembles a chromosome with the value of each gene being either the 'don't care' symbol, #, or a value from the set of alleles of that gene. Chromosomes that belong to the set defined by a schema are called instances or representatives of that schema. A schema has six properties: its base, its defining length, its order, the number of instances it defines, the number of instances it defines within a population, and the average fitness of the instances it defines within a population.

The base of a schema is the cardinality of the largest domain of values (or set of alleles) for a variable (or gene) of a candidate solution (or chromosome). For example, schemata represented by binary coded chromosomes are base 2 schemata. This means that each position of a base 2 schema can take on one of 3 values which make up its alphabet: the 2 values in the set of alleles for each gene and the "don't care" symbol, #. In general, base n schemata represent chromosomes where the cardinality of the set of alleles for each gene is n. This means that each position of a base n schema can take on one of $(n+1)$ values which make up its alphabet.

To illustrate the other five properties of schemata, let $H = \#1\#\#10$ be a base 2 schema with the alphabet $\{\#, 0, 1\}$. The defining length of H, $\delta(H)$, is the distance between the outermost nonwildcard values. The defining length of H is $6-2=4$ because the outermost nonwildcard values correspond to the second and the sixth positions when counting from left to right.

The order of a schema, $o(H)$, is the number of nonwildcard symbols in the schema. Therefore $o(H)$ is equal to 3. The number of instances of a schema is a function of its defining length and order. A schema, H, represents exactly $2^{l-o(H)}$ binary coded instances where l is the length of each chromosome. The number of instances of a schema, denoted $m(H,t)$, and the average fitness of those instances, denoted $f(H,t)$, for any population t are two properties of a schema that are dynamic. As individuals of a population die and are replaced, the number of instances that some schemata represent within a population may increase or decrease. As schemata gain and lose instances, their average fitness will tend to fluctuate as well.

To demonstrate how SGAs search based on the Schema Theorem, let $f_{avg}(t)$ represent the average fitness of all schemata with at least one instance in generation t, let p_χ represent the rate that single-point crossover is used, and let p_μ represent the mutation rate. Also, assume that better individuals have larger fitness values assigned to them.

A selection algorithm that selects binary coded individuals based on fitness also implicitly selects schemata based on their average fitness. If an individual has an above average fitness, then that individual has an above average chance of being selected to reproduce.   Similarly, if a schema has a "better than average" fitness then it has a "better than average" chance of being present in the next generation. For now let us envision an SGA with no genetic operators. The selection algorithm of this SGA selects a new population from the old population. For example, if the population size were 20 then the 20 parents selected would form the next population. Since SGAs allow duplicates, a population would eventually evolve where every member is a duplicate. For this SGA, we can predict the number of instances of schema H there will be in a population at generation $t+1$ by the following equation:

$$m(H,t+1) = m(H,t)\, f(H,t)\, /\, f_{avg}(t). \qquad (17.2)$$

The fraction $f(H,t)\,/\,f_{avg}(t)$ represents the probability that an instance of H will be selected to be a parent.

Since better than average individuals have a better than average chance of being selected to be a parent, let us suppose that at generation t there is a schema, H, whose average fitness exceeds the average fitness of all schemata within the population by some constant c (where $c>0$). We can rewrite the previous equation as:

$$m(H,t+1)=m(H,t)\, (f_{avg}(t)+cf_{avg}(t))\,/\,f_{avg}(t). \qquad (17.3)$$

By factoring out $f_{avg}(t)$ we can reduce Equation 17.2 to:

$$m(H,t+1)=m(H,t)(1+c). \qquad (17.4)$$

Now suppose that there is at least one instance of *H* at generation *t=0*. Then *m(H,0)>0*, and we can rewrite Equation 17.4 in the following manner:

$$m(H,t)=m(H,0)(1+c)^t. \tag{17.5}$$

Equation 17.5 shows that a selection algorithm that selects individuals in proportion to their fitness actually allocates exponentially increasing trials to above average schemata. In GAs, this process of allocating exponentially increasing trials to above average schemata is done for a large number of schemata at the same time and is referred to as implicit parallelism [24, 25].

GAs need to use genetic operators in order to create new individuals; however, genetic operators can disrupt schemata. A disrupted schema is one that loses instances due to the application of genetic operators. In order to predict how many instances of a schema will be present in a population at generation *t+1*, the probability that a schema does not get disrupted must be figured into Equation 17.2. Let p_χ represent the crossover rate and let p_μ represent the mutation rate.

The probability that schema *H* does not get disrupted by using single point crossover depends on the number of cut points within an individual's chromosome, $\delta(H)$, and p_χ. A chromosome with *l* genes has *l-1* cut points. A schema is disrupted via single point crossover when a cut point is generated between its two defining positions. The probability of *H* surviving single point crossover, $S_\chi(H)$, is

$$S_\chi(H) =1- [p_\chi \delta (H)/(l-1)]. \tag{17.6}$$

Mutation can also disrupt a schema when it changes the value of a nonwildcard symbol within the schema. This depends on *o(H)* and p_μ. The probability that *H* will survive mutation, $S_\mu(H)$, is

$$S_\mu(H) = (1- p_\mu)^{o(H)}. \tag{17.7}$$

By figuring the probability that a schema will not be disrupted by single point crossover and mutation into Equation 17.2, we can predict a lower bound on the number of instances of a schema that will be in the population at generation *t+1* by the following equation:

$$m(H,t+1) \geq m(H,t) \, f(H,t)/f(t) \, S_\chi(H) \, S_\mu(H) . \tag{17.8}$$

Equation 17.8 is called the Schema Theorem. It is also known as the Fundamental Theorem of Genetic Algorithms. By observing Equation 17.8 more closely one can see that some schemata have a greater probability of losing instances while others have a greater probability of gaining instances. In fact, one can see that the schemata with the greatest probability of gaining instances are those schemata that have a short defining length, a low order, and have an above average fitness. This observation forms the basis of what is known as the building block hypothesis. This hypothesis says that GAs converge upon solutions by actually building them from the bottom up. The Fundamental Theorem of Genetic Algorithms shows how building blocks of a particular problem can be placed together to build larger building blocks ultimately resultingin the GA's development of a solution.

17.4 GENETIC PROGRAMMING

Genetic programming is an attempt to apply the given notion, "How can computer programs learn to solve problems without being explicitly programmed?" [26]. According to Koza [5], founder of genetic programming, allowing computers to seek solutions in the form of programs is the basis for achieving nonexplicit programming.

17.4.1   Structure Representation

The structures undergoing adaptation in genetic programming are noted as hierarchically formed programs (individuals represented in parse tree form) which dynamically change size and shape. The set of possible structures in genetic programming is primarily based on the set of all possible valid compositions that can be constructed from the set of n problem dependent functions from $F = \{f_1, f_2, ..., f_n\}$ and the set of n terminals from $T = \{t_1, t_2, ..., t_n\}$.

Arithmetic operations, conditional operators, mathematical and Boolean operations, or any defined functions specific to the problem may describe functions within the function set. They may also refer to standard programming operations. Each respective function takes a prespecified number of arguments, primarily based upon its operability, which can be either terminals or other functions.

Terminals within the terminal set may represent a variety of atoms that are generally problem dependent. These atoms are either represented in constant or variable form. Generally, terminals can be viewed as the inputs to the as-yet-undiscovered computer program.

Consider the following function (F) and terminal (T) sets: $F = \{+, -, sqrt\}$, and $T = \{A, B, C, \pi\}$. The representation of a possible structure that may be generated from these sets is shown in Figure 17.9. We refer to this type of representation as a rooted point labeled tree with ordered branches. Note that the internal points in the tree are denoted by functions, and terminals denote the external points (leaves).

17.4.2   Closure and Sufficiency

When determining the function and terminal sets, the satisfaction of the closure and sufficiency properties is desirable. The closure property requires that each element being a member of the chosen function set be able to accept any function or terminal in their respective sets. More specifically, each function should be well defined. In practical problems, however, this property is difficult to satisfy. Thus special approaches or provisions are frequently used to preserve closure. Consider a problem where the division operator is present in the function set. One case, when zero is randomly chosen as the divisor, is considered undefined. The use of a protected division operator is a simple approach that effectively guarantees closure. Consequently, when zero is encountered, the use of the operator automatically returns a value of one.

The second desirable property that should be satisfied is sufficiency. The terminal and function sets chosen for a problem being capable of generating a solution to the problem characterizes sufficiency. Expert knowledge of a particular problem generally allows the user's chosen sets to satisfy this requirement.

17.4.3 Fitness Evaluation

The fitness measure for a given application can be described as the driving mechanism for the evolutionary process. This measure basically determines the probability of an individual surviving to the age of reproduction and successfully reproducing. The nature of the fitness measure varies with the problem. Considering that it is fully defined, the fitness measure should be capable of evaluating any individual that it encounters within a run. Usually fitness is evaluated over a set of fitness cases that is generally chosen to sufficiently represent the domain space. This serves as the basis for generalizing evolved individuals to the entire search space.

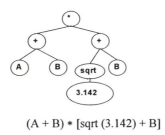

(A + B) * [sqrt (3.142) + B]

Figure 17.9: Illustration of Parse Tree Structure.

17.4.4 Genetic Operators

The initial populations in genetic programming are produced by randomly generating computer programs composed of functions and terminals appropriate to the problem domain. Thus, the initial population is a blind search of the search space of the problem represented as computer programs. However, breeding of successive generations is done by using three primary genetic operators: Darwinian reproduction, crossover (sexual recombination), and mutation. The use of Darwinian reproduction increases the probability of stronger individuals (programs having higher fitness) receiving multiple copies in the next generation while the weaker individuals receive fewer copies and eventually become extinct. The use of crossover provides variation in the population by producing offspring that are essentially a product of genetic material taken from its two parents. Figure 17.10 shows an example of this type of recombination using the same function and terminal sets used in Figure 17.9.

Note that, unlike the ordinary genetic algorithm, genetic programming allows the flexibility of mating individuals to cross material at different points. This provides greater flexibility in sampling the search space as well as enhances the opportunity for genetic programming to deliver some counterintuitive solutions.

Finally, to reintroduce diversity in a population that may tend to converge prematurely, the mutation operator is introduced. Mutation is implemented by performing random alterations in the program structures.

17.5   SUMMARY

In this chapter we provided a brief introduction to the field of EC and an overview of GAs. We also presented the fundamental theorem of GAs, the Schema Theorem, which describes the behavior of genetic search. Finally, we provided a brief overview of another EC technique for which research interest is rapidly growing, genetic programming.

Presently, there is a promising trend underway. Researchers are now combining evolutionary, neural, and fuzzy computing techniques to form hybrid systems that are even more efficient and robust [27, 28]. In the chapters to follow, the reader will be introduced to a number of these exciting new hybrids.

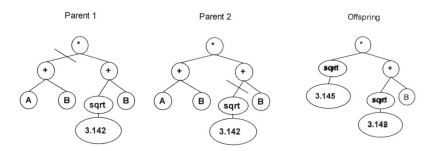

Figure 17.10: Crossover in Genetic Programming.

ACKNOWLEDGMENTS

This work is partially funded by grants from NASA Autonomous Control Engineering Center (ACE) at North Carolina A&T SU under grant # NAG2-1196 and NASA Dryden Flight Research Center under grant # NAG4-131. The authors wish to thank the ACE Center and NASA Dryden for their financial support. A portion of the research described in this chapter was performed at the Jet Propulsion Laboratory, California Institute of Technology, under contract with the National Aeronautics and Space Administration.

REFERENCES

1.   Bäck, T., Hammel, U., and Schwefel, H.P., Evolutionary Computation: Comments on the History and Current State., *IEEE Trans. on Evolutionary Computation*, Vol. 1, No. 1, 3-17, 1997.

2. Davis, Lawrence, *Handbook of Genetic Algorithms*, Van Nostrand Reinhold, New York, 1991.
3. Fogel, D. B., *Evolutionary Computation: Toward a New Philosophy of Machince Intelligence*, IEEE Press, 1995.
4. Holland, J. H., *Adaptation in Natural and Artificial Systems*, University of Michigan Press, Ann Arbor, MI, 1975.
5. Koza, J.R., *Genetic Programming: On the Programming of Computers by Natural Selection*, MIT Press, Cambridge, MA, 1992.
6. Michalewicz, Z., *Genetic Algorithms + Data Structures = Evolution Programs*, 2nd ed., Artificial Intelligence Series, Springer-Verlag, Berlin, 1994.
7. Spears, W.M., De Jong, K.A., Bäck, T., Fogel, D.B., and De Garis, H., An Overview of Evolutionary Computation, *Proc. of the 1993 European Conf. on Machine Learning*, 442-459, 1993.
8. Bäck, T., Hoffmeister, F., and Schwefel, H.-P., A Survey of Evolution Strategies, *Proc. of the Fourth Int. Conf. on Genetic Algorithms*, 2-9, Morgan Kaufmann Publishers, San Francisco, CA, 1991.
9. Fogel, L.J., Owens, A.J., and Walsh, M.J., *Artificial Intelligence Through Simulated Evolution*, John Wiley & Sons, NY, 1966.
10. De Jong, K. and Spears, K., On the State of Evolutionary Computation, *Proc. of the Fifth Int. Conf. on Genetic Algorithms*, 618-623, Morgan Kaufmann Publishers, San Francisco, CA, 1993.
11. Syswerda, G., Uniform Crossover in Genetic Algorithms. *Proc. of the Third Int. Conf. on Genetic Algorithms*, 2-9, Morgan Kaufmann Publishers, San Francisco, CA, 1989.
12. De Garis, H., Genetic Programming: Modular Evolution of Darwin Machines, *Proc. of 1990 Int. Joint Conf. on Neural Networks,* 194-197, 1990.
13. Reynolds, R.G., An Introduction to Cultural Algorithms, *Proc. of Evolutionary Program, (EP-94)*, 131-139, San Diego, CA, 1994.
14. Adleman, L.M., Molecular Computation of Solutions to Combinatorial Problems, *Science*, Vol. 266, 1021-1024, 1994.
15. Kennedy, J. and Eberhart, R., Particle Swarm Optimization, *Proc. of the 1995 IEEE Int. Conf. on Neural Networks*, 1942-1948, 1995.
16. Dorigo, M. and Gambardella, L.M., Ant Colony System: A Cooperative Learning Approach to the Traveling Salesman Problem, *IEEE Trans. on Evolutionary Computation,* Vol. 1, No. 1, 53-66, 1997.
17. Goldberg, D. E., *Genetic Algorithms in Search, Optimization & Machine Learning*, Addison-Wesley Publishing Company, Inc., Reading, MA, 1989.
18. Eshelman, L. J. and Schaffer, J. D., Real-Coded Genetic Algorithms and Interval Schemata, in *Foundations of Genetic Algorithms II*, Whitley, L.D., (ed.) Morgan Kaufman Publishers, San Francisco, CA, 1993.

19.   Schaffer, J. D., Caruna, R., A., Eshelman, L.A., and Das, R., A Study of Control Parameters Affecting Online Performance of Genetic Algorithms for Function Optimization, *Proc. of the Third Int. Conf. on Genetic Algorithms*, 51-60, Morgan Kaufmann Publishers, San Francisco, CA, 1989.

20.   Baker, J. E., Reducing Bias and Inefficiency in the Selection Algorithm, *Proc. of the Second Int. Conf. on Genetic Algorithms and Their Appl.*, 14-21, Erlbaum, Cambridge, MA, 1987.

21.   Whitley, D., The Genitor Algorithm and Selection Pressure: Why Rank-Based Allocation of Reproductive Trials is Best, *Proc. of the Third Int. Conf. on Genetic Algorithms*, 116-121, Morgan Kaufmann Publishers, San Francisco, CA, 1989.

22.   Grefenstette, J.J., Optimization of Control Parameters for Genetic Algorithms, *IEEE Trans. on Sys., Man & Cybern. SMC-16*, Vol. 1, 122-128, 1986.

23.   Bridges, C. L. and Goldberg, D. E., An Analysis of Reproduction and Crossover in a Binary-Coded Genetic Algorithm, *Proc. of the Second Int. Conf. on Genetic Algorithms*, 9-13, Erlbaum, Cambridge, MA, 1987.

24.   Grefenstette, J. J. and Baker,L. How Genetic Algorithms Work: A Critical Look at Implicit Parallelism, *Proc. of the Third International Conference on Genetic Algorithms*, 20-27, Morgan Kaufmann Publishers, San Francisco, CA, 1989.

25.   Vose, M.D., Generalizing the Notion of Schema in Genetic Algorithms, *Artif. Intell.*, 50, 385-396, 1991.

26.   Samuel, A., Some Studies in Machine Learning Using the Game of Checkers, *IBM J. of Res. and Develop.*, 3(3), 210-229, 1959.

27.   Homaifar, A. and McCormick, E., Simultaneous Design of Membership Functions and Rule Sets for Fuzzy Controllers Using Genetic Algorithms, *IEEE Trans. on Fuzzy Systems*, Vol. 3, No. 2, 129-139, 1995.

28.   Yao, X., Evolving Artificial Neural Networks, *Proc. of the IEEE*, 87(9), 1423-1447, September, 1999.

18   EVOLUTIONARY CONCEPTS FOR IMAGE PROCESSING APPLICATIONS

Madjid Fathi and Lars Hildebrand

18.1   INTRODUCTION

This chapter describes how evolutionary concepts can be used to improve the performance of operators for image processing applications. The evolutionary concepts are based on evolution strategies, which are almost unknown outside Europe and explained in more detail here. Evolution strategies provide a good alternative to genetic algorithms if real valued problems have to be solved. This chapter gives an overview of evolution strategies. These principles are described using mathematical formulas as well as two- and three-dimensional example diagrams. The second part of this chapter consists of two applications examples in which evolutionary concepts are applied.

18.2   OPTIMIZATION TECHNIQUES

18.2.1   Basic Types of Optimization Methods

The term "optimization" is understood as the process by which the most favorable parameters for a system are chosen with respect to the system's objective function while taking the restrictions into account. If the structure of such a system can be formulated in the form of a mathematical model and a quantitative description of the optimal state can be given, then optimization procedures can be developed to determine the values of the parameters needed to reach the optimal state of the system. If the optimal state is defined by a single value or a set of individual values, then the optimization procedure used is called parameter optimization. If the optimum is defined by a function, then the procedure used is called functional optimization. If, in the following discussion, the word optimization is used, it will always refer to parameter optimization.

The optimization is either a linear or a nonlinear optimization depending on the kinds of restrictions. Nonlinear optimizations can be further divided into two classes: convex and concave. For a characterization of these classes, see [1].

18.2.2 Deterministic Optimization Methods

All optimization methods in which the changes in the variables during an optimization phase are determined in a deterministic fashion are united under the general term "deterministic optimization methods". This means that for an initial state Z_0 of an optimization problem, the state Z_j can always be reached in the same manner. The path from the initial state Z_0 to the optimal state Z_{opt} is always reached through the same intermediate states Z_i. A complete optimization is uniquely characterized by the optimization method used, the corresponding control parameters, and the states Z_0 and Z_{opt}. In the following text, a few examples of deterministic methods are given [2].

18.2.2.1 *Minimization in the Direction of the Coordinates*

Given is the function $F(\hat{x})$ and a starting point $\hat{x}_0$. Starting at this point, the minimum $\hat{x}_1$ is searched for in a specific direction. From this new point, a new minimum $\hat{x}_2$ is searched for in a new direction, etc. Any direction at all can be chosen, but it must be ensured that the sequence of directions chosen includes all dimensions of the search space. One possible sequence of search directions is given in [3]: The unit vectors of the coordinate system are chosen one after the other as the direction for the search. If the search has gone through all elements of the base vector, then one starts again with the first element chosen. This procedure ends when the distance between two vectors found, $\hat{x}_{i-1}$ and $\hat{x}_i$, does not exceed a certain length ε, i.e. $\|\hat{x}_{i-1} - \hat{x}_i\| < \varepsilon$ where $\|\hat{x}_{i-1} - \hat{x}_i\|$ is the distance between the two points $\hat{x}_{i-1}$ and $\hat{x}_i$ in the n-dimensional space.

18.2.2.2 *Minimization in the Direction of the Steepest Slope*

Again, the function $F(\hat{x})$ and a starting point $\hat{x}_0$ are given. To reach the minimum of the function $F(\hat{x})$ from a point, one only has to follow the negative gradient $\hat{b} = -\nabla F(\hat{x})$. The minimum $\hat{x}_1$ in the direction of $\nabla F(\hat{x}_0)$ is determined, and from this point the minimum $\hat{x}_2$ in the direction of $\nabla F(\hat{x}_1)$ is determined, etc. These steps are repeated until a point $\hat{x}_{opt}$ is reached which corresponds to the minimum within a certain tolerance range.

One characteristic of this method is that the search directions are always perpendicular to each other because the search for the minimum near $\hat{x}_i$ in the direction $\hat{b}_{i-1}$ means that the derivative of the function in the direction of $\hat{b}_{i-1}$ is 0 at $\hat{x}_i$, i.e. $\hat{b}_{i-1} \nabla F(\hat{x}_i) = 0$ is true, meaning that the gradient is perpendicular to $\hat{b}_{i-1}$. A disadvantage of this method is that the direction of the search cannot be adjusted according to the function, resulting in the fact that the method converges to the minimum slowly [3]. Path of optimization is in the direction of the steepest slope.

18.2.2.3 *Simplex Minimization*

The EVOP (**E**volutionary **Op**eration) optimization method has been expanded by Spendley, Hext, and Himswoth to become the so called simplex method [4] - [6]. When applying the simplex method in an n-dimensional space, the simplest geometric figure that can be made with $n + 1$ vertices is drawn, although this figure cannot be drawn in an $n - 1$-dimensional space anymore. A simplex in a two- dimensional space is a triangle, a simplex in a three-dimensional space is a tetrahedron, and for higher dimensional spaces, there is no visual representation of the corresponding simplex anymore. A regular simplex is a simplex for which all vertices of the figure are equidistant from each other.

Starting with a randomly chosen regular simplex, the values of the function at the $n + 1$ vertices of the simplex are evaluated. The point with the worst value of the function is thrown out, and a new vertex point is determined which corresponds to the projection of this point through the center of gravity of the surface defined by the remaining n points. In the course of the optimization process, it can occur that the reflected point is the worst point of the new simplex. In this case, the process will oscillate. In order to avoid this, the second worst point will be reflected instead of the worst point. If the process approaches the optimum, then the succession of new simplexes will rotate around one vertex. If the succession of rotating simplexes reaches the initial simplex of the rotation, then the length of the edges of the last simplex are halved and the process continues until a certain break criteria is reached.

A disadvantage of this optimization method is the fact that the size of the simplexes can only be reduced, thereby reducing the step size when approaching the optimum. An enlargement of the simplex is not taken into account, and if the size of the initial simplex is too small or one of the simplexes created during a local optimization lies in an area of steeply sloped gradients, then the process will converge slowly.

18.2.3 **Probabilistic Optimization Methods**

In contrast to the deterministic optimization methods, probabilistic optimization methods contain at least one random component or a corresponding calculation step [7] - [11]. At this point, it is important not to assume that the term random means without having any specific plan in mind. The term probabilistic is not used to identify methods that randomly evaluate points in the search space without following a certain strategy or without learning after each step. An example of such a method is the blind search method thought out by Brooks. This method will be explained in a little more detail here in order to differentiate it from probabilistic optimization methods [11].

The starting point is an n-dimensional vector which is restricted by n intervals $a_i \leq \hat{x}^i \leq b_i$, where $a_i \neq b_i$ and $i \in \{1, ..., n\}$. The minimum of

$F(\dot{x})$ within the bounds of this interval is to be determined. A few points are chosen at random and evaluated to do this. The probability density for points outside the interval bordered by $a_i \le \dot{x}^i \le b_i$ are set to 0, and inside this area the densities are evenly distributed so that the following is true:

$$p(\dot{x}) = \begin{cases} 1/V, & \text{for all } a_i \le \dot{x}^i \le b_i \\ 0, & \text{else} \end{cases} , \tag{18.1}$$

whereby V is the volume of the n-dimensional space created by the interval. This is calculated using

$$V = \prod_{i=1}^{n} (b_i - a_i) . \tag{18.2}$$

The point with the smallest value of the function is considered to be the optimum. Due to the random control characteristic of this optimization method, only one probability can be given which describes whether or not the optimum will be found within N attempts. To determine this, a new volume v is selected which contains all points in the search space that fulfill the conditions for the optimum. The probability that the optimum can be found within N attempts is then determined by

$$p = 1 - (1 - v/V)^N \tag{18.3}$$

For low dimensional search spaces, this method appears to be promising. For a desired determination probability p of 0.99 and a target volume v where $v = V/100$, the number of attempts N can be determined using

$$N = \frac{\ln(1-p)}{\ln\left(1 - \dfrac{v}{V}\right)} \tag{18.4}$$

resulting in $N = 459$. In addition, the effect of a reduction of the volume on the length of the interval can be determined with

$$\frac{d}{D} = \sqrt[n]{\left(\frac{v}{V}\right)} \tag{18.5}$$

For $v = V/100$ and $n = 2$, this results in a value of 0.1 for the reduced interval length d. This means that when we assume that the optimal values correspond to 1% of the two-dimensional search space, and that these are to be found with a probability of 0.99, then 459 random values within the search space must be evaluated. The borders of the interval which contains the optimum are reduced to 10% of the original interval area. 90% of the defined area per interval can be excluded from the search. Unfortunately, this method fails when optimization problems of higher order dimensions are attempted to be solved. For $n = 10$ and assuming the same conditions as above, this results in the same number of attempts N, but the interval area is not reduced to 10% of the original interval, rather it is now 63%. In order to achieve a 90% reduction in the interval area, we would have to reduce the target volume by a factor of 10^{-10} of the search space. The number of attempts can be determined using (18.4) and becomes then $N \approx 4{,}6 \cdot 10^{10}$.

Schwefel has shown that a blind search for all probabilities $p > p_{critical}$ with $p_{critical} \approx 0.63$ is a less effective search method than the grid method, which has an optimal behavior for non directional probabilistic methods [11]. In spite of this, methods such as the blind search are still useful as all search steps can be carried out at the same time, and, when processed in parallel, this method results in a good run time behavior.

If one removes the requirement that all attempts must be made independent of each other, then one has a directional probabilistic method. These methods are able to "learn" in which direction to search by allowing previous attempts to influence the current attempt. A few of the methods from this field will be presented in the following section: $(1 + 1)$-, $(\mu + 1)$-, and (μ, λ)- evolution strategies.

18.3 EVOLUTION STRATEGIES

Evolution strategies are based on the fact that biological evolution represents an almost perfect method to adapt an individual to the environment. The fundamental evolutionary concepts were transferred to the technical field of optimization methods by Rechenberg in 1964 [12, 13]. These fundamental concepts will be explained in this section.

18.3.1 Biological Evolution

In 1859, Darwin produced the theory that all creatures have developed from more primitive forms over a long period of time [14]. He observed three basic principles of nature:

1. Nature creates a potential overpopulation of life forms. In spite of this, the size of a population does not generally change very much.

2. No life form is an identical copy of another life form. There are always at least a few minor differences.

3. Changes in a life form that have proven to be favorable can often be found in their offspring.

Darwin derived the following from his observations. There exist life forms which have been able to survive better in their environment than others. The reason for this can be found in the minute differences between individual life forms. These differences can be passed on to later generations.

Through Mendel's research, the basic principles of inheritance were recognized. In 1865, he formulated three laws:

1. Reciprocity:
   When two pure breed (F_0 generation) individuals, i.e., individuals which are not the result of cross breeding and which differ in at least one aspect, are crossed, then the members of the following generation (F_1 generation) will have a uniform, equal appearance.

2. Division:
   The second filial generation F_2 is not uniform. The appearance of individual, inherited characteristics occurs according to a numerical model discovered by Mendel. The possible ratios are 3:1, 1:2:1 and 9:3:3:1, whereby the individual numbers stand for the various combinations of characteristics in the offspring.

3. Recombination:
   Mendel's third law states that genes can be assembled in new combinations. This recombination is extremely significant with respect to the variety of forms which a given life form may take and is therefore the basis for evolution.

The formulation of these laws raised many questions concerning the encoding and replication of genetic information. Through the research conducted by Avery (1944) and Watson and Crick (1953), the significance of nucleic acids, as well as the assembly and structure of the two most important representatives of the nucleic acids (deoxyribonucleic acid - DNA and ribonucleic acid - RNA), were discovered. With only a few exceptions, DNA consists of a two-stranded winding chain (double helix) of deoxyribose and phosphate, as well as the purine bases adenine (A), cystosine (C), guanine (G) and thymine (T). The combination of these purine bases in groups of 3 results in about 20 amino acid codes.

These amino acids then constitute the basis for the polypeptides and therefore for the synthesis of proteins. Chromosomes constitute the carrier for the DNA, and a complete set of chromosomes is called a genome. The basic mechanism for inheritance is the division of the DNA double helix into individual strands, the interpretation of these strands, and the synthesis of the encoded amino acids. At the same time, the DNA double helix can duplicate itself due to the complimentary character of the double helix. A single strand of DNA can complete itself and become a DNA molecule again. These complex processes have a tendency to go wrong, i.e., a mutation can occur. Possible mutation forms are genome mutations (a change in the number of chromosomes), chromosome mutations (a change in the structure of the chromosome), and gene mutations (a change in the nucleic acids). These basic principles, such as the encoding of all information concerning a life form, duplication, mutation, and selection have been carried over into the field of technical optimization in the form of evolution strategies.

18.3.2 Mechanisms of Evolution Strategy

In this section, the transfer of the basic biological evolutionary principles to the field of optimization will be explained. This transfer is a direct transfer in many cases, but it will be shown that only a portion of the principles of biological evolution are applicable to evolution strategies. In the field of evolution strategies, single potential solutions of an optimization problem are regarded as individuals. A set of individuals which belong to the same optimization step in the optimization process builds a population. In order to differentiate between the populations, each population is classified as being a generation. The various operations, for example recombination, mutation, and selection, are carried out during the transition from one generation to the next. As evolution strategies are probabilistic optimization methods, a few terms from probability theory have to be used, see [15].

The encoding of an individual serves as the basis for all further examinations. Given is an n-dimensional optimization problem in $\mathbb{R}^n$. An individual $\vec{I}$ is described as a point in an $(n + s + a)$-dimensional space by:

$$\vec{I} = (\vec{x}, \vec{\sigma}, \vec{\alpha})^T \in \mathbb{R}^n \times \mathbb{R}^s \times [-\pi, \pi]^a, \qquad (18.6)$$

for $s \in (1, ..., n)$ and $a \in (0, ..., (n(n-1))/2)$, where $\vec{x} = (x_1, ..., x_n)^T \in \mathbb{R}^n$ is the object variable and $\vec{\sigma} = (\sigma_1, ..., \sigma_s)^T \in \mathbb{R}^s$ and $\vec{\alpha} = (\alpha_1, ..., \alpha_a)^T \in [-\pi, \pi]^a$ are the strategy variables.

The object variable $\vec{x}$ gives the position of the individual $\vec{I}$ in the search space. In addition to the position of the individual, there are two other pieces of information belonging to each individual $\vec{I}$ which are called the strategy components. The component $\vec{\sigma}$ gives the step size in each of the coordinate directions of the search space, and using the component $\vec{\alpha}$, the angle between the various step directions can be set. The importance of the strategy components will be explained using an example after having defined the step size and the mutation operation.

An individual $\vec{I}$ must be assigned a fitness value $\Psi(\vec{I}) = \Psi(\vec{x}) \in \mathbb{R}$ which states the "goodness" of an individual regarding the optimization problem, and which therefore has a direct influence on the individual's chance for survival. The function Ψ is only determined using the object variable $\vec{x}$. The influence of $\vec{\sigma}$ and $\vec{\alpha}$ on the fitness of an individual can only be indirectly determined through the fitness of the object variable $\vec{x}$. A new individual $\vec{I}'$

can be created by mutating the individual $\vec{I}$. The following is true for the mutation:

$$
\begin{aligned}
mut(\vec{I}) &= mut(\vec{x}, \vec{\sigma}, \vec{\alpha}) \\
&= (mut(\vec{x}), mut(\vec{\sigma}), mut(\vec{\alpha}))
\end{aligned} \tag{18.7}
$$

Equation (18.7) states that an individual $\vec{I}$ is mutated by mutating each of its components, the object variable $\vec{x}$, and the strategy variables $\vec{\sigma}$ and $\vec{\alpha}$. Each of these components is mutated differently.

The step vector $\vec{\sigma}$ is mutated by mutating all coordinates $\vec{\sigma}^i$ according to

$$
\vec{\sigma}^{i} = \vec{\sigma}^{i} e^{(\tau'N(0,\,1) + \tau N_{i}(0,\,1))}, \tag{18.8}
$$

where $i \in (1, ..., s)$. The term $\tau'N(0, 1)$ is determined once for the individual, and the term $\tau N_i(0, 1)$ is calculated anew for each coordinate σ_i. Schwefel recommends the following to determine the values of the parameters τ and τ' [11]:

$$
\tau \approx \frac{1}{\sqrt{2\sqrt{n}}}, \quad \tau' \approx \frac{1}{\sqrt{2n}}. \tag{18.9}
$$

The angle vector $\vec{\alpha}$ is mutated by mutating the elements $\vec{\alpha}^i$ according to

$$
\vec{\alpha}^{i} = \vec{\alpha}^{i} + \beta N_{i}(0, 1), \tag{18.10}
$$

where $i \in (1, ..., a)$. The term $N_i(0, 1)$ is determined for each coordinate $\vec{\alpha}^i$. Schwefel also recommends a value for the parameter β [11]:

$$
\beta = 0,873(rad) \cong 5°. \tag{18.11}
$$

After having introduced the mutation functions for the strategy parameters, the mutation of the object variables $\vec{x}$ can be defined using them. The object vector $\vec{x}$ is mutated using

$$
\vec{x}' = \vec{x} + N(\vec{0}, C(\vec{\sigma}', \vec{\alpha}')) \tag{18.12}
$$

whereby $(\vec{\sigma}', \vec{\alpha}')$ specifies the covariance of an n-dimensional normal distribution. In the following, a two-dimensional example will be given which demonstrates the effect of each of the mutation steps of an individual $\vec{I}$ on an offspring $\vec{I}'$. The section of the search space shown in the diagrams ranges over the interval $[-4,4] \times [-4,4] \in \mathbb{R}^2$ on both the x- and the y-axes (the x- and y-axes are shown without any specific units). The probability $p(\vec{I}')$ that the offspring $\vec{I}'$ will be assigned a particular location in this section of the search space is represented in the diagram for $p(\vec{I}') > 0,01$ by a variation in the gray scale in the area near $\vec{I}$. In Figure 18.1, $\vec{\sigma}^i = 1$ and $\vec{\alpha}^j = 0$ for

$i \in (1, ..., n)$ and $j \in (1, ..., n(n-1)/2)$.

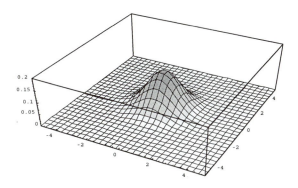

Figure 18.1: Mutation with $\vec{\sigma} = (1, 1)^T$ and $\vec{\alpha} = (0)$.

By changing the strategy parameters $\vec{\sigma}$ and $\vec{\alpha}$, the probability $p(\vec{I}')$ of a possible new individual $\vec{I}'$ at a certain location can be influenced. The values of the coordinates $\vec{\sigma}^i$ and $\vec{\sigma}$ contain the step sizes in the direction of the corresponding coordinate i, and the ratio $\vec{\sigma}^1/\vec{\sigma}^2$ determines the shape of the ellipse around $\vec{I}$. Figure 18.2 shows $p(\vec{I}')$ for $\vec{\sigma} = (2, 1)^T$.

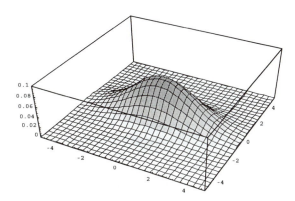

Figure 18.2: Mutation with $\vec{\sigma} = (2, 1)^T$ and $\vec{\alpha} = (0)$.

Additionally, if one allows the ellipse described by $\vec{I}$ and $\vec{\sigma}$ to be rotated by defining a rotation angle $\vec{\alpha}$, then the shape and location of the ellipse can be adjusted as shown in Figure 18.3.

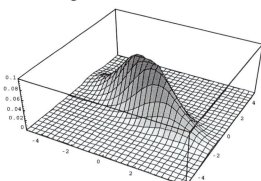

Figure 18.3: Mutation with $\vec{\sigma} = (2, 1)^{T}$ and $\vec{\alpha} = (-0.7)$.

In summary, a change in the step size vector $\vec{\sigma}$ changes the shape of the ellipse, and a change in the orientation vector $\vec{\alpha}$ rotates the ellipse around the origin of the coordinate system. The ellipse determines the probability of a given position of the new individual $\vec{I}'$ for a given initial position of the individual $\vec{I}$.

If one applies this concept to an n -dimensional space, then $\vec{\sigma}$ describes a hyperellipse in which each element $\vec{\sigma}^{i}$ describes the length of an axis of the ellipse, and in which each element $\vec{\alpha}^{j}$ of $\vec{\alpha}$ describes the angle of rotation between two coordinate axes. Figure 18.4 shows the control possibilities for the two-dimensional case once again.

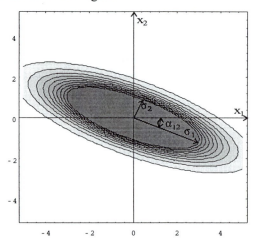

Figure 18.4: Meaning of the Strategy Parameters $\vec{\sigma}$ and $\vec{\alpha}$ in $\mathbb{R}^{2}$.

The position of an individual in the search space is changed by the mutation. The step size for this change in position is the standard deviation. At the beginning of an optimization, the individuals should be well spread out throughout the search space so that possible optima can be quickly found.

The mutation operation is not able to find these optima even in its optimum form as the limited standard deviation decreases the possibility that large step sizes can be used, and a mutation of the step size to obtain a larger value can require the computation of several generations. For this reason, the recombination operator was introduced. Similar to chromosome mutation in biology, the information inherited from the previous generations is also mutated, whereby inheritance information from two individuals is exchanged between the individuals. As with mutation, strategy and object variables can be recombined in many ways.

A new individual $\vec{I}'$ can be created by recombination using at least two individuals $\vec{I}_1$ and $\vec{I}_2$ from a parent generation $P(t)$ with μ individuals. The following is true for the recombination:

$$\begin{aligned} &= rec(P(t)) \\ &= rec(\vec{I}_1, ..., \vec{I}_\mu) \\ &= (rec(\vec{x}_1, ..., \vec{x}_\mu), rec(\vec{\sigma}_1, ..., \vec{\sigma}_\mu), rec(\vec{\alpha}_1, ..., \vec{\alpha}_\mu) \end{aligned} \qquad (18.13)$$

Before the individual recombination operations are described, an extra function used to index the individuals will be introduced. The function G_n produces a value in the interval $[1, n] \in \mathbb{N}$ for a given $n > 0$. The choice of this value is determined by a discrete, constant distribution over the range $[1, n]$. In contrast to the various mutation operations, the recombination operations are very similar to each other as each operator can be applied to each component of an individual. Four recombination methods are generally used:

1. Discrete recombination: Two parents $\vec{I}_1$ and $\vec{I}_2$ are chosen from the set of all parents. An offspring $\vec{I}'$ is created by selecting the corresponding element from either $\vec{I}_1$ or $\vec{I}_2$ for all elements of $\vec{I}'$:

$$f(\vec{I}_1, \vec{I}_2) = f \begin{pmatrix} I_1^1, I_2^1 \\ I_1^2, I_2^2 \\ ..., ... \\ I_1^n, I_2^n \end{pmatrix} = \begin{pmatrix} I_{G_2}^1 \\ I_{G_2}^2 \\ ... \\ I_{G_2}^n \end{pmatrix}, \qquad (18.14)$$

where $\vec{I}_1 = \vec{I}_{G_\mu}$ and $\vec{I}_2 = \vec{I}_{G_\mu}$.

2. Global discrete recombination:
   For each element of the offspring $\hat{I}'$, one parent individual $\hat{I}$ is chosen from the set of all parents. In contrast to discrete recombination, the set of possible parents is not limited to two individuals:

$$f(\hat{I}_1, ..., \hat{I}_\mu) = f\begin{pmatrix} \hat{I}_1^1 & , ..., & \hat{I}_\mu^1 \\ \hat{I}_1^2 & , ..., & \hat{I}_\mu^2 \\ ..., & ..., & ... \\ \hat{I}_1^n & , ..., & \hat{I}_\mu^n \end{pmatrix} = \begin{pmatrix} \hat{I}_{G_\mu}^1 \\ \hat{I}_{G_\mu}^2 \\ ... \\ \hat{I}_{G_\mu}^n \end{pmatrix} \tag{18.15}$$

3. Intermediate recombination:
   Two parents $\hat{I}_1$ and $\hat{I}_2$ are chosen from the set of all parents. An offspring $\hat{I}'$ is created by selecting for all elements of $\hat{I}'$ the average values of the elements from $\hat{I}_1$ and $\hat{I}_2$:

$$f(\hat{I}_1, \hat{I}_2) = f\begin{pmatrix} \hat{I}_1^1 & , \hat{I}_2^1 \\ \hat{I}_1^2 & , \hat{I}_2^2 \\ ..., & ... \\ \hat{I}_1^n & , \hat{I}_2^n \end{pmatrix} = \begin{pmatrix} (\hat{I}_1^1 + \hat{I}_2^1)/2 \\ (\hat{I}_1^2 + \hat{I}_2^2)/2 \\ ... \\ (\hat{I}_1^n + \hat{I}_2^n)/2 \end{pmatrix}, \tag{18.16}$$

   where $\hat{I}_1 = \hat{I}_{G_\mu}$ and $\hat{I}_2 = \hat{I}_{G_\mu}$.

4. Global intermediate recombination:
   For each element of the offspring $\hat{I}'$, two parent individuals are chosen from the set of all parents. An element of $\hat{I}'$ is then the average value of the corresponding elements from $\hat{I}_1$ and $\hat{I}_2$:

$$f(\hat{I}_1, ..., \hat{I}_\mu) = f\begin{pmatrix} \hat{I}_1^1 & , ..., & \hat{I}_\mu^1 \\ \hat{I}_1^2 & , ..., & \hat{I}_\mu^2 \\ ..., & ..., & ... \\ \hat{I}_1^n & , ..., & \hat{I}_\mu^n \end{pmatrix} = \begin{pmatrix} (\hat{I}_{G_\mu}^1 + \hat{I}_{G_\mu}^1)/2 \\ (\hat{I}_{G_\mu}^2 + \hat{I}_{G_\mu}^2)/2 \\ ... \\ (\hat{I}_{G_\mu}^n + \hat{I}_{G_\mu}^n)/2 \end{pmatrix} \tag{18.17}$$

The difference between discrete and intermediate recombination is shown in Figures 18.5 and 18.6. In addition to the recombination, the influence of $N(0, 1)$ mutations is also shown.

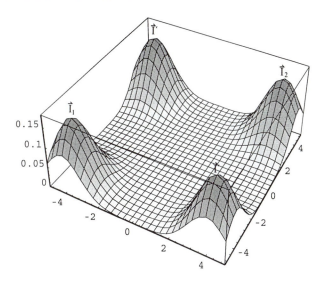

Figure 18.5: Probabilities for the Position of a New Individual.

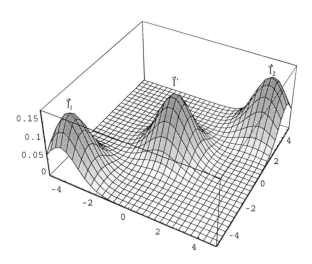

Figure 18.6: Probabilities for the Position of a New Individual Through Mutation and Intermediate Recombination.

By choosing a parameter $\gamma \neq 1/2$ (see Equation (18.18)), it is possible to transfer the individuals recombined using intermediate recombination to a single parent. The influence of each parent is then determined by:

$$\vec{I}' = \gamma \vec{I}_1^i + (1 - \gamma)\vec{I}_2^i \text{ , where } \gamma \in [0,1] \subset \mathbb{R}. \tag{18.18}$$

Up to now, only methods that alter single individuals have been introduced. In the next section, three different types of evolution strategies will be discussed. Schwefel introduced the following notation to differentiate among the three types [11]. An evolution strategy with μ parent individuals and λ offspring is designated with $(\mu \dot{+} \lambda)$. If the parents and the offspring are taken into consideration during the selection process, then it is a $(\mu + \lambda)$ strategy. When only the offspring are taken into consideration, then it is a (μ, λ) strategy. This notation has been extended by Schöneburg [16] and Bäck/Hoffmeister [17] in order to reflect further characteristics of the natural selection process. These extensions will not be discussed in this paper, though, as they serve more to make evolution strategies better reflect natural evolution than to improve the speed and probability of convergence.

18.3.3   The (1+1) Evolution Strategy

As the name says, this variation of an evolution strategy deals with one parent and one offspring. An individual $\vec{I} \in \mathbb{R}^n \times \mathbb{R}_+$ consists of the components $\hat{x}$ and $\vec{\sigma}$. The object component x describes the position of the individual in the search space, and the strategy component $\vec{\sigma}$ consists of a single element which contains the standard deviation, and therefore the step size, for each element $\hat{x}^i$ of $\hat{x}$ for an $N(0, \vec{\sigma})$ normal distribution. The following algorithm represents an implementation of such a strategy.

```
 1   t:=0;
 2   initialize  (t):= {(x̂, σ⃗)};
 3   evaluate  Ψ(P(t))= Ψ(x̂) ;
 4   while termination_criteria  T(P(t))  not fulfilled
 5           x̂', σ⃗') = mut((x̂, σ⃗));
 6           evaluate Ψ(x̂') ;
 7           if Ψ(x̂') ≤ Ψ(x̂)
 8               then  (t + 1):= {(x̂', σ⃗')};
 9               else  P(t + 1)=P(t) ;
10           t:=t+1;
11   end
```

The mutation operation (line 5) is carried out in two steps:

$$mut = mut_{\vec{x}} \circ mut_{\vec{\sigma}}. \tag{18.19}$$

In the first step, a new standard deviation is computed using

$$\vec{\sigma}' = mut_{\vec{\sigma}}(\vec{\sigma}) = mut_{\sigma}(\sigma) = \begin{cases} \sigma/c \,, \text{ if } p > 1/5 \\ \sigma \cdot c \,, \text{ if } p < 1/5 \\ \sigma \,, \text{ if } p = 1/5 \end{cases} \tag{18.20}$$

Schwefel recommends $c = 0.817$ [11], which he derived from the optimal standard deviation for the sphere model, as the value of the constant c. The parameter p must be reevaluated for each generation, and gives ratio of successful mutations to the total number of mutations. In the second step, the object component $\vec{x}$ is mutated using the new standard deviation by determining a new element $\vec{x}'^i$ for all elements $\vec{x}^i$ from $\vec{x}$ according to

$$\vec{x}'^i = \vec{x}^i + N(0, \vec{\sigma}'). \tag{18.21}$$

The selection operation (lines 7 - 9) selects the individual with the best fitness Ψ value from the parents and offspring. As this selection is a so called + -selection, it is possible that a parent with a high fitness value Ψ will survive over many generations. At regular intervals, the survival behavior of the parents and offspring are checked (18.20). If the fitness value of the offspring is too high ($p > 1/5$) then the search space is enlarged. It is assumed here that the reason for the large number of successful mutations is that we are far away from the optimum. If, however, a parent individual survives too many generations $p < 1/5$, then the step size is decreased in order to reduce the size of the search space. This "1/5 success rule" has been empirically determined by Rechenberg [12]. One disadvantage of this rule is that for optimization problems which do not achieve a reasonable success rate due to their topology, the step size will constantly decrease upon each application of the deterministic operation $mut_{\vec{\sigma}}$, so that the evolution strategy will eventually stagnate at a local optimum.

18.3.4 The (μ+1) Evolution Strategy

As the (1+1) evolution strategies only contain one parent per generation, the recombination concept taken from biological evolution cannot be applied. In order to apply recombination, Rechenberg developed the (μ+1) evolution strategy where $\mu > 1$ [12]. This variation has μ parents and produces one offspring from these parents. Due to the +-selection, one parent (usually the parent with the lowest fitness value Ψ) is replaced by the offspring when at least one of the parent individuals has a lower fitness value than the offspring. If

one compares this method to the extended simplex method, then one realizes
that both methods operate according to the principle of "throw the worst point
out". Rechenberg did not find a satisfactory extension of his "1/5 success rule"
for the mutation operation, and therefore only the recombination operation
will be introduced in this section. A combination of mutation and recombination operations in connection with a high quality control of the strategy parameters can be found in the following section. The following algorithm
represents an implementation of such a strategy.

1   t:=0;
2   initialize $P(t) := \{\dot{x}_1, ..., \dot{x}_\mu\}$;
3   evaluate $\Psi(P(t)) = (\Psi(\dot{x}_1), ..., \Psi(\dot{x}_\mu))$;
4   **while** termination_criteria $T(P(t))$ **not** fulfilled
5           $\dot{x}' = \text{rec}(P(t))$;
6   evaluate $\Psi(\dot{x}')$;
7           $P(t+1) := \text{sel}_\mu^{\mu+1}(\{\dot{x}'\} \cup P(t))$;
8   t:=t+1;
9   **end**

The recombination operation (line 5) creates one offspring from μ parent
individuals. This is done by applying the recombination operations (18.14) to
(18.17). The selection operation $\text{sel}_\mu^{\mu+1}$ selects the best μ from the population $(\{\dot{x}'\} \cup P(t))$ which contains $\mu + 1$ individuals. The discussion of the
$(\mu+1)$ evolution strategy ends here, as it has only served to present the intermediate step between the (1+1)- and the (μ,λ) evolution strategies.

18.3.5   The (μ,λ) Evolution Strategy

An individual $\vec{I}$ consists of the components $\dot{x}$, $\vec{\sigma}$ and $\vec{\alpha}$. The start population $P(0)$ is initialized by assigning random vectors from the search space to
the object component $\dot{x}$, $(1, ..., 1)^T$ to the strategy component $\vec{\sigma}$ and
$(0, ..., 0)^T$ to the strategy component $\vec{\alpha}$ (line 2). The evaluation of an individual is done through the evaluation of the object component $\Psi(\dot{x})$ (lines 4
and 9). The recombination operation (line 6) is carried out λ times in order to
produce λ offspring from μ parents (see Equation (18.13)). All of the recombination types given in Equation (18.14) through Equation (18.17) may also
be applied. Schwefel recommends a global intermediate strategy for the strategy components and a discrete recombination for the object components [11].
This recommendation should be reevaluated according to the specific application, however. A general statement as to which recombination type is suited to
which component does not exist. After producing λ offspring through the

recombination, the offspring are mutated according to (18.7). Mutations carried out according to Equation (18.8) through Equation (18.12) can occur depending on the complexity of the strategy components. The number of individuals in a population is not changed by the mutation, so that the selection operation (line 10) selects μ individuals from λ individuals. These μ individuals will become the parent individuals in the next generation. The following algorithm represents an implementation of such a strategy.

```
1   t:=0;
2   initialize  P(t):= {Ì₁(t), ...,Ìμ(t)} ;
3              // where Ìᵢ ∈ ℝⁿ × ℝˢ × [−π,π]ᵅ for all i = (1, ..., μ) /
                /
4   evaluate Ψ(P(t)) = (Ψ(x̌₁(t)), ..., Ψ(x̌μ(t))) ;
5   while termination_criteria T(P(t)) not fulfilled
6          ∀(k ∈ {1, .., λ} ):Ì'ₖ(t):= rec(P(t)) ;
7          ∀(k ∈ {1, .., λ} ):Ì'ₖ(t):= mut(Ì'ₖ(t)) ;
8          P"(t):= {Ì"₁(t), ..,Ì"λ(t)} ;
9          evaluate  Ψ(P"(t)) = (Ψ(x"₁(t)), ..., Ψ(x̌"λ(t))) ;
10         P(t+1):= selλμ (P"(t)) ;
11         t:= t + 1 ;
12 end
```

The (μ,λ) evolution strategy is the only strategy of those presented in this paper in which one is able to comprehensively adapt the strategy parameters of the optimization problem, and therefore adapt the strategy to the topology of the optimization problem. The basis for this is the ratio of the number of elements contained in the strategy components $\vec{\sigma}$ and $\vec{\alpha}$ to the number of parent individuals μ and offspring λ. In order to be able to adapt the strategy components, at least $(s + a)/10$ offspring should be produced per generation. One problem is the quadratic growth of the number of orientation angles $\vec{\alpha}^i$ with $i = n(n - 1)/2$ for optimization problems of higher dimensions. This fact has lead to the exclusion of the use of the orientation angles for optimization problems with approximately ten dimensions or more. For these optimization problems, the only adaptation carried out is the adaptation using various step sizes.

18.4   IMAGE PROCESSING APPLICATIONS

The authors have shown that the use of single methods from the field of computational intelligence, as well as the use of combinations of methods, can lead to powerful applications. Examples can be found for the design of composite materials [18, 19], the optimization of fuzzy rule-based systems [20] -

[22], or the combined application of neural networks, evolution strategies, and fuzzy logic [23] - [25]. The following two examples are image processing applications, in which the combined use of evolution strategies and fuzzy logic allows a powerful extension of the image processing operators [26, 27].

18.4.1   Generating Fuzzy Sets for Linguistic Color Processing

This section describes how fuzzy based color processing can benefit from optimization techniques. The example is part of mechanical engineering application, in which the quality of welding spots has to be determined [27] - [30].

18.4.1.1   Resistance Spot Welding

Resistance spot welding is a welding process that uses the inherent resistance of metal workpieces to join two sheets of metal by the flow of electrical current. Typical areas are the automotive, aerospace, and engineering industry. Resistance spot welding can easily be automated and in most cases only assistant helpers or robots are needed to supply the material. This fact has lead to the economical success of resistant spot welding as well as to the need for quality testing systems. Most of the quality testing systems have to destroy the welding joint in order to obtain quality measures like longitudinal and transverse tensile strength, bend strength, or hardness. Microscopic and macroscopic examination of the joint also require destructive operations. Nondestructive tests often need large and expensive equipment, like gamma or X-ray tubes, or are too sensitive to be used directly at the welding machine, like most ultrasonic sensors [31].

Human experts are able to check the quality of a welding spots using optical criteria. Color is one of the most important criteria. Typical areas that carry quality information are a blue or red inner and outer spot, and the impact zone with its color. Standard techniques of image processing are able to detect these different areas in the images. Figure 18.7 shows two examples, one for good quality and one for poor quality [27].

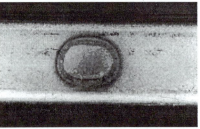

Figure 18.7: Example of a Good (Left) and Poor (Right) Welding Spot.

It can be seen that color carries a lot of information, but processing of color information is a complex task. One approach is the modeling of color information using sophisticated fuzzy sets to hold linguistic color names [27].

18.4.1.2 Linguistic Color Processing

Many color representations are based on technical demands. Examples are the red, green, and blue division of colors for televisions or the cyan, magenta, and yellow division for the printing media. These two representations reflect the additive and subtractive mixing of a few base colors to obtain a large set of displayable or printable colors. Technical representations are suitable for displaying colors, but fail if a deeper understanding of color is needed.

Human description of colors is not based on the additive or subtractive mixing of base colors [27, 32]. It is more oriented by characteristics like hue, brightness or lightness. Apart from technical color models, another class of color models exists that fulfills these human demands.

The HSI-model is well suited for the linguistic processing of color because colors that are similar for humans are grouped together and there is a clear distinction of colors and grays [33] - [35]. The 3-dimensional model can be easily reduced to a 2-dimensional one, by simply dropping the intensity coordinate when the pure and light colors are important, or by dropping the saturation coordinate when pure and dark colors are important. These operations result in the HS- and HI-sub models that are used throughout approach.

A fuzzy set over the HS-color model is defined by eight points. Each point contains two components to represent the angle (hue) and the radius (intensity or saturation). This results in an extension of the polar coordinate system of the HS-color space towards a cylindrical coordinate system. Each point represents one corner of the fuzzy set. This shape is chosen to reduce the computational demands and to accelerate computation time. A typical shape of such a fuzzy set is shown in Figure 18.8 [27]:

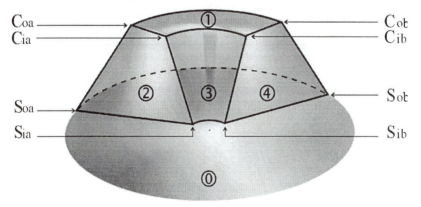

Figure 18.8: Fuzzy Set for Linguistic Color Processing.

If all significant colors are labelled, one can build a fuzzy rule system that allows a linguistic oriented way of expressing quality assessments. Below is an extract of such a rule system [22].

IF	COLOR_OF_INNER_SPOT	IS	BLUE
THEN	QUALITY	IS	GOOD
IF	COLOR_OF_OUTER_SPOT	IS	LIGHT_BLUE
THEN	QUALITY	IS	GOOD
IF	COLOR_OF_IMPACT_ZONE	IS	DARK_RED
THEN	QUALITY	IS	GOOD
IF	COLOR_OF_INNER_SPOT	IS	LIGHT_RED
THEN	QUALITY	IS	POOR
IF	COLOR_OF_OUTER_SPOT	IS	BLUE
THEN	QUALITY	IS	POOR
IF	COLOR_OF_OUTER_SPOT	IS	RED
THEN	QUALITY	IS	POOR
IF	COLOR_OF_IMPACT_ZONE	IS	LIGHT_RED
THEN	QUALITY	IS	POOR

This set of rules can be evaluated using standard fuzzy techniques in combination with the described methods to calculate the membership values of colors. Any fuzzy system that allows addition of new functions for the calculation of membership values can be extended to benefit from fuzzy color processing. The colors that exist in a sample image, as well as their frequency, can be expressed using a frequency distribution. Two examples of frequency distributions are shown in Figures 18.8 and 18.10 [36, 37]:

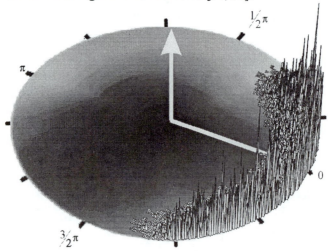

Figure 18.9: Typical Frequency Distribution of a Good Sample Point.

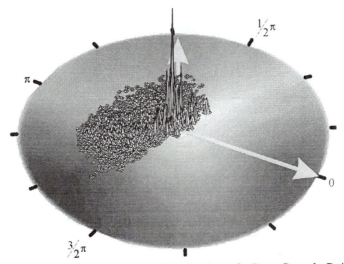

Figure 18.10: Typical Frequency Distribution of a Poor Sample Point.

To generate proper fuzzy sets an evolution strategy can be used. The object parameters are the definition points of the fuzzy set. The whole object vector consists of eight two-dimensional coordinates, yielding a 16-figure vector. The dimension is average so different step sizes for each figure in the vector can be used. The fitness function calculates the difference of the frequency distribution and the surface of the optimized fuzzy set. The smaller the difference, the better is the fitness value. As a result, fuzzy sets are created which are a good approximation of the sampled color frequency, as shown in Figures 18.11 and 18.12.

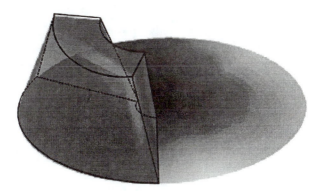

Figure 18.11: Dark Color as Fuzzy Set for Linguistic Color Processing.

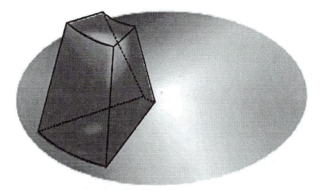

Figure 18.12: Light Color as Fuzzy Set for Linguistic Color Processing.

18.4.2   Developing Specialized Digital Filters

A common task during image processing is the enhancement of certain image features. This task is performed if a direct analyzing of the image is not possible (arrow in Figure 18.13). Well known are the techniques for edge detection, or the use of digital filters to enhance vertical or horizontal lines [38, 39]. These filters yield only poor results, however, if more complex features have to be detected. The following paragraph shows how digital filters in combination with evolution strategies can be used to generate image enhancement methods that are able to detect circular features.

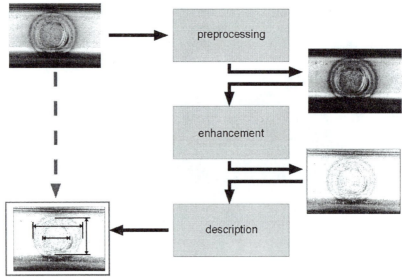

Figure 18.13: Image Processing.

18.4.2.1   Digital Image Filters

Digital image filters use a matrix of filter coefficients that is applied to each pixel in the input picture. The pixel and its neighbors are used to calculate the value of the output pixel. The whole process is described by the coefficients of the filter matrix.

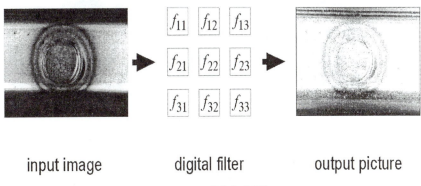

input image   digital filter   output picture

Figure 18.14: Digital Filter.

Some matrix filters are well known, the Frei/Chen-, Sobel-, Prewitt-, or Kirsch filter [38, 39]. The result of these filters applied to a welding spot image is shown in the next figures.

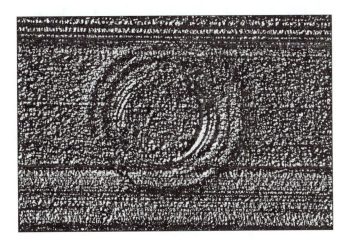

Figure 18.15: Frei/Chen Digital Filter.

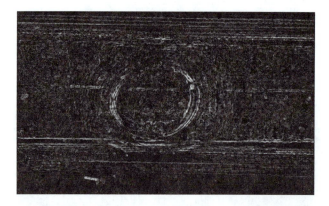

Figure 18.16: Frei/Chen Digital Filter.

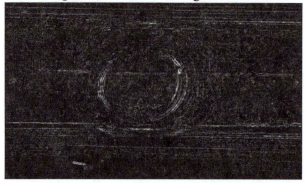

Figure 18.17: Prewitt Digital Filter.

Figure 18.18: Kirsch Digital Filter.

It can be seen that all four filters are not able to enhance the circular structure, that is necessary for the image description. Due to the fact, that the filter matrix can be expressed as a real valued vector, the use of evolution strategies to optimize the filter matrix is possible and allows a faster optimization compared to other types of evolutionary algorithms, e. g. genetic algorithms.

18.4.2.2 Optimization of Digital Filters

The matrix can be rearranged to build a vector, as shown in Figure 18.19. The low dimensionality allows the use of all types of self adaptation and evolution strategy variants. The fitness function compares the filtered image with a control image, in which the relevant features are enhanced by hand.

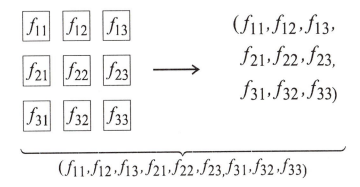

Figure 18.19: Rearranging a Matrix to a Vector.

After the optimization a new filter matrix is created that enhances circular features as shown in Figure 18.20. The values for the amtrix elements after the optimization took place are :(7.77, -11.31, 0.13, -8.32, 10.77, 3.52, -6.11, 14.41, -5.35).

18.5 CONCLUSION

Image processing techniques can be optimized and enhanced in many ways. This chapter demonstrates the use of specialized digital filters and the use of fuzzy logic for linguistic oriented techniques. The use of fuzzy logic is a powerful extension and allows the use of human-like feature descriptions, as shown for linguistic color processing. Fuzzy sets that are used can be generated and optimized using evolutionary concepts, in this case the use of evolution strategies.

If the numerical values of some image processing operators can be altered, evolutionary concepts can be used to find optimal values for these operators. The example shown uses this technique to generate high specialized digital filters for feature detection in images.

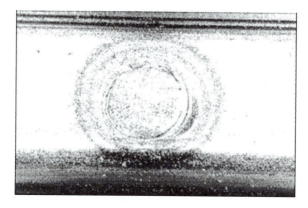

Figure 18.20: Optimized Filter for Detection of Circular Features.

REFERENCES

1. Bronstein, I. N., *Taschenbuch der Mathematik (handbook of mathematics)*, ergänzende Kapitel (additional chapters), 6. edition, BSB Teubner, Leipzig, 1990.
2. Vanderplaats, G. N., *Numerical Optimization Techniques for Engineering Design*, McGraw-Hill, New York, 1984.
3. Brandt, S., *Datenanalyse*, 3. edition, BI Wissenschaftsverlag, Mannheim, 1992.
4. Nelder, J. A. and Mead, R., A Simplex Method for Function Minimization, *Computer*, 6, pp. 308 - 313, 1965.
5. Spendley, W., Hext, G. R., and Himsworth, F. R., Sequential Application of Simplex Design in Optimization and Evolutionary Operation, *Technometrics*, 4, 1962.
6. Box, G. E. P., Evolutionary Operation - A Method for Increasing Industrial Productivity, *Appl. Stat.*, 6, 1957.
7. DeJong, K., An Analysis of the Behavior of a Class of Genetic Adaption Systems, Ph.D. Thesis, University of Michigan, MI, 1975.
8. Fogel, L. J., Owens, A. J., and Walsh, M. J., *Artificial Intelligence Through Simulated Evolution*, John Wiley & Sons, New York, 1966.
9. Goldberg, D. E., *Genetic Algorithms in Search, Optimization and Machine Learning*, Addison-Wesley, Reading, PA, 1989.
10. Holland, J. H., *Adaption in Natural and Artificial Systems*, University of Michigan Press, Ann Arbor, 1975.
11. Schwefel, H.-P., *Evolution and Optimum Seeking*, John Wiley & Sons, New York, 1994.

12. Rechenberg, I., *Cybernetic Solution Path of an Experimental Problem*, Royal Aircraft Establishment, Farnborough, 1965.
13. Rechenberg, I., *Evolutionsstrategie: Optimierung technischer Systeme nach Prinzipien der biologischen Evolution*, Frommann-Holzboog, Stuttgart, 1973.
14. Darwin, C., *On the Origin of Species by Means of Natural Selection*, 1859.
15. Bronstein, I. N., *Taschenbuch der Mathematik*, 24. edition, BSB Teubner, Leipzig, 1989.
16. Schöneburg, E., and Heinzmann, F. and Feddersen, S., *Genetische Algorithmen und Evolutionsstrategien*, Addison-Wesley, Bonn, 1995.
17. Bäck, T. and Hoffmeister, F., Extended Selection Mechanisms in Genetic Algorithms, in Belew, R. K. and Booker, L. B., *Proc. of the Fourth Int. Conf. on Genetic Algorithms and their Applic.*, Morgan Kaufmann, San Mateo, CA, 1991.
18. Hildebrand, L. and Fathi, M., *Multi Agent Based Design of Composite Materials*, in: *Proc. Int. Conf. on Composite Eng.* ICCE/6 '99, Hawaii, USA, pp. 213 - 215, 1999.
19. Fathi, M. and Hildebrand, L., *Intelligent Design Methods for Smart Materials*, in: *Proc. IPMM'99 Int. Conf. on Intell. Proc. and Manufacturing of Materials*, Honolulu, Hawaii, pp. 1011 - 1015, 1999.
20. Fathi, M. and Hildebrand, L., *Intelligent Methods for the Optimization of Composite Materials*, in: *Proc.* IPMM97, *Australasia-Pacific Forum on Intell. Proc. & Manufacturing of Materials*, Gold Coast, Australia, 1997.
21. Fathi, M. and Hildebrand, L., Evolution Strategies for the Optimization of Fuzzy Rules, Proc. of the "IPMU", Paris, 1994.
22. Fathi, M. and Hildebrand, L., The Application of Evolution Strategies to the Problem of Parameter Optimization in Fuzzy Rulebased System, Proc. of the *IEEE Int. Conf. on Evolutionary Computing*, Perth, 1995.
23. Hildebrand, L., Jäger, M. and Fathi, M., Learning of Linguistic and Numerical Knowledge - Application of Neural Networks and Evolutionary Algorithms, in: *Proc. Eng. of Intell. Syst. EIS'98*, La Laguna, Spain, 1998.
24. Hildebrand, L. and Fathi, M., Evolutionary Design of Screw Rotor Profiles, in: *Proc. Int. Conf. on Composite Eng. ICCE/5 '98*, Las Vegas, 1998.
25. Fathi, M. and Hildebrand, L., Complex System Analysis Using CI-Methods, in: *Proc. AeroSense - SPIE 13th Ann. Int. Symp. on Areospace/ Defense Sensing, Simulation, and Controls*, Orlando, FL, 1999.
26. Hildebrand, L. and Fathi, M., Soft Computing as a Methodology for Color Processing, in: *Proc. Eusflat - Estylf Joint Conf.*, Palma de Mallorca, Spain, 1999.

27. Hildebrand, L. and Reusch, B., Fuzzy Color Processing, in: Kerre, E.
     and Nachtegael, M., (Ed.) *Fuzzy Techniques in Image Processing*,
     Springer-Verlag, Heidelberg, Germany, pp. 267 - 286, 2000.
28. Hildebrand, L. and Fathi, M., Probabilistic Optimization - Evolution
     Strategies, in Real World Applications of Intelligent Technologies, in:
     National Institute for Reasearch and Development in Microtechnologies,
     Bucharest, Romania, 1997.
29. Hildebrand, L. and Fathi, M., Vision Systems for the Inspection of Resis-
     tance Welding Joints, in: *Proc. Electron. Imaging 2000, IS&T/SPIE 12th
     Inter. Symp.*, San Jose, CA, 2000.
30. Hildebrand, L. and Fathi, M., Detection of 3-Dimensional Image Fea-
     tures Using a Single Camera, in: *Proc. Inter. Forum on Multimedia and
     Image Process.*, World Automation Congress, Hawaii, 2000.
31. Waschkies, E., Process-Integrated Resistance Spot Welding Testing
     Using Ultrasonic Techniques, *Welding in the World* 39, No. 6, 1997.
32. Silverstein, L. D., *Human Factors for Color CRT Displays*, Soc. for Inf.
     Displays, 1982.
33. Teichner, W. H., Color and Information Coding, *Proc. of the Soc. for Inf.
     Displays*, Vol. 20, 1970.
34. Watt, A. H., *Fundamentals of Three-Dimensional Computer Graphics*,
     Addison-Wesley, Wokingham, England, 1989.
35. Wyszecki, G. and Siles, W. S., *Color Science*, 2nd Edition, John Wiley.
     New York, 1982.
36. Hildebrand, L. and Fathi, M., Linguistic Color Processing for Human-
     Like Vision Systems, in: *Proc. Electron. Imaging 2000, IS&T/SPIE 12th
     Int. Symp.*, San Jose, CA, 2000.
37. Reusch, B., and Fathi, M. and Hildebrand, L., Fuzzy Color Processing
     for Quality Improvement, in: *Proc. Int. Forum on Multimedia and Image
     Process.*, World Automation Congress, Anchorage, Alaska, pp. 841 -
     848,1998.
38. Rosenfeld, A. and Kak, A. C., *Digital Picture Processing*, Vol. 1 and 2,
     2nd Edition, Academic Press, San Diego, CA, 1982.
39. Foley, J., VanDam, A., and Feiner, S., *Computer Graphics: Principles
     and Practice*, 2nd Edition, Addison-Wesley, Reading, MA, 1990.

19 EVOLUTIONARY FUZZY SYSTEMS

Mohammad -R. Akbarzadeh-T. and A. -H. Meghdadi

19.1   INTRODUCTION

Fuzzy logic has been described as a practical, robust, economical, and intelligent alternative for modeling and control of complex systems. This long list of superior traits, however, can only be realized if quality expert knowledge exists and is made available to the control engineer. This is while, in conventional applications of fuzzy logic, there is not yet a systematic way of acquiring such expert knowledge. Quality expert knowledge often does not exist, for example, in remote environments that humans have not experienced such as the surface of Mars or the hazardous environment of an underground nuclear waste storage tank. Even when such expert knowledge does exist, it is not clear whether the expert would be able to objectively present his knowledge in terms of a constrained set of rules and membership functions, and whether such expertise would indeed be optimal. In solving this paradox, fuzzy logic has been complemented by various strategies such as neural networks, fuzzy clustering, gradient based methods, and evolutionary optimization algorithms.

Evolutionary optimization algorithms have been particularly appealing various scientific circles, primarily because such algorithms allow autonomous adaptation/optimization of fuzzy systems without human intervention, are not easily trapped in locally optimal solutions, allow parallel exploitation of an optimization parameter space, and do not require gradient evaluation of the objective function. Evolutionary fuzzy systems are hybrid fuzzy systems in which evolutionary optimization algorithms are used to optimize/adapt fuzzy expert knowledge. The evolutionary optimization algorithms operate by representing the optimization parameters via a gene-like structure and subsequently utilizing the basic mechanisms of Darwinian natural selection to find a population of superior parameters.

There are various approaches to evolutionary optimization algorithms including evolution strategies, evolutionary programming, genetic programming and genetic algorithms. These various algorithms are similar in their basic concepts of evolution and differ mainly in their approach to parameter representation [1]. Genetic algorithms (GA), in particular, is an evolutionary method which has performed well in noisy, nonlinear, and uncertain optimization landscapes typical of fuzzy systems. In this chapter, we will explore further why and how GA is used for optimization of fuzzy systems and, in particular, fuzzy controllers. Various issues such as determining the set of parameters, designing the transformation function for representing the parameter space in genetic domain, creating the initial population, and determining the fitness function will be discussed. Finally, application of GA will be illustrated in optimizing fuzzy control of a DC motor.

19.1.1   The Problem Statement and Design Outline

Based on the universal approximation theorem [2], we know that, for every continuous and nonlinear/linear function g, a fuzzy system f exists which can approximate g to any desirable degree of accuracy. In other words, if there exists a desirable nonlinear function g which meets a given system's performance criteria, there also exists a fuzzy function f closely approximating g. Fuzzy controller design can therefore be viewed as a complex optimization problem in the search space of all possible nonlinear controllers. The type and characteristics of this search space determines the best optimization method to automate this design process. These characteristics include a *large* parameter space, a *non-differentiable* objective function involving *uncertainty and noise*, and, finally, *multi modality* and *deceptiveness*, i.e., similar sets of membership functions and rule sets may perform quite differently.

Genetic algorithms are particularly suitable for such optimization and hence automating the above design process. As we will see later in this chapter, genetic algorithms can easily encode a large number of parameters, are based on function evaluation as compared with gradient evaluation, always keep and combine a population of potential solutions in parallel and hence can easily avoid local optima, and, due to their guided evolutionary mechanism, are more computationally efficient when compared with random search [3].

In many ways, GA can be likened to a piece of sculpting wax, its designer to a sculptor, and the whole design process of a GA to an art. Like a sculptor, a GA designer has great many choices which, when combined, create a unique GA. As a sculptor patiently forms the base, the body, the hands, and finally the head of his creation, the GA designer has to complete several stages of design before a genetic algorithm is completely defined and capable of optimizing a particular system.

As illustrated in Figure 19.1, the first step involves identifying the parameters that need to be optimized. Reducing the number of free parameters usually reduces the complexity of the optimization task, thereby achieving a faster convergence of the optimization algorithm. However, genetic algorithms can typically handle a large number of parameters efficiently. Also, by constricting too many of the parameters, we might just be eliminating the optimal solution set from the GA search landscape. So a careful trade off exists between complexity of the optimization task and convergence of the genetic algorithm.

Genetic algorithms operate on populations of individuals that are usually binary strings. Since in most applications a phenotype (a solution in problem parameter space) consists of real numbers, an encoding function is required to map the phenotype to its representation in GA space (genotype). Step II is, therefore, to determine this encoding (interpretation) function. In designing this interpretation function, a higher number of bits per real number produces a higher degree of accuracy in the representation, but also a longer GA string (increased complexity). Additionally, an interpretation function should be designed to minimize competing conventions and deception. The problem of competing conventions arises when two or more completely different genotypes (individuals in GA domain) represent one phenotype (an individual in problem domain). In such cases, crossover of such individuals is not likely to yield improved individuals. Due to a high number of parameters and their interaction in fuzzy logic systems, the interpretation function and design of fitness function can significantly affect performance of a genetic algorithm.

Step III is creation of the initial population, or the starting points of the

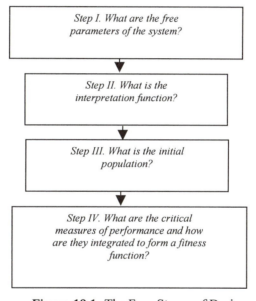

Figure 19.1: The Four Stages of Design.

optimization process. As with any optimization task, GA can be expected to perform better when provided with a fitter initial population. As we will see, however, this issue is not as trivial as it may seem. In fact, there are occasions when a GA performs poorly even with a highly fit initial population.

Step IV is defining the fitness function (objective function). Since the theme of GA is the "survival of the fittest," GA is inherently an optimization algorithm (as compared with a minimization algorithm). So, the improved individuals in a given population are assigned higher fitness values. Every candidate solution for a problem is evaluated to determine the degree of fitness for that solution. Since the type of the fitness function determines the shape of the search space and since there is a great degree of freedom in choosing a fitness function, design of fitness function has a large impact on the performance of the algorithm.

As we will see in this chapter, the inherent flexibility of the evolution based optimization algorithms and the large number of the free parameters in a fuzzy system have created a large diversity and variety in how these two complementary approaches are coupled. Different methods vary in their answers to the above questions. For the rest of this chapter, we will use GA as the optimization algorithm and name the resulting hybrid system as GA-fuzzy system. GA-fuzzy systems are in fact the most common evolution based fuzzy system.

19.2   FREE PARAMETERS

Fuzzy expert knowledge can be divided into two basic components: *Domain knowledge* and *Meta Knowledge*. The Domain knowledge is generally the conscious operating knowledge about a particular system such as the membership functions and the fuzzy rule set. The Meta knowledge is the unconscious knowledge that is also needed to completely define a fuzzy system such as the mechanism of executing the fuzzy rules, methods of implication, rule aggregation, and defuzzification.

Most of the existing methods in evolutionary fuzzy systems attempt to optimize parameters of the domain knowledge only (namely membership functions and rule set) while ignoring the effect of meta knowledge. Consequently, there are 4 basic methods of optimization as follows

1) Automatic optimization of membership functions while there is a fixed and known rule set;
2) Automatic selection of the rule set with fixed membership functions;
3) Optimization of both the membership functions and rule set in two steps. First selecting the optimal rule set with fixed known membership functions and then tuning the membership functions with the resulting rule set; and
4) Simultaneous optimization of fuzzy rule set and membership functions.

Note that the number of membership functions or rules can also be optimized in the algorithm. There may be various reasons for a method to be selected. Some of those advantages and disadvantage are mentioned below:

1) Since the rule set and membership functions are codependent, they should be defined simultaneously. This can lead to more optimal solutions. [4][5];
2) Since the performance of a fuzzy system is more dependent on fuzzy rules rather than membership functions, fine tuning of the fuzzy system is better possible by tuning of membership functions. So it seems that it is better first to select the optimal rule set (coarse tuning) and then tune the membership functions (third method);
3) Even though various methods exist to encode both the rule base and membership functions, such encoding can have several potential difficulties. In addition to the level of complexity and large number of

optimization parameters, the problem of *competing conventions* may arise and the landscape may unnecessarily become multi-modal.

19.2.1   Competing Conventions

Competing conventions means that there are different chromosomes representing the same exact nonlinear function in the evaluation space. In other words, there is more than one string in GA domain (genotype), which corresponds to only one solution in problem domain (phenotype). To illustrate this, consider the following example from Akbarzadeh[3].

Example 19.1:
Consider two fuzzy rules related to temperature control of a room:

- Individual A says: "If temperature is *hot,* turn on the cooler"
- Individual B says: "If temperature is *cold,* turn on the cooler"

Under normal circumstances, these two rules are expected to be contradictory. If evaluated using same membership functions, one will result in proper control compensation and the other will result in an uncomfortably cold room. However, consider if parameters defining the membership functions for the fuzzy sets *cold* and *hot* temperatures are interchanged for individual B, which may happen if both rules and membership functions are optimized simultaneously. Then both of these rules are essentially the same nonlinear function with same control action for same input variables.

As will be illustrated in the example, even though the GA might produce two highly fit individuals (with two competing conventions), the genetic operators, such as crossover, will not yield fitter individuals if both membership functions and rules are to be evaluated under the same string structure. Consider the following two individuals:

- Individual A says: "If temperature is *hot,* turn on the cooler"
- Individual B says: "If temperature is *cold,* turn on the heater"

Both of these individuals are expected to perform well in an evaluation. Now let us perform a crossover operator by interchanging part of the genetic code corresponding to the output as follows:

- Individual A says: "If temperature is *hot,* turn on the heater"
- Individual B says: "If temperature is *cold,* turn on the cooler"

Obviously, these two individuals will not fare well in a performance evaluation. As is illustrated here, the design of the transformation function can significantly alter the behavior of GA.

19.3   DESIGN OF INTERPRETATION (ENCODING) FUNCTION

In this section, we will explore various possible ways that genetic algorithms can optimize membership functions and rules of a fuzzy expert system.

19.3.1   Membership Functions (MF)

Fuzzy partitioning is the process of partitioning a variable's universe of discourse $[u-, u+]$ by defining fuzzy sets. Figure 19.2 shows a fuzzy partitioning with five fuzzy sets.

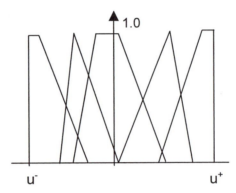

Figure19.2: Fuzzy Partitioning of a Variable's Universe of Discourse by Fuzzy Sets.

One may choose to partition a fuzzy variable's universe of discourse with any desirable number of fuzzy sets. Membership functions can be all or some part of a genetic representation (chromosome) of the system. Their genetic representation is named MFC (membership function chromosome).

Every fuzzy set in a fuzzy partitioning is defined by its type and shape as shown below;

- Type of the membership functions: triangular, trapezoidal, gaussian, ...
- Shape of the membership function: important points and parameters of a membership function such as left base, center, base width, etc.

Thus the encoding problem is divided into two parts:

1. Selection of free parameters:

Selecting free parameters is in fact a compromise between more optimal solutions and less complex spaces. Higher numbers of free parameters may yield a higher fit final solution, but also yield a more complex landscape with higher multimodality and more difficulty in finding the optimal parameters in the landscape. Consequently, the GA designer has to decide which parameters to fix and which parameters to tune. For example, we can assume only triangular membership functions with fixed base width and tune the center of the membership functions. Triangular membership functions are widely used in evolutionary fuzzy systems. So we discuss them separately.

2. *Encoding of Chosen Parameters:*
Several methods exist for encoding MF parameters; among them, binary string encoding is the most common.

19.3.1.1 Triangular Membership Functions

There are different types of coding methods for triangular membership functions as discussed below,

1. In this method a triangular membership function is defined by its three parameters: left base, center, and right base. A binary string MFC is developed as shown in the figure below where each parameter is encoded as a binary string.
2. Symmetric triangular membership functions are assumed here; thus two

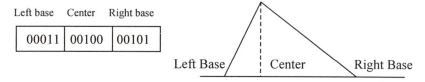

Figure 19.3: Binary Encoded Triangular Membership
Function Chromosome.

parameters are sufficient to define a membership function, left base (starting point) and right base (ending point).

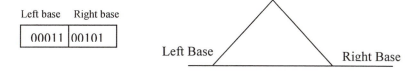

Figure 19.4: Symmetric Triangular Membership Function and Its MFC.

3. In this method, triangular membership functions are symmetric and have fixed centers, only their base widths are tuned. Thus for every membership function, there is only one encoded parameter.

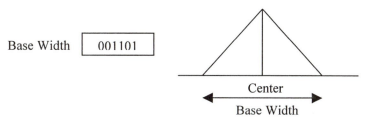

Figure 19.5: Genetic Representation of Symmetric Triangular
Membership Function with Fixed Center.

4. In this method, triangular membership functions with fixed base width are assumed and only their centers are encoded and tuned. Thus there is only one free parameter.

Center
000101
Center

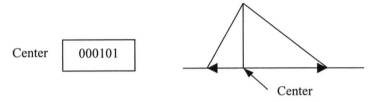

Figure 19.6: Genetic Representation of Triangular Membership Functions with Fixed Base Width.

5. In this method, symmetric triangular membership functions are assumed while their centers and widths are encoded and tuned, yielding two free parameters.

Center
Base Width
001110
000101
Base Width
Center

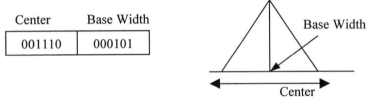

Figure 19.7: Genetic Representation of Symmetric Triangular Membership Functions by Its Center and Base Width.

Here, the MFs are assumed to be normalized, i.e., y-axis is fixed.

19.3.1.2 Non-triangular Membership Functions

To use other types of membership functions, another parameter in an MFC is needed in order to completely define the membership functions. This coding, for example, includes an index referring to the available types of

001	Triangular membership function
010	Trapezoidal membership function
011	Gaussian membership function
100	Sigmoidal membership function
...	...

Starting Point
Ending Point
010
00011
00101
Type
Starting Point
Ending Point

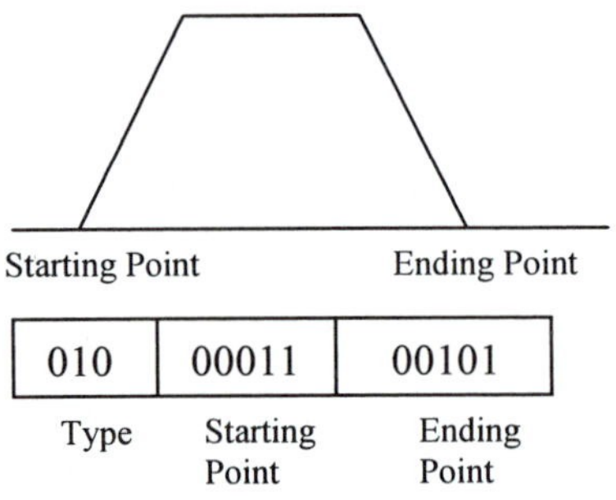

Figure 19.8: Non-triangular Membership Function.

membership functions. To simplify the problem we may use only symmetric membership functions and thus encode every membership function with three parameters, the type of the function, starting point and ending point, with a fixed ratio of points in between starting and ending points.

19.3.1.3 General Method of MF Encoding:

To define and encode any unknown membership function, a method presented in [6] is presented here. In this method all the membership functions in a domain of a variable are encoded together in a matrix form. Every column of this matrix is a gene and is associated with a real value x in the domain X. The gene is a vector having n elements, where n is the number of the membership functions in that partition. Every element is the membership value of all membership functions at x. In practice, a finite number of the points in the partition are considered and thus this method is a discrete representation of the membership functions. Thus, if p points in the domain are considered and there are n membership functions, $n*p$ parameters are encoded. The figure below is a genetic representation of the domain where a and b are the starting and ending points of the domain. Although this method is very general and can be implemented for every membership function in the domain, it has the disadvantage that the number of the encoded parameters can be very large and thus enlarge the search space.

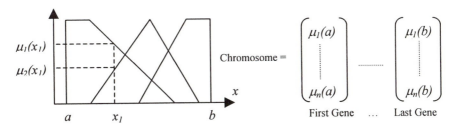

Figure 19.9: Genetic Representation of a Fuzzy Partitioning.

19.3.2 Rule Encoding

Rule set encoding can be more complicated than membership function encoding. An important problem is the simultaneous cooperation and competition of the fuzzy rules in a fuzzy rule set. This means that, although each rule is in competition with others for being selected in the rule set, the impact of each rule in the system is dependent on other rules that concurrently exist in the rule set. For example, two fuzzy rules may be highly fitted if they both exist in the rule set while neither of the two rules may be desirable separately. Rule set optimization is widely used in fuzzy classification problems, which can be generally categorized as either the Michigan approach or the Pittsburgh approach.

The Michigan approach:

In this approach every individual in the GA is a fuzzy rule encoded as a string with fixed length. The GA operates on the individual rules and more fit rules are combined together via genetic operators to create the next population of rules. The fitness function is designed so as to show the fitness of one rule. The method was first introduced by Holland and Retain in 1983. The most important disadvantage of this method is the problem of competing convention.

The Pittsburgh approach:

In this approach, every individual in the GA is a fuzzy rule set encoded as a string with variable length. Fitness function, therefore, operates on the rule sets and higher fit rule sets are combined via genetic operators to produce rule sets with higher fitness. This method is more desirable because of the competing convention problem. This method was first developed in 1980 by Smith and named LS-1. For the rest of this chapter, we will use this approach.

19.3.2.1 A Control System Problem Formulation:

Consider a fuzzy system with *n* inputs and *1* output, referral to as multi input single output (MISO). Let i^{th} input have m_i fuzzy sets as input membership functions and the only output variable to have *p* fuzzy sets as output membership functions. The MISO can be easily generalized to multi input multi output (MIMO).

Figure 19.10: MISO Control System.

Thus we have *R* fuzzy rules where the maximum number of the rules in the system will be:

$$R \le R_{\max} = \prod_{i=1}^{n} m_i \qquad (19.1)$$

R could be either a fixed or free number. Each rule is associated with a selection of the *n*-input membership functions and *1*-output membership functions from the possible MFs. Thus parameters of the fuzzy rules are indices rather than real variables. The indices specify which membership functions are to be selected for the antecedent and consequent parts of the fuzzy rules.

Now let us define the following for the MISO fuzzy system:

x_i : i^{th} input to the system.
m_i: the number of the fuzzy variables (membership functions) for the ith input.
y : output of the system (single output).

$IMF(i,j)=IMF_{ij}$: the membership function of the fuzzy variable of the ith input in the jth rule in a rule set.

OMF_j: the membership function of the output fuzzy variable in the jth rule.

n: the number of the inputs to the system.

q: the number of the fuzzy variables of the output.

R: the number of rules.

And:

$ISET(i)$: ordered set of the membership functions corresponding to fuzzy variables of the ith input.

$$ISET(i)=\{IMF(i,1),IMF(i,2), \dots , IMF(i,m_i)\} , i=1,\dots,n \qquad (19.2)$$

$OSET$: ordered set of the membership functions corresponding to fuzzy variables of the output.

$$OSET=\{OMF (1), OMF (2)\dots OMF (q)\} \qquad (19.3)$$

Thus fuzzy rule set of the system can be shown as below:

$$Rule \ 1, Rule \ 2\dots Rule \ j\dots Rule \ R$$

where the jth rule can be represented as:

$$If (x_1 \ is \ IMF_{1j} \ \& \ x_2 \ is \ IMF_{2j} \ \& \ \dots \ \& \ x_n \ is \ IMF_{nj}) \ Then \ y \ is \ OMF_k$$

where

$$IMF_{ij} \in ISET \ (i), OMF_k \in OSET, and \ 1 \le k \le q , \qquad (19.4)$$

For the case of two-input, one-output system, the fuzzy rule set–also known as fuzzy associative memory (FAM)–can be shown graphically in a table every cell shows the output membership function of a fuzzy rule with known input membership functions.

Example 19.2

Referring to Figure 19.11, the fuzzy if-then rules of the rule set are

$If (x1 \ is \ IMF_{11} \ \& \ x_2 \ is \ IMF_{21})$	*then*	$y \ is \ OMF_2$
$If (x1 \ is \ IMF_{11} \ \& \ x_2 \ is \ IMF_{22})$	*then*	$y \ is \ OMF_{10}$
$If (x1 \ is \ IMF_{12} \ \& \ x_2 \ is \ IMF_{22})$	*then*	$y \ is \ OMF_7$

$\vdots$

	IMF_{11}	IMF_{12}	IMF_{13}	IMF_{14}
IMF_{21}	OMF_2	OMF_1	OMF_{13}	OMF_{14}
IMF_{22}	OMF_{10}	OMF_7	OMF_9	OMF_5
IMF_{23}	OMF_3	OMF_{11}	OMF_8	OMF_4
IMF_{24}	OMF_{16}	OMF_6	OMF_{12}	OMF_{15}

Figure 19.11 A Sample Fuzzy Rule Set Table
for $m_2=m_1=4$.

If $R<R_{max}$, some of the cells in the table are *don't care* and can be shown with a *0* or * in the cell. In the case of more input variables, the table can be extended to higher dimensional arrays.

Regarding the above table, every fuzzy rule set is defined with R_{max} free parameters, where R_{max} is defined previously. These parameters are indices that represent an output membership function among q membership functions in *OSET*. We can use only indices of the membership functions and build a matrix of indices named P.

p_{15}	p_{14}	p_{13}	p_{12}	p_{11}
p_{25}	p_{24}	p_{23}	p_{22}	p_{21}
p_{35}	p_{34}	p_{33}	p_{32}	p_{31}
p_{45}	p_{44}	p_{43}	p_{42}	p_{41}
p_{55}	p_{54}	p_{53}	p_{52}	p_{51}

Figure 19.12: Two-dimensional Array of Indices, P.

$$P=[p_{kl}]=[p(k,l)], \ p_{kl}\in Z, \ 0\le p_{kl}\le q \qquad (19.5)$$

$$k\in[1,m_2],l\in[1,m_1]$$

Similarly, in the case of three inputs, we have a three dimensional array P such that:

$$P=Array[p_{skl}]=p(s,k,l)$$
$$P_{skl}\in Z \ \ ,0\le p_{skl}\le q \qquad (19.6)$$

$$s\in[1,m_3] \ , \ k\in[1,m_2] \ , \ l\in[1,m_1]$$

Genetic Representation

Having the indices array of the rule set, P, it is possible to use either string or array representation of the rule set. Naturally, in the case of nonstring representation, genetic operators should be modified so as to be useful for that representation. One of the simple array representations of the rule set is the matrix representation in two input systems. We will discuss the matrix representation, which has been introduced in Kinzel et al.[6].

a) String representation:

String representation of the rule set table can be obtained in two steps:

Step 1: encoding all the elements of matrix P (defined previously). Binary encoding is a common method to encode the parameters. S is the resulting matrix after encoding.

$$S_{kl}=Decimal\_to\_Binary(p_{kl}) \qquad (19.7)$$

$$S=[s_{kl}] \quad , \quad k \in [1,m_2], \quad l \in [1,m_1] \tag{19.8}$$

Step 2: Obtaining the string representation of the table using the rows of the **S** matrix as follows:

$$Chromosome = s_{11}s_{12}s_{13}...s_{1m1}\,s_{21}\,s_{22}\,s_{23}...s_{2m1}... \quad ... \quad ... \quad s_{m21}s_{m22}\,...s_{m2m1} \tag{19.9}$$

Number of the bits in every chromosome will be

$$N = (\prod_{i=1}^{n} m_i) * K \tag{19.10}$$

where K is the number of the bits in every element of S (s_{kl}), and is the greatest integer in the following inequality: (19.11)

$$K \geq \log_2(q+1)$$

where q is the number of the output fuzzy sets.

The genetic operators in this type of representation can be the same as the standard genetic algorithms.

b) Matrix representation:

Because of the matrix nature of the rule set, a matrix representation seems to be more efficient. A matrix chromosome and a set of genetic operators are needed to operate on this chromosome. We will discuss the method presented in reference [6] for two input systems. Kinzel et al. used the previously defined array of indices as the chromosome (Figure 19.12).

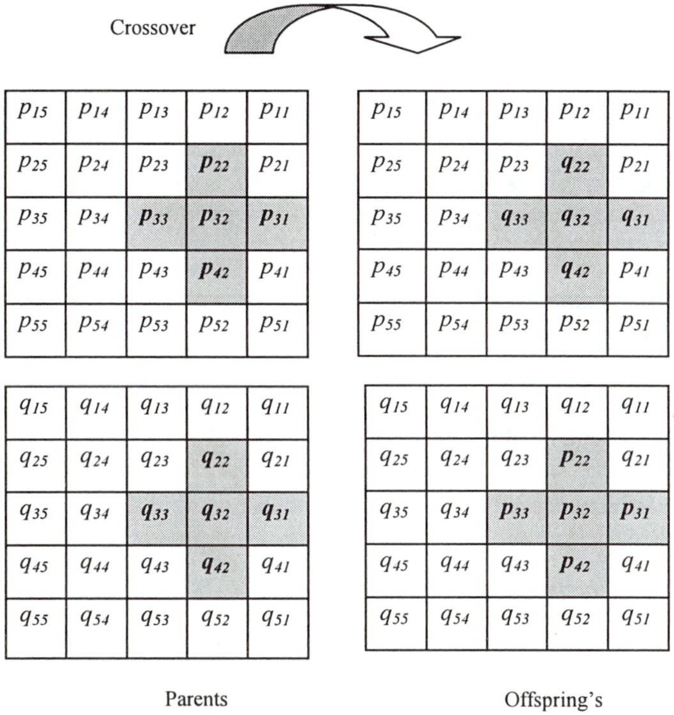

Figure 19.13: Point-Radius Crossover.

In this method, a Point-Radius crossover is used where each crossover is determined with a circle with known center and radius. The region in the table that is surrounded by the circle is exchanged with the similar region in the other chromosome (Figure 19.13).

Example 19.3:
Consider a two input system where the set of output membership functions is *OSET={A, B, C, D, E}*. Notice that it is not necessary to use integers for the indices. Here, every output fuzzy set is determined with an alphabet between *A* and *E*. Figure 19.13 shows two sample chromosomes and the crossover operation between them.

Mutation in this method is simply the alternation of one index in the table with a different index in *OSET*. Figure 19.14 is an example of the mutation operator in the matrix representation.

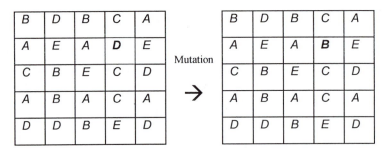

Figure 19.14: Mutation.

Note: Genetic programming (GP) also seems to be a good alternative to GA for optimization of fuzzy rules. This is because optimization of fuzzy rules has a symbolic nature as compared with optimization of numeric information (such as in membership functions). Here, we want to find the optimum arrangement of a few objects in a table just like making a puzzle.

Rule Firing Strength
Although it is common to consider only the antecedent and consequent parts of a fuzzy rule as the free parameters of optimization, it is possible to introduce other parameters as well. Lee and Esbensen [7] introduced degree of hedging as another free parameter. Let $\mu(x_1, x_2, \ldots, x_n)$ be the rule firing strength for a fuzzy rule. A free parameter named p is used to increase or decrease the rule firing strength and is named degree of hedging. The modified firing strength is obtained as below:

$$\text{Modified Firing Strength} = \mu^p(x_1, x_2, \ldots, x_n).\qquad(19.12)$$

19.4   THE INITIAL POPULATION

In many applications of genetic algorithms, initial population is chosen purely randomly. Many times this is an obligation because there is no initial knowledge about the system. However, since human knowledge is often available and also applicable in fuzzy system, it may be reasonable to include such knowledge in an initial population in order to decrease the time needed to reach the optimal solution, even though such knowledge may not be optimal. But nonrandom initial population is not *always* better than random initial population even though such a population may initially exhibit a higher average fitness. The following example by Lee and Takagi[8] shows how and whether a random initial population can be more desirable than nonrandom initial populations.

Example 19.4

Lee and Takagi used a genetic fuzzy controller for an inverted pendulum. Inverted pendulum is a system consisting of an inverted beam (pole) on a moving cart. The task of the controller is to stabilize the pole angle θ and the cart position x by applying the force F to the cart. The inverted pendulum is a nonlinear system with an unstable equilibrium point and thus is a common platform for testing in control systems technology.

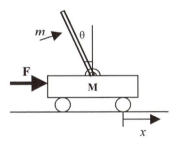

Figure 19.15 Inverted Pendulum System.

Simulation results in four different cases are sketched in Figure 19.16

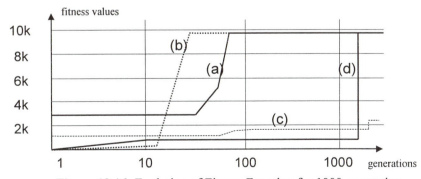

Figure 19.16: Evolution of Fitness Function for 1000 generation.

(a) Symmetrical rules (even number of rules) – human knowledge initial population.
(b) Symmetrical rules (even number of rules) – random initial population.
(c) Non-symmetrical rules - human knowledge initial population.
(d) Non-symmetrical rules - random initial population.

It can be seen that GA with random initial population in case (b) reaches its optimum value in less time than case (a) where initial population is selected based on human knowledge. Also, symmetrical rules are more preferable.

In the above example, the reason for the faster convergence of the random initial population is because of the lack of diversity in initial population. Due to the stochastic nature of the GA, it is very important to have sufficient diversity in the initial population such that GA can exploit the landscape properly and efficiently. Therefore, incorporating *a priori* expert knowledge in optimization process needs be done with consideration for diversity. Akbarzadeh and Jamshidi in 1998 proposed a method for maintaining diversity in initial population while utilizing a priori expert knowledge. The following section from Akbarzadeh [3] discusses the grandparenting method.

19.4.1   Grandparenting: A Method of Incorporating a *priori* Expert Knowledge

The method presented here is based on a grandparenting scheme where the grandparent is the genotype representation of one expert's control strategy in the form of a fuzzy controller.

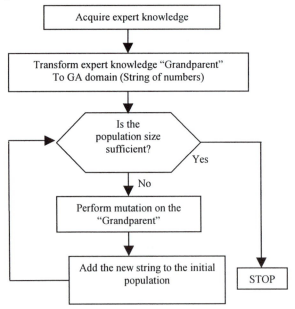

Figure 19.17: The Grandparenting Technique in Creating Initial Population.

All members of the initial population are binary mutations of the grandparent. Figure 19.17 illustrates the process of creating the initial population. The mutation rate is a significant factor since, as the mutation rate increases, diversity among members of the initial population increases as well.

One important concern about this approach is the diversity issue. If all members of the initial population are to be derived from one individual (the grandparent), will there be enough diversity among them such that the GA may exploit the landscape properly? In fact, this is one of the reasons why one might ignore the initial knowledge altogether and start with a totally random population. Let us look at two extreme situations.

• One extreme: The mutation rate is set to zero. Hence all members of population are exactly alike (no diversity). The initial population will consequently have a high average fitness and very low diversity. Individuals in the initial population cannot recombine to create more fit individuals unless the mutation rate in the algorithm is set to a high value. This is because all members are alike and therefore crossover is not a productive operator. This sort of initial population often leads to premature convergence.

• Second extreme: The mutation rate is set to one. In this case, a totally random initial population is created with a low average fitness and a high diversity. This type of initial population will in general result in a faster learning rate, but since the initial fitness is low, it will take a long time to converge.

From above, it is clear that in most cases, mutation rate should be set to a value between zero and one. The lower mutation rate indicates a higher degree of confidence that the optimal string is in close proximity/similarity to the grandparent's string. In other words, if the expert knowledge is already performing well and we only require fine tuning, exploiting the whole parameter vector space may not be necessary. In exchange, by setting mutation rate to a low value, we ensure faster convergence, which helps the implementation of GA in a real time system environment.

In contrast, the higher mutation rate indicates a low degree of confidence in the expert and the need for exploring the rest of the representation space more fully and with a higher diversity. Depending on the complexity of the problem, this means a poorly fit initial population and longer convergence time. However, it may be the only feasible alternative if expert knowledge is not available. In short, the grandparenting technique adds a control variable, the mutation rate, as a new parameter by which the GA designer can weight diversity vs. convergence and average fitness of the initial population.

The following example from Akbarzadeh [3] illustrates the mechanism of the grandparenting method and its benefits in a higher fit initial condition and faster convergence.

Example 19.5

Determine the parameters, b_i, such that the following fitness function, $f(B)$, is maximized:

$$f(B) = B.B^T = \sum_{i=1}^{8} b_i \tag{19.13}$$

where, $B = [b_1, b_2, ..., b_i, ..., b_8]$ is a an 8-bit binary row vector. Also given is an expert opinion about the possible values for the optimal solution B^*, $B_{expert}=[1,1,1,0,1,1,1,1]$.

Solution: Intuitively, the solution may be clear to the reader: $B^*=[1,1,1,1,1,1,1,1]$. However, it is interesting to see how GA finds the optimal solution automatically, and furthermore how the grandparenting method enhances the GA performance in contrast with the method of random initial population. Using the standard random initial population and from the law of averages, it can be concluded that the average fitness of a random initial population is $f_{initial}=4$ for the above problem. Now let us compare the above with the grandparenting method. The grandparenting method requires an expert opinion. In our situation, the expert (or the grandparent) offers the following as a possible solution: $B_{expert} = [1,1,1,0,1,1,1,1]$. The fitness of this grandparent is seven, which is, as expected, higher than average fitness of random initial population. Figure 19.18 illustrates the evolution of an initial population generated through the proposed grandparenting technique.

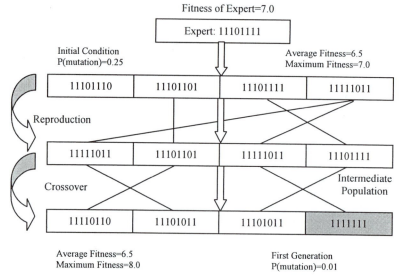

Figure 19.18: An Example of the Grandparenting Method.

The initial population developed through the "grandparenting" technique has an average fitness of $f_{initial} = 6.5$ which is significantly higher than average fitness of a random initial population. In this example, the $P_{mutation}=0.25$ is used for creating the initial population. Furthermore, the optimal solution B^* is found in only one generation as compared to several generations required if starting with a random initial population. Note that the intermediate population is not counted since it is an "intermediate" step in creating the new population and its fitness is not evaluated.

The above example illustrates the method by which grandparenting technique utilizes *a priori* expert knowledge to improve the fitness of individuals within the initial population while keeping diversity in the population, hence improving the performance of GA. Moreover, in many control systems, there is usually access to more than one expert. Often, views and opinions of various experts are different. This difference in opinion, although complicating the process of knowledge acquisition, is not a weakness; is, in fact, the strength of biological systems. The evolution in nature is not limited to manipulating genes and chromosomes. In fact, the diversity in human minds is indeed the strength of man where each human may differ in perspective and opinion if faced with the same constraints and criteria. It is therefore no surprise that, if several experts are interviewed in regard to a control system, each would give us differing heuristics. The grandparenting idea provides ability to combine multiple experts' opinions in the above process by repeating the process for each expert. Hence, the resulting population will consist of variations of multiple expert systems competing with each other for the right to survival.

19.5   FITNESS FUNCTION

In any genetic algorithm, fitness function plays an important rule because GA depends on fitness function to guide the direction of its search. There is no general way to define a fitness function for a problem; however, it is often designed such that the more desirable solutions correspond to higher fitness. So, the GA optimization is usually regarded as a search for the parameters which maximize the fitness function. Obviously, a fitness function needs to include all the pertinent parameters which need to be optimized (free parameters). One of the objectives of the optimization problem is to find the systems with higher performance, so it is common to include some of the performance measures of the system into the fitness function. Other requirements can also be considered. For example in Example 19.6, the number of rules is included in the fitness function as a measure of complexity to reduce the number of rules in the system.

Example 19.6:

In a simple control system, the shape of the step response can be a good measure of the controller operation. A fast and accurate response with lower overshoot is often desired. So a fitness function can be of the form below:

$$fitness = \int_{t_i}^{t_f} \frac{1}{e^2 + \gamma^2 + 1} dt \qquad (19.14)$$

where e represents the error between desired and system response and γ represents overshoot. t_i and t_f are the starting and ending time of simulation.
A more general form for a normalized fitness function regarding time domain response has been proposed in Akbarzadeh [3]:

$$f(p_1, p_2, ..., p_n) = \frac{\gamma}{t_f - t_i} \int_{t_i}^{t_f} \frac{1}{k_1 p_1^2 + k_2 p_2^2 + ... + k_n p_n^2 + \gamma} \, dt \tag{19.15}$$

where p_i, $i=1,...,n$ are time varying system parameters, k_i are positive constant multipliers as weighting functions to emphasize or de-emphasize a parameter significance, and γ is a positive constant which sets the slope of the fitness function near its optimal solution. A common problem with many fitness functions is that once the algorithm nears the neighborhood of the optimal solution, it cannot reach the optimal point because it cannot accurately differentiate between near-optimal and optimal solutions. The nearer the slope of the fitness function is to its optimal solution, the more it can distinguish between optimal and near-optimal solutions. The proposed fitness function has the advantage that this slope can be controlled by γ. Smaller values of γ correspond to greater slopes.

Example 19.7:

Lee and Takagi [8] introduced a fitness function for the inverted pendulum control problem (Example 19.4). The objective of controlling the inverted pendulum is to balance the system in the shortest amount of time for a specific range of initial conditions. The fitness function for a trial is evaluated in two steps:

Step1: first define a score function of ending time (t_{end}) for the trial, based on the termination conditions. It is possible to consider three different termination conditions. Figure 19.19 illustrates how the scoring function evaluates the score for a trial.

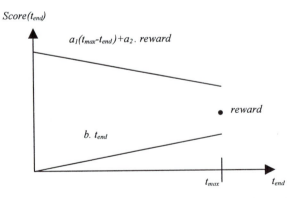

Figure 19.19: Scoring Function for Inverted Pendulum.

Condition 1: the system balanced the pole before time is expired.
$$(|\theta| \leq \varepsilon^{\circ} \quad t_{end} < t < t_{max})$$
(19.16)

Condition 2: time is expired and the system could not balance the pole.
$$(\varepsilon^{\circ} \leq |\theta| \leq 90^{\circ} \quad t < t_{end} = t_{max})$$
(19.17)

Condition 3: the pole fell over before ending time. (19.18)
$$(|\theta| \geq 90^{\circ} \quad t_{end} < t < t_{max})$$

The score function has been defined as:

$$
\begin{array}{ll}
score(t_{end}) = a_1(t_{max} - t_{end}) + a_2 . \ reward & if \ \ condition \ 1 \\
score(t_{end}) = reward & if \ \ condition \ 2 \\
score(t_{end}) = b . \ t_{end} & if \ \ condition \ 3
\end{array}
$$
(19.19)

where θ is the angle of the pole, ε is a real number defined so that for $|\theta| \leq \varepsilon^{\circ}$ the system is considered as stable, and a_1, a_2, b and *reward* are constants. (Figure 19.19)

The general idea is that if the system balances the pole (*condition 1*), a shorter time for t_{end} is ranked higher, but if the pole falls over, a longer time for t_{end} means that it is kept from falling for a longer period of time and thus is ranked higher.

Step 2: The fitness function is not only a function of the *score*. To consider steady state error and to penalize systems according to the number of the rules in the system, fitness function is defined as below:

$$fitness = \frac{(score(t_{end}) + c \sum_{0}^{t_{end}} |\theta_t|)}{number \ of \ rules + offset_{rules}}$$
(19.20)

The steady state error is a summation of the pole angle displacement weighted with constant c. Offset$_{rules}$ is a parameter which controls the degree of penalty for the number of rules.

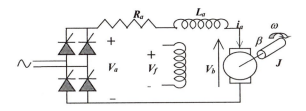

Figure 19.20: The Drive System of the Separately Excited DC Motor.

19.6   SPEED REGULATION OF A DC MOTOR

In this section, we will examine one of the recent successful applications of GA-fuzzy systems in detail. The following example is adapted from Akbarzadeh et al. [16].

In this example, a GA fuzzy system is simulated for velocity control of a DC motor. Due to its excellent speed control characteristic, the DC motor has been widely used in industry even though its maintenance costs are higher than the induction motor. As a result, speed control of the DC motor has attracted considerable research and several methods have evolved. Proportional-integral (PI) and Proportional-integral-derivative (PID) controllers have been widely used for speed control of the DC motor. Kim and Baek [17] surveyed the current state of the PI, PID and command matching controllers for speed regulation of DC motors. To reduce the loading effect and minimize time delay, they added a feed-forward controller to the PID controller. In Iracleos and Alexandridis[18], the feedback gains of a PI controller are first nominally determined and thereafter tuned using fuzzy logic. Yousef [19] determined a fuzzy logic based controller with superior performance over a DC motor PI controller. Yousef controlled both speed and current variables. In Fisher et al. [20] three different intelligent control architectures are considered. There, a feedforward/feedback control strategy is used to ensure effective, high performance tracking of reference speed trajectories.

The above works indicate successful utilization of fuzzy logic over nonfuzzy PI and PID controllers in regulating DC motor drive systems. Yet, for *best* response, the above approaches have no capability to search for an *optimal* knowledge base. Here, an automatic way of searching for optimal knowledge is proposed and applied to the speed regulation problem.

This section is organized as follows. First, the system model of a DC motor is formulated. Next, the simulation results of the corresponding system are compared with nonfuzzy PID and nonoptimized fuzzy PID controllers.

Let us now consider a separately excited DC motor as is shown in Figure 19.20. Where $\omega(t)$ is rotational speed, $i_a(t)$ armature circuit current, $T_1(t)$ constant torque-type load, $R_a(t)$ armature circuit resistance, β coefficient of viscous-friction, k torque coefficient, J moment of inertia, and L_a armature circuit inductance. In state space form, if we let

$$x_1(t) = i_a(t), x_2(t) = \omega(t), u(t) = V_a(t), d(t) = T_1(t) \qquad (19.21)$$

be our choice of state and control variables, then the state space model of the system can be represented by the following:

$$X(t) = [x_1(t), x_2(t)]^T \qquad (19.22)$$

$$\dot{X}(t) = AX(t) + Bu(t) + Ed(t),$$
$$y(t) = CX(t) \qquad (19.23)$$

and:

$$A = \begin{bmatrix} \dfrac{-R_a}{L_a} & \dfrac{-k}{L_a} \\ \dfrac{k}{J} & \dfrac{-\beta}{J} \end{bmatrix}, B = \begin{bmatrix} \dfrac{1}{L_a} \\ 0 \end{bmatrix}, C = \begin{bmatrix} 0 & 1 \end{bmatrix}, E = \begin{bmatrix} 0 \\ \dfrac{-1}{J} \end{bmatrix} \qquad (19.24)$$

where the load torque is considered as disturbance input.

Numerical Values

The DC motor under study has the following specifications and parameters:
a) Specifications: *1 hp, 220 Volts, 4.8 Amperes, 1500 rpm*
b) Parameters:

$$R_a = 2.25\Omega, L_a = 46.5mH, J = 0.07kg.m^2, \beta = 0.002N.m..\frac{\sec}{rad}, k = 1.1\frac{V}{\sec.rad}$$

19.6.1 The Control Architecture

The control architecture used here is the standard application of genetic algorithms on fuzzy controllers with proportional, integral, and derivative variable inputs, with the difference that the idea of grandparenting is used to shape the initial population. Through the GA, various combinations of candidate solutions are evaluated and the best, fittest, solution is chosen to control the actual system. The GA has the capability to alter the shape of the membership functions of individual inputs. Figure 19.21 illustrates a block diagram of the closed loop control system.

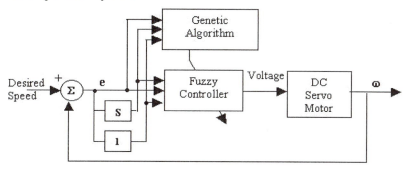

Figure 19.21: GA Optimized Fuzzy PID Control Architecture for a DC Motor.

19.6.2 Results

Here, three different controllers are simulated and compared. The first simulation involves a model based PID controller as discussed by Kim and Baek [17]. The corresponding control law is as follows:

$$u(t) = -\frac{P}{\omega(t)} + Q\int_{t_0}^{t_f}(\omega_r - \omega)dt - R\frac{d\omega(t)}{dt} \tag{19.25}$$

where $\omega_r = 10.0\frac{rad}{\sec}$ is the reference input, $P = 1.1712$, $Q = 13.236$, $R = 0.03$. The second simulation is a fuzzy PID control law $u(t)$ based on a *nonoptimized a priori* expert knowledge.

$$u(t) = f(e, \dot{e}, \int e) \tag{19.26}$$

where f is a nonlinear function determined by the fuzzy associative memory and parameters of input and output membership functions, $e = \omega_r - \omega$, $\dot{e} = \dot{\omega}_r - \dot{\omega}$, and:

$$\int e = \int_0^t (\omega_r - \omega)dt \tag{19.27}$$

In the third simulation, GA is used to optimize parameters of the above fuzzy controller. In order to minimize the parameter set, GA is applied to optimize *only* the input membership parameters of the fuzzy controller as is shown in Figure 19.21. Other parameters in the knowledge base are not allowed to vary. This will reduce simulation processing time and will still demonstrate the potential utility of GA. The following fitness function was used to evaluate various individuals within a population of potential solutions:

$$fitness = \frac{1}{t_f - t_i} \int_{t_i}^{t_f} \frac{1}{\left(k_1 e^2 + k_2 \dot{e}^2 + k_3 \gamma^2 + 1\right)} dt \tag{19.28}$$

where e and $\dot{e}$ represent the errors in angular position and velocities, γ represents overshoot, and k_1, k_2, and K_3 are design parameters. Consequently, a fitter individual is an individual with a lower overshoot and a lower overall error (shorter rise time) in its time response. The above fitness function is normalized such that a fitness of 1 represents a perfectly fit individual with zero error and overshoot. Similarly, a divergent response receives a fitness of zero. In this simulation, the following values were used

$$k_1 = 25 \; , \; k_2 = 150 \; , \; k_3 = 1 \; \cdot$$

Figure 19.22 shows the maximum and average fitness of each GA generation. A total of 40 generations were simulated; each generation included 100 individuals. The performance measure never reaches steady state since it is constantly trying out new directions of search through mutation. As is shown in Figure 19.22, the curve for the maximum fitness converges very quickly, i.e., within the first two generations. However, the fitness of the whole population converges within 20 generations. The mutation rate for creating the initial population was set to 0.1. Thereafter, the mutation rate was set to 0.033. The probability of crossover was set to 0.6.

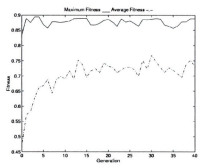

Figure 19.22: Plot of Maximum and Average Fitness Values of GA for a DC Motor.

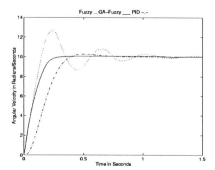

Figure 19.23: Comparison of PID, Fuzzy PID, and GA-Optimized Fuzzy PID Controllers for a DC Motor.

Figure 19.23 illustrates the three controllers' time responses. The GA optimized fuzzy controller is a significant improvement over the initial fuzzy controller based on crude expert knowledge. While keeping the same rise time, the GA optimized controller has no oscillations and almost no overshoot. When compared with the model based PID controller, the GA-optimized fuzzy controller also shows significant improvement. In this respect, the overshoot and rise time are reduced by over 50%.

19.7   CURRENT PROBLEMS AND CHALLENGES

Real time implementations
The capabilities of evolutionary algorithms will be highly increased if they can be implemented online. Even with the aid of new computers with high computational speeds, real time GA implementation is still a challenging problem. Such real time evolutionary algorithms can have many applications such as in design of more intelligent robots with a higher degree of autonomy.

For example, robots used for hazardous material handling or space robots commonly operate in unstructured and noisy environments. Such robots often need to be locally controlled with an intelligent controller [21] and with little or no contact with a human supervisor. Online evolutionary methods can enable these robots to learn from past experiences and evolve their performance and adapt themselves to the environment while in operation. Therefore, one of the important research activities is to develop and implement genetic algorithms with higher levels of computational efficiency.

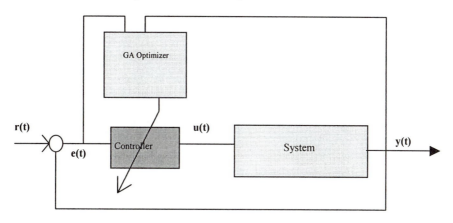

Figure 19.24: Online Genetic Optimizer for Control System.

Side effects of evolutionary fuzzy systems
G.V. Tan and X. Hu [10] introduced some of the undesired side effects that can surface when using evolutionary fuzzy systems. The main problem is that evolutionary optimization algorithms may drastically alter the control system

architecture (acting autonomously) to the extent that the resulting system would
no longer be identifiable by a human operator. This is particularly true for fuzzy
systems where an intuitive human understanding of the control system is a
significant aspect of a control system. Indeed, the meaningful relation between
membership functions and associated fuzzy rules may be lost from a human
point of view. Figure 19.25 shows a fuzzy partition after optimization that is
meaningless for humans.

To overcome this problem, some constraints should be included in the
optimization algorithm such as limiting the system to a symmetric rule set and
uniformly distributed membership functions for a more meaningful fuzzy
system.

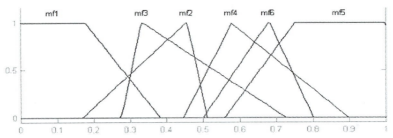

Figure 19.25: An Example of Optimized Fuzzy Partitioning.

19.8   SUMMARY AND RESULTS

As we enter the era of more complex systems, the need for more intelligent and
ultimately autonomous controllers arises. This need is currently being addressed
by applying fuzzy logic to bridge the gap between machine number processing
capability and human thinking. Even though the power of original thinking and
innovation is what we look for in intelligent and autonomous systems, fuzzy
logic, in its conventional form, does not provide that power. That is why we
equip fuzzy logic with nature based evolutionary algorithms in search of
machine self innovation. Through genetic operations such as mutation and
crossover, GA is able to invent and recombine new search paths.

In this chapter, several design aspects of such fuzzy and GA improved fuzzy
controllers were discussed. Following summary remarks can be made:

- Genetic algorithms fill the existing gap between fuzzy logic and its
  application to complex systems by removing the need for optimal human
  expert knowledge, and instead allowing machine self innovation.
- GA is a relatively robust alternative for automatic design and optimization
  of fuzzy systems. Successful applications of such hybrid GA fuzzy
  architecture are abundant in the literature.

- There is not a unique approach to implementation of GA fuzzy systems. Various approaches often differ in their selection of free parameters, genetic representation of the parameters, and fitness functions.
- Selection of free parameters for optimization is a compromise between more complex and larger search spaces and less optimal solutions.
- The main parameters of a fuzzy system that are usually selected for optimization are membership function parameters and fuzzy if-then rule parameters. They can be optimized simultaneously or in steps.
- After selection of the free parameters and the way they are optimized, the most important parts of the optimization are
  - *Genetic representation of the parameters*
  - *Definition of the genetic operators*
  - *Definition of the fitness function for evaluation of the solutions.*
  - *Selection of the initial population.*
- The time needed for optimization and the computational time in every generation is an important property of an optimization algorithm because real time implementation of the algorithm would be possible if the computation delay were small enough.

ACKNOWLEDGMENT

The authors gratefully acknowledge the support of the Center for Autonomous Control Engineering, The University of New Mexico, in writing this chapter.

REFERENCES

1. Fogel, L. J., *Intelligence Through Simulated Evolution*, John Wiley & Sons, NY, 1999.
2. Wang, L. X. and Mandel, J. M., Fuzzy Basis Functions, Universal Approximation, and Orthogonal Least- Square Learning, *IEEE Trans. on Neural Networks*, Vol. 3, No 5, September, 1992.
3. Akbarzadeh-T, M. R., Fuzzy Control and Evolutionary Optimization of Complex Systems, Ph.D. thesis, The University of New Mexico, 1998.
4. Homaifar A. and McCormick, E., Simultaneous Design of Membership Functions and Rule Sets for Fuzzy Controllers Using Genetic Algorithms, *IEEE Trans. on Fuzzy Systems*, Vol. 3, No 2, 129–139, May 1995.
5. Shi, Y. et al., Implementation of Evolutionary Fuzzy Systems, *IEEE Trans. on Fuzzy Systems*, Vol. 7, No 2, 109–119, April 1999.
6. Kinzel, J. et al., Modification of Genetic Algorithms for Design and Optimizing Fuzzy Controllers, *IEEE Int. Conf. on Fuzzy Systems*, 28–33, 1994.
7. Lee, M. and Esbensen, H., Evolutionary Algorithms Based Multi objective Optimization Techniques for Intelligent Systems Design, FUZZ-IEEE'96, 360–364, 1996.

8.   Lee, M.A. and Takagi, H. , Integrating Design Stages of Fuzzy Systems using Genetic Algorithms, in *Proc. 2nd IEEE Int. Conf. Fuzzy Systems*, 612-617, San Francisco, 1993.

9.   Karr, C.L. and Gentry, E. J. , Fuzzy Control of pH Using Genetic Algorithms, *IEEE Trans. on Fuzzy Systems*, Vol. 1, No 1, 46-53, February 1993.

10.   Tan, G. V. and Hu, X., On Designing Fuzzy Controllers using Genetic Algorithms, *IEEE Int. Conf. on Fuzzy Systems*, 905–911, 1996.

11.   Cooper, M. G. and Vidal, J. , Genetic Design of Fuzzy Controllers: The Cart and Jointed -Pole Problem, *IEEE Int. Conf. on Fuzzy Systems*, 1332–1337, 1994.

12.   Xu, H. Y. and Vukovich, G. , Fuzzy Evolutionary Algorithms and Automatic Robot Trajectory Generation, FUZZ-IEEE'94, 595-600, 1994.

13.   Moon, S.Y. and Kwon, W.H., Genetic-based Fuzzy Control for Automotive Active Suspensions, FUZZ-IEEE'96, 923–929, 1996.

14.   Wang, C.H. et al., Integrating Fuzzy Knowledge by Genetic Algorithms, *IEEE Trans. on Evolutionary Computations*, Vol. 2, No 4, 138–149, Nov 1998.

15.   Park, Y.J., Lee, S.Y., and Cho, H. S., A Genetic Algorithm-Based fuzzy Control of an Electro-Hydraulic Fin Position Servo System, *1999 IEEE Int. Fuzzy Systems Conf. Proc.*, Aug. 22–25, Seoul, Korea, 1999.

16.   Akbarzadeh-T, M. R. et al. , Evolutionary Fuzzy Speed Regulation for a DC Motor, *The 29th Southeastern Symp. on Syst. Theory*, Cookeville, TN, March 1997.

17.   Kim, Y.T. and Baek, S.H., The Speed Regulation of a DC Motor Drive System with a PI, PID, and Command Matching Controllers, *Dongguk Journal*, (in Korean), Vol. 29, 1990, 525–541.

18.   Iracleous, D. P. and Alexandridis, A. T. Fuzzy Tuned PI controllers for Series Connected DC Motor Drives*, Proc. of the IEEE Int. Symp. on Industrial Electronics*, Athens, Greece, 495–499, 1995.

19.   Yousef , H. and Khalil, H. M. , Fuzzy Logic-based Control of Series DC Motor Drives, *Proc. Of the IEEE Int. Symp. on Industrial Electronics*, Athens, Greece, 1995.

20.   Fisher, M.E., Ghosh, A. and Sharaf, A.M., Intelligent Control Strategies for Permanent Magnet DC Motor Drives, *Proc. of the IEEE Int. Conf. on Power Electron.*, New Delhi, India, 1996.

21.   Akbarzadeh-T, M. R., et al., Genetic Algorithms and Genetic Programming: Combining Strengths in One Evolutionary Strategy, *Proc. of the 1998 Joint Conf. on the Environment*, Albuquerque, NM, March-April 1998.

20 GENETIC AND EVOLUTIONARY METHODS FOR MOBILE ROBOT MOTION CONTROL AND PATH PLANNING

Abdollah Homaifar, Edward Tunstel, Gerry Dozier, and Darryl Battle

20.1   INTRODUCTION

A variety of evolutionary algorithms, operating according to Darwinian concepts, have been proposed to solve problems of common engineering applications.   Applications often involve automatic learning of nonlinear mappings that govern the behavior of control systems, as well as parallel search strategies for solving multiobjective optimization problems.   In many cases, hybrid applications of soft computing methods have proven to be effective in designing intelligent control systems. This chapter presents two instances of such hybrid applications to problems of mobile robot control. In particular, evolutionary computation and fuzzy logic are combined to solve robot motion control and path planning problems.   The first part of the chapter describes a methodology for applying genetic programming (GP) to design a fuzzy logic steering controller for mobile robot path tracking. Genetic programming is employed to learn the rules and membership functions of the fuzzy logic controller, and also to handle selection of fuzzy set intersection operators (t-norms).   The second part of the chapter describes an application of fuzzy logic to enhance the performance of an evolutionary robot path planning system.   In this case, fuzzy logic is employed in the selection phase of the simulated evolution process.

20.2 GENETIC PROGRAMMING FOR PATH TRACKING CONTROL

In applications of genetic and evolutionary methods, the data structures of individuals being evolved are different depending upon the specific type of evolutionary algorithm employed. Genetic programming is a method of program induction introduced by Koza [1]. It has been demonstrated to be useful as an approach to learning fuzzy logic rules for mobile robot control and navigation [2, 3]. It has also proven useful for the classical cart centering control problem [4]. The data structures undergoing adaptation in GP are noted as hierarchically formed programs of a given host programming language. In the host language, individuals are represented as parse trees, which dynamically change size and structure during simulated evolution.

The set of possible structures produced by GP is primarily based on the set of all possible valid compositions that can be constructed from the set of n problem dependent functions defined in a function set, $F = \{f_1, f_2, ..., f_n\}$, and the set of m

terminals (function arguments, variables, and/or constants) defined in a terminal set, $T = \{t_1, t_2, ..., t_m\}$.

In order to appreciate the utilization of GP for the design of fuzzy logic controllers, we introduce the steering control problem next and follow it by discussing some of the important implementation issues to be considered for the application of GP.

20.2.1 Path Tracking Formulation

The first of two control problems examined in this chapter is a path tracking problem, which was formulated by Hemami et al. [5, 6] for a class of low speed (less than 2 m/s) tricycle-model vehicles. Essentially, the control objective is to successfully navigate a mobile robot or automated guided vehicle along a desired path in a two-dimensional environment. We wish to automatically design a multiple input, single output fuzzy controller that will achieve this objective. The inputs consist of a measurable position error, ε_d, and a measurable orientation error, ε_θ, associated with path following in the plane (see Figure 20.1). The output is the steering angle, δ, which is the corrective control action that would cause the errors to approach zero and, thus, force the robot to follow the desired path. The position error is taken as the deviation of the center of gravity, C, or any other desired point of the robot from the nearest point on the path. The orientation error is the angular deviation of the robot from the tangent of the desired path.

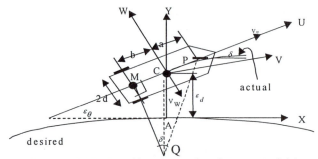

Figure 20.1: Tracking Control and Error Variables.

Hemami et al. derived a state-space kinematic model for this robot where the state vector comprised the pose errors described. The resulting kinematic model is repeated herein for clarity in the discussion that follows. The reader is referred to reference [5] or [6] for details of the derivation, which culminates in the following:

$$\begin{bmatrix} \dot{\varepsilon}_d \\ \dot{\varepsilon}_\theta \end{bmatrix} = \begin{bmatrix} 0 & V_u \\ 0 & 0 \end{bmatrix} \begin{bmatrix} \varepsilon_d \\ \varepsilon_\theta \end{bmatrix} + \begin{bmatrix} MC/MP \\ 1/MP \end{bmatrix} V_u \tan \delta \pm \begin{bmatrix} \dot{\eta}_d \\ \dot{\eta}_\theta \end{bmatrix} \tag{20.1}$$

where V_u is forward linear velocity of the robot, and $\dot{\eta}_d$ and $\dot{\eta}_\theta$ are rates of change of the effects of path curvature. In Hemami et al. [6] it is concluded (based on dynamic analysis of the same vehicle) that for small steering angle, δ (tan $\delta = \delta$), Equation (20.1) approximates the slow dynamics of the vehicle when its forward velocity is low. In the simulations presented later, we have simplified the robot kinematic model by taking this small steering angle approximation into account. Furthermore, we apply the controller to straight line path following and, therefore, neglect the model effects of path curvature. Such a simplification does not preclude autonomous tracking of reasonably complicated paths since multisegment paths can be defined to be piecewise linear.

To allow for control of the mobile robot, some means of measuring the input information is needed to feed into the system in order to generate a desired output. Thus the system under control is assumed to have some suitable sensory apparatus. For our implementation, we assume that the robot has dead-reckoning/odometry sensors that provide access to the error states at all times, or permit calculations thereof. This sensory input data is then mapped to control outputs according to the desired control policy. In path following simulations, the position and orientation errors in Equation (20.1) are updated using the fourth-order Runge-Kutta method, which is widely used in computer solutions to differential equations [7].

20.2.2 GP Solution

The path tracker to be learned by GP is a two input, single output fuzzy controller that will map the error states into a proper steering angle at each time step. A population of candidate solutions is created from which a solution will emerge. The allowance for rule bases of various sizes enhances the diversity of the population. That is, the GP system creates individuals in the initial population that each have possibly different numbers of rules within a finite range (15-30) specified before a run. In the process of learning fuzzy control rules and membership functions, GP manipulates the linguistic variables directly associated with the controller. Given a desired motion behavior, the search space is contained in the set of all possible rule bases that can be composed recursively from a set of functions and a set of terminals. The function set consists of membership function definitions (describing controller inputs), components of the generic fuzzy *if-then* rule, and common fuzzy logic connectives. More specifically, these include functions for fuzzy sets, rule antecedents and consequents, fuzzy set intersection and union, and fuzzy inference. The terminal set is made up of the input and output linguistic variables and the corresponding membership functions associated with the problem.

Selection of appropriate t-norms is automated, thereby giving the GP system greater control of the evolutionary design. That is, the influence of GP is extended to include selection of the type of t-norm employed to compute the conjunction of fuzzy propositions in the antecedent of a rule. The two most commonly used t-norms for fuzzy control are Mamdani's *min* and Larsen's *product* [8]. T-norms for each conjunctive rule are selected at random by GP for rule bases in the initial population, and are carried along based on fitness in successive generations.

To achieve the goal of evolving membership functions and rules for FLCs, the GP system must conform to strong syntactic constraints when breeding individuals. Special rules of construction were introduced in Tunstel and Jamshidi [2] and later extended using algorithms described in references [9] and [10]. We refer the reader to Homaifar et al. [11] for a detailed description of the resulting syntactic rules, the full design algorithm, and other GP implementation issues related to fuzzy controller design.

20.2.2.1 Controller Fitness Evaluation

Each rule base in the current population is evaluated to determine its fitness value for steering the robot from initial locations near the desired path to final locations on the path such that steady state and final pose errors are minimized. This evaluation involves frequent simulation of the robot's motion from each of a finite number of initial conditions until either the goal state is achieved or the allotted time expires. The initial conditions are referred to as *fitness cases* in the GP community. For this problem we use eight different initial conditions, which is a logical choice given the pair-wise symmetry of the possible error categories illustrated in Figure 20.2. Consider error category (d), which represents a case where the robot is located on the left of the desired path with a negative heading orientation. There also exists a symmetric case where the robot is located on the right of the desired path with a positive heading orientation. These symmetric cases are each represented by error category (d). The same holds for categories (a), (b) and (c) illustrated in the figure, yielding a total of eight fitness cases that fully describe the possible combinations of errors with respect to the path.

The fitness function is a measure of performance used to rank each individual relative to others in the population. We compute path tracking performance by summing the Euclidean norms (normalized) of the final error states plus the average control effort ($\bar{\delta}$) over all eight fitness cases. Thus, the following fitness function drives the evolution process

$$Raw\ Fitness = \sum_{i=1}^{8} \sqrt{(\varepsilon_d^2 + \varepsilon_\theta^2 + \bar{\delta}^2)_i} \qquad (20.2)$$

where ε_d and ε_θ are the position error and orientation error existing at the end of each fitness case simulation. The objective of this fitness function is to minimize final path tracking errors as well as the control effort expended. As such, a perfect fitness score is zero and, in general, lower fitness values are associated with better controllers.

Simulations show that adding average control effort as part of the path tracking metric significantly reduces undesired steering oscillations. Fitness functions based solely on final error states sometimes yielded impractical controllers that exhibited rapid oscillations in the steering control signal, which would cause damage to the steering mechanism of a real mobile robot.

The path tracking success of an individual in the population is also based on its ability to minimize the error states to within the following specified tolerances, $|\varepsilon_d| < 0.15$m and $|\varepsilon_\theta| < 0.26$ radians, for each fitness case. A fitness case simulation in which these tolerances are satisfied is considered a hit, or successful trial. Thus, each individual has the potential of receiving a total of eight hits during fitness evaluation for this path tracking problem.

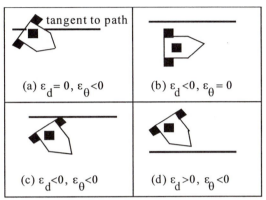

(a) $\varepsilon_d = 0$, $\varepsilon_\theta < 0$ (b) $\varepsilon_d < 0$, $\varepsilon_\theta = 0$

(c) $\varepsilon_d < 0$, $\varepsilon_\theta < 0$ (d) $\varepsilon_d > 0$, $\varepsilon_\theta < 0$

Figure 20.2. Error Categories for Path Tracking Control Problem.

20.3   PATH TRACKING SIMULATION RESULT

In this section, we present representative results of simulated path tracking performance for an evolved controller. Results are presented, in particular, for a fuzzy controller designed with t-norms selected randomly during co-evolution of rules and membership functions. Selection of appropriate t-norms is one of several design decisions that could lengthen the manual trial-and-error procedure typically used by FLC designers. We elect to automate this decision and thereby give the GP system greater control of the evolutionary design. To achieve this, the GP system is allowed to choose at random between the two common t-norms mentioned above. T-norms for each conjunctive rule are selected at random by GP for rule bases in the initial population, and are thus carried along based on fitness in successive generations.

The simulated robot is based on Hemami's kinematic model with dimensions taken from the Heathkit Hero-1 mobile robot. The Hero-1 has a tricycle wheel configuration in which the front wheel is driven by a DC motor and steered by a stepper motor. Its two rear wheels are passive. Dimensions employed are 0.3 m for the wheelbase, and 0.2 m for the offset from the rear axle to the front wheel. These dimensions correspond to the constant lengths $2d$ and MP of Figure 20.1, respectively. All simulations were conducted assuming a controller sampling-rate of 20 Hz and run for a maximum of 10 seconds. In each case, the robot travels at a constant nominal forward speed of 1.5 m/s unless otherwise stated.

All GP runs for the path tracking problem were executed on a 260 MHz MIPS DECstation using a restructured version of the simple genetic programming in C (SGPC) system [12]. Five consecutive runs (initialized using different random number generator seeds) were executed using the GP control parameters listed in Table 20.1. About one hour of computation time is required for a run of this magnitude.

Table 20.1: GP Control Parameters.

Parameter	Value
Population size	200
Number of generations	50
Mutation probability	0.001
Crossover probability	0.600
Maximum mutation depth	4
Reproduction probability	0.399
Maximum new tree depth	5
Maximum depth after crossover	7

A rule base of 25 rules emerged as the fittest among all five runs. This rule base used five conjunctive rules, three employing the Mamdani t-norm and two employing the Larsen t-norm. The evolved input membership functions associated with the best rule base are shown in the left half of Figure 20.3.

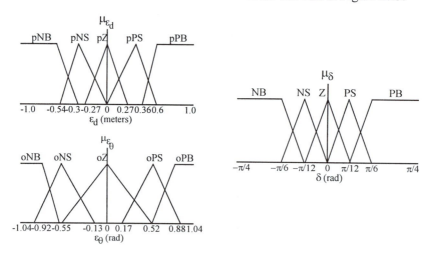

Figure 20.3: Co-Evolved Input Membership Functions and Fixed Output Membership Functions.

Co-evolved rules are listed in Table 20.2, where the notations *NB, NS, Z, PS,* and *PB* represent fuzzy linguistic terms of "negative big," "negative small," "zero," "positive small," and "positive big," respectively. Terms describing the inputs, ε_d and ε_θ, are preceded with the prefixes "p" and "o," respectively. The fixed output membership functions are shown in the right half of Figure 20.3, where the linguistic terms are labeled without prefixes.

Table 20.2: Best Evolved Rule Base.

1	IF oZ THEN NS
2	IF pPB THEN Z
3	IF pNB THEN Z
4	IF pPS THEN NB
5	IF pNS and oPS THEN NS (Mamdani's *min*)
6	IF pNB THEN PB
7	IF oNS THEN Z
8	IF oNB THEN PS
9	IF pNS THEN NS
10	IF pNS and oZ THEN PB (Larsen's *prod*)
11	IF oPB THEN NB
12	IF pNS and oPB THEN NB (Larsen's *prod*)
13	IF pPS THEN NS
14	IF oNS THEN PB
15	IF pPB THEN NB
16	IF oZ THEN PS
17	IF oNB THEN PB
18	IF pNS and oNS THEN PB (Mamdani's *min*)
19	IF pNS THEN Z
20	IF oPS THEN NB
21	IF pZ THEN PS
22	IF pPB and oZ THEN Z (Mamdani's *min*)
23	IF pPB THEN PS
24	IF oPS THEN PS
25	IF oNS THEN PS

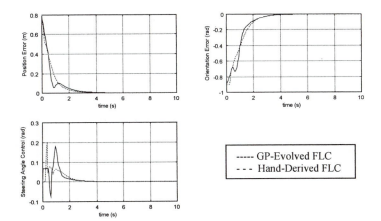

Figure 20.4: Co-evolved FLC Path Tracking Performance.

The evolved controller received a raw fitness of 0.1091 with 8 hits. In Tunstel and Jamshidi [2], an FLC designed manually, through a lengthy process of trial and error, is presented which also used 25 rules. Hours of iterative refinement of membership functions and rules were invested before arriving at a suitable design. In comparison, the hand derived FLC received a comparable raw fitness (0.08 with 8 hits) for the identical tracking problem. Figure 20.4 shows the temporal responses of position error, orientation error, and control effort for the evolved controller and for the hand derived controller. This result corresponds to error category (d) of Figure 20.2, with initial conditions of ε_d = 0.8 m and ε_θ = -0.9 rad. As was shown in Hemami et al.[6], this error category is the most general for studying path tracking by tricycle-type vehicles, in the sense that corrective vehicle steering from states in other error categories ultimately leads to vehicle error status in category (d) or its counterpair. The evolved controller achieved comparable response characteristics to those of the hand derived controller using an equivalent number of rules.

20.3.1   Evolved Controller Robustness

Given the capability to evolve FLCs that can effectively follow paths, an important next step is to examine their robustness to practical perturbations. To test the noise robustness of the evolved controller, simulations were performed with the imposition of a noise signal upon the sensor measurement related to heading (orientation). We assume that the error states are derived from sensor measurements which, due to their imperfect nature, introduce an additive sinusoidal noise signature of small amplitude and low frequency (relative to the controller sampling frequency) that corrupts the orientation error. For this investigation we impose the sensor noise signal, $n(t) = 0.15\cos(3t)$ with $t = kT$, where k=1,2,3,... is the sampling instant, and T is the sampling period. Thus, the noise amplitude is bounded by 0.15 radians (10°), and at any sampling instant the corrupted orientation error signal lies in the range of ($\varepsilon_\theta \pm 0.15$) radians.

In addition to the additive noise, we also increased the constant nominal forward speed of the robot by 20%, which resulted in a simulated speed of 1.8 m/s. A typical result is shown in Figure 20.5, which illustrates the performance of both the evolved controller and the hand derived controller when induced with noise and an increased vehicle speed. While the oscillatory effects of the added noise are clearly evident in the steady state response, the controller successfully navigates the robot onto the path and maintains the steady state errors within the tolerances specified earlier. Thus, this evolved fuzzy controller exhibits path tracking robustness to the imposed perturbations. This result is representative of temporal responses for each of the remaining fitness cases. In simulations completed thus far, the most robust fuzzy controllers were those evolved when GP was allowed to randomly select t-norms.

The performance assessment of the evolved controller with regard to robustness is based upon the assumption that low frequency oscillations within the control signal of amplitude less than 0.026 radians (1.5°) are practical. In light of this assumption, the results indicate that the evolved FLC was able to navigate the robot along the desired path with the imposed perturbation of sensor noise and the increase in the robot's nominal speed.

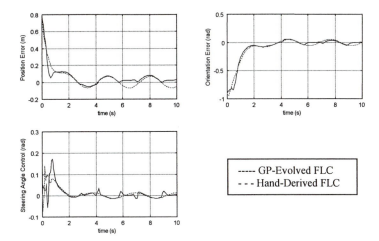

Figure 20.5: Co-evolved FLC Response to Sensor Noise and Increased Forward Speed.

20.4   EVOLUTIONARY PATH PLANNING

Thus far, we have discussed genetic programming techniques for solving a mobile robot tracking problem. We now move on to a related application of evolutionary methods for mobile robot motion planning. Mobile robots that are capable of tracking paths can be used effectively in mapped environments where specific paths from location to location can be designated. If autonomy is desired, the robot software should be capable of using map-based information to plan suitable paths in the operating environment. In the very least, a facility for offline path planning should be available to generate suitable paths for the robot. What constitutes a suitable path depends on the specific features of the application and robot functionality. A single objective, or multiple objectives, may be imposed to define suitable paths for a given application. Many path planning systems consider shortest paths as the primary criterion. However, the shortest path may not always be the most efficient. For example, path smoothness could be of considerable importance. In general, the path planning problem can often be posed as a multiobjective optimization problem. Depending on the nature of the objectives to be met for suitable paths, the formulation of an effective closed-form, multiobjective function to be optimized could be quite difficult. As a way to circumvent such difficulties, a multiobjective *selection* method has been developed for use with an evolutionary path planning system.

The remainder of this chapter presents an approach to path planning that employs evolutionary methods to find suitable paths in a robot's operating environment. The main attributes and evolutionary mechanisms of the path planner are described. In addition, a technique for enhancing path planning performance using fuzzy logic in the evolutionary process is presented.

20.4.1   Evolutionary Path Planning System

In this section, we present the salient attributes of an evolutionary path planning system called GEPOA (global evolutionary planning and obstacle avoidance system), which has been applied to robot planning problems [13]. GEPOA uses steady state reproduction, flat crossover [14] with Gaussian mutation, and uniform mutation in an effort to develop feasible paths. In each generation of path evolution, two parents are selected using tournament selection with a tournament size of two. If the first parent selected represents an infeasible path, it is repaired 50 percent of the time using a method called visibility-based repair (VBR), described below. If the first parent selected is feasible then the two parents create one offspring, which replaces the worst individual in the population. The following attributes of GEPOA will be briefly described: environment and path representation, visibility-based repair, path evaluation and selection functions, and evolutionary operators.

20.4.1.1 Environment and Path Representation

An obstacle within the robot's environment is represented as a set of intersecting line segments. Each line segment connects two distinct vertices. Associated with each vertex within the environment is a value, which represents the number of obstacles that contain it. This value is referred to as the *containment value* (CV) of a vertex. If a vertex lies along the boundary of an environment its CV is assigned a value of infinity. Figure 20.6 provides an example of how obstacles are represented in GEPOA. Notice that the four-sided obstacle (Obstacle 1) is represented by only two lines in GEPOA.

An individual in the evolving population of candidate paths (CPs) contains four fields. The first field is a chromosome, which contains a gene corresponding to the Cartesian coordinates of each node of the path (where nodes of a path are connected by a straight-line segment). The second field is called the seed. The seed of an individual is the gene that will be crossed or mutated to created an offspring. Initially, an individual will have only three genes: the start gene, the seed gene, and the destination gene. Repair genes are inserted into the chromosome by the VBR algorithm each time a straight-line segment of an individual is found to pass through an obstacle. The third field is a value referred to as the violation distance. The violation distance represents the Euclidean distance of the CP, which cuts through one or more obstacles. The fourth field records the Euclidean distance of the path from the start to destination genes.

20.4.1.2. Visibility-Based Repair of Candidate Paths

VBR facilitates construction of valid paths through free space and is performed as follows. When an obstacle, o_i, lies along a straight-line segment between two nodes P and Q, each line of o_i is checked to see if it is intersected by PQ. If a line of o_i is intersected by PQ, then a repair node is created using the following set of rules:

   1.   if the CVs of a line's vertices are both equal to one, then the repair node

is selected to be a point along an extension (the distance outside an obstacle at which a repair node is placed is a user-specified parameter);

2. if the CVs of a line's vertices are different, then the repair node is selected to be a point just outside the vertex which has the lower CV;

3. if the CVs of a line's vertices are greater than one and equal, then the repair node is selected to be a point just outside of the vertex which is farther from the point of intersection.

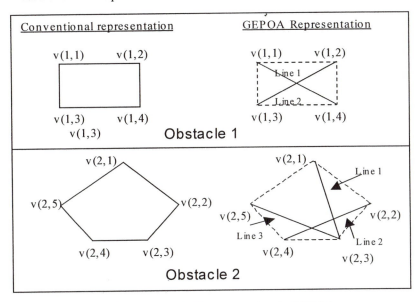

Figure 20.6: Obstacle Representation in GEPOA.

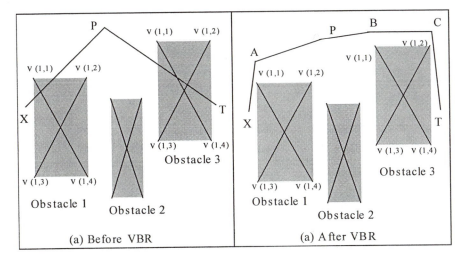

Figure 20.7: Visibility Based Repair of Paths.

Figure 20.7 shows an example of how VBR can be used to transform an infeasible path into one that is feasible. In Figure 20.7a, an infeasible path *XPT* is shown. The path *XPT* is infeasible because the line segment *XP* passes through Obstacle 1 and the line segment *PT* passes through Obstacle 3. Before proceeding further, notice that each vertex in the environment shown in Figure 20.7a has a CV of one.

Using VBR, the line segment *XP* can be repaired to *XAP*. Since *XP* intersects Line 1 of Obstacle 1, a repair node corresponding to a point just outside either v(1,1) or v(1,3) must be selected. By applying Rule 1, Node *A*, which corresponds to a point just outside vertex v(1,1), is selected as the repair node. Similarly, the line segment *PT* can be repaired to *PBCT*. Again Rule 1 must be applied to Line 1 and Line 2 of Obstacle 3. The repair node that results from the intersection of *PT* and Line 1 is Node *B*. The repair node that results from intersection of *PT* and Line 2 is Node *C*. Figure 20.7b shows the result of using VBR on *XPT*. The repaired, feasible version of *XPT* is *XAPBCT*.

Given a candidate path, the VBR algorithm used by GEPOA works as follows. Each obstacle within the environment is checked with each straight line segment from the start gene to the destination gene of the candidate path until a segment is found that passes through the obstacle. The infeasible segment is repaired via VBR and the process is repeated using the next obstacle. As an example of how this repair algorithm works, notice once again Figure 20.7. When given the path *XPT* the algorithm works as follows. Obstacle 1 is checked to see if it is violated by segment *XP*. Since it is, a repair gene (Node *A*) is generated and Obstacle 2 is then considered. Obstacle 2 is checked to see if it is "cut" by segment *XA*. Since it is not cut by segment *XA*, Obstacle 2 is checked with segment *AP* then segment *PT*. Since there are no more segments to inspect, Obstacle 3 is considered. Obstacle 3 is checked to see if it is cut by segments *XA*, and *AP*. Finally, Obstacle 3 is checked to see if it is cut by *PT*. Since it is, two repair genes are generated (Nodes *B* and *C*) and the algorithm terminates.

Since this repair algorithm considers an obstacle only once, it is possible for a repair gene to be generated that creates a line segment that cuts through a previously considered obstacle. Therefore, a candidate path may need to be repaired by this VBR algorithm more than once.

20.4.1.3. *Path Evaluation, Selection, and Evolutionary Operators*

The evaluation function computes the Euclidean distance of each straight line segment of the path that an individual represents as well as the violation distance. GEPOA uses a modified version of tournament selection, with a tournament size of two, to select individuals to become parents. The selection process is as follows. Two individuals are randomly selected from the current population. If the violation distances of the two are different, then the individual with the smaller violation distance is selected to be a parent. If the violation distances are the same then the individual with the smaller overall distance is selected to be a parent.

GEPOA uses two operators along with VBR to create and/or refine individuals. The two operators are as follows: (1) a version of Radcliffe's flat crossover [14], which we refer to as seed crossover and (2) a version of uniform mutation, which we refer to as uniform seed mutation. Seed crossover proceeds as follows. Given two seed genes $s_1 = (x_1,y_1)$ and $s_2 = (x_2,y_2)$, a seed gene for an offspring is created as follows:

$$s_{off} = (rnd(x_1,x_2) + N(0,4.0)\ rnd(y_1,y_2) + N(0,4.0)) \qquad (20.3)$$

where *rnd* is a uniform random number generator and *N(0,4.0)* is a Gaussian random number with zero mean and a standard deviation of 4.0. The resulting offspring has a chromosome containing three genes: a gene corresponding to the node representing the current position of $R(X)$, the seed node, and the destination node. The offspring then undergoes VBR and may have additional repair genes added by the VBR algorithm. In uniform seed mutation, either the x or y coordinate of a parent is mutated using uniform mutation to create a seed gene for an offspring. A resultant offspring created by seed mutation is similar to one created by seed crossover in that it also has a chromosome containing three genes. Once again the offspring undergoes VBR and may have additional repair genes added by the VBR algorithm.

20.5   PATH EVOLUTION WITH FUZZY SELECTION

During evolution of candidate paths the selection of the parent paths that undergo reproduction is based on several objective criteria. This section describes a tournament selection procedure that employs fuzzy logical inference to enhance the performance of the GEPOA system. The fuzzy tournament selection algorithm (FTSA) selects CPs to be parents and undergo reproduction based on:
1. the Euclidean distance of a path from the origin to its destination,
2. the sum of the changes in the slope of a path,
3. the average change in the slope of a path.

As such, the overall objective of the FTSA is to allow evolutionary path planners to evolve CPs that feature minimal distances from start to destination, minimal sums of the changes in slope (SCS), and minimal average changes in slope (ACS). Given two candidate paths (CP_1 and CP_2) that are randomly chosen from the current population, the FTSA takes six inputs – the path distances, the SCS, and the ACS. It returns one output in the continuous interval [-1, 1], which corresponds to the CP that should be selected to be a parent. Any output less than zero indicates that CP_1 is to be selected, while any output greater than zero indicates that CP_2 is to be selected.

20.5.1 Fuzzy Inference System

Let (d_1, s_1, a_1) and (d_2, s_2, a_2) denote the distance, SCS, and ACS for CP_1 and CP_2, respectively. These six inputs are converted into three derived parameters,

d, s, and a, whose computed values lie in [-1,1] according to the following expressions:

$$\left(d = \frac{d_1 - d_2}{d_1 + d_2}, s = \frac{s_1 - s_2}{s_1 + s_2}, a = \frac{a_1 - a_2}{a_1 + a_2} \right) \qquad (20.4)$$

Note that for values of d, s, and a, which are less than zero, the more desirable attribute belongs to CP_1 and vice versa for CP_2. Each of the derived inputs has a domain partitioned by three fuzzy subsets defined using overlapping membership functions. Figure 20.8 shows the three membership functions, where x is d, s, or a. If the value of x is nonpositive then it is a member of the fuzzy set "less than," LT, which represents the set of all tuples (x_1, x_2) such that $x_1 < x_2$. Similarly, values of x that are nonnegative are members of the fuzzy set "greater than," GT, and represent the set of tuples (x_1, x_2) for which $x_1 > x_2$. All values of $|x| < X$ are members of the fuzzy set "equal," EQ, representing the set of all tuples (x_1, x_2) for which x_1 and x_2 are approximately equal. By varying the value of X, the FTSA has the ability to focus on optimizing a particular objective. In the sequel, X is D, S, or A.

The fuzzy rules are formulated as listed below. For each of the seven rules, P represents the singleton consequent of the rule. If the consequent of a rule is ($P = -1$), then the rule has specified that CP_1 should be selected to be a parent. Similarly if a rule's consequent is ($P = 1$) then it has specified that CP_2 should be selected. The defuzzification method used is the Mean of Maxima.

- IF d is LT THEN $P = -1$
- IF d is $EQ(D)$ AND s is LT THEN $P = -1$
- IF d is $EQ(D)$ AND s is $EQ(S)$ AND a is LT THEN $P = -1$
- IF d is $EQ(D)$ AND s is $EQ(S)$ AND a is $EQ(A)$ THEN $P = 0$
- IF d is $EQ(D)$ AND s is $EQ(S)$ AND a is GT THEN $P = 1$
- IF d is $EQ(D)$ AND s is GT THEN $P = 1$
- IF d is GT THEN $P = 1$

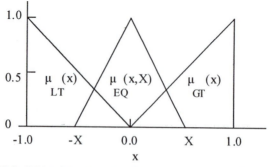

Figure 20.8. FTSA Membership Functions for LT, $EQ(X)$, and GT.

20.5.2 Experimental Example

Here, we present an illustrative example of the utility of the FTSA for enhancing the GEPOA path planning system described above. For fuzzy tournament selection to be effectively used in this type of system, it must be able to adequately rank individuals of a population. Hereafter, let GEPOA+FTS denote GEPOA with fuzzy tournament selection.

Using a hypothetical environment consisting of obstacles distributed throughout enclosed free space, we compared paths evolved by GEPOA and GEPOA+FTS. The parameters for each of these algorithms were as follows: the population size was 20, the flat crossover rate with Gaussian mutation (standard deviation = 4.0) was 0.66, and the uniform mutation rate was 0.34. After the initial population was created, both algorithms were allowed to run for 500 generations, thus, creating a total of 520 individuals. For GEPOA+FTS, we set $D = 0.15$, $S = 0.15$, and $A = 0.15$.

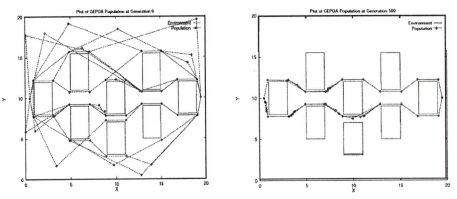

Figure 20.9: Path Population of GEPOA: Generation 0 (left); after Generation 500 (right).

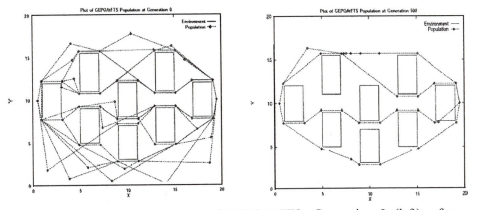

Figure 20.10: Path Population of GEPOA+FTS: Generation 0 (left); after Generation 500 (right).

The left halves of Figures 20.9 and 20.10 show the initial populations that were randomly generated by GEPOA and GEPOA+FTS, respectively. Since GEPOA and GEPOA+FTS use a visibility based algorithm to repair infeasible paths, it is not uncommon for feasible (but suboptimal) paths to appear in the initial population.

The right half of Figure 20.9 shows the population of paths developed by GEPOA after 500 steady state generations. Notice that GEPOA has converged upon the two equal and shortest paths; however, these paths are quite rugged. By contrast, the right half of Figure 20.10 shows the population of paths developed by GEPOA+FTS after 500 steady state generations. First of all notice that GEPOA+FTS has converged upon a number of good paths. Notice also that among the paths evolved with GEPOA+FTS, the shortest path is still represented. The fact that it is infeasible is not a major concern because it has a chance of being repaired! Not only does fuzzy tournament selection allow evolutionary search to converge upon the best path, but it also allows for a great deal of valuable and much needed diversity.

20.6   SUMMARY AND CONCLUSIONS

GP was successfully applied to discover FLCs capable of navigating a mobile robot to track straight-line paths in the plane. The overall performance of the best evolved rule bases was comparable to that of a manually designed rule base that utilized more rules in most instances. Instances of simultaneous evolution of membership functions and rules showed that GP was capable of evolving a FLC that demonstrated satisfactory responsiveness to various initial conditions while utilizing minimal human interface. Suboptimal solutions with respect to the employed fitness function were consistently found, demonstrating that GP performs well as a global adaptive search method. Further automatic improvement towards optimal solutions can be made by synthesizing a hybrid between GP and a localized search method such as hill climbing [15, 16].

GP was also applied to larger population sizes facilitated by the dramatic speed increase of our coding implementation in C vs. the previously investigated LISP implementation. The 82% increase in speed of evolution alone serves as a strong basis for practical application of GP in the controller design process. The approach provides a means for expeditious design of FLCs that can be directly applied to a physical system. Alternatively, human experts can use the rapidly evolved FLCs as design starting points for further manual refinement [4]. To assess the practicality of the GP solutions, robustness characteristics of evolved FLCs were examined. The controllers evolved with random selection of fuzzy t-norms were particularly robust when subject to imposed perturbations of sensor noise and an increase in nominal robot speed. The results support the notion that a genetically evolved fuzzy logic controller can have practical utility.

Fundamental features of the GP system include manipulation of linguistic variables directly associated with the fuzzy system (as opposed to numerical encoding/decoding), a syntactic structure that provides context preservation via structure preserving genetic operators, and provision for evolving rule bases of various sizes in a single population. These features were inherited from our previous implementation in LISP. Beyond these, the implementation proposed herein provides several improvements and extensions that make GP a more powerful tool for FLC design. Namely, we have dramatically improved upon the required speed of evolution and extended the system to handle full FLC design, including evolution of the most appropriate t-norms for the controlled system.

In the second part of the chapter we presented an evolutionary algorithmic approach to robot path planning. It serves as an effective means of circumventing the difficulties associated with formulating complex multiobjective functions for suitable paths. A multiobjective selection method based on fuzzy logic was applied with an evolutionary path planning system. The fuzzy tournament selection algorithm can be used for multiobjective path planning by almost any evolutionary based motion planning system. Despite the simple nature of the fuzzy inference system employed, the FTSA exhibits complex behavior. The approach allows evolutionary search to converge upon a diversity of optimal and/or near optimal paths. The availability of alternative feasible paths is important in the event that a local navigation system cannot traverse a particular global path. This can happen, for example, when unfavorable conditions are sensed locally, replanning becomes necessary, or task constraints intervene.

ACKNOWLEDGMENTS

This work is partially funded by grants from NASA Autonomous Control Engineering Center (ACE) at North Carolina A&T SU under grant # NAG2-1196 and NASA Dryden Flight Research Center under grant # NAG4-131. The authors wish to thank the ACE Center and NASA Dryden for their financial support. A portion of the research described in this chapter was performed at the Jet Propulsion Laboratory, California Institute of Technology, under contract with the National Aeronautics and Space Administration.

REFERENCES

1.   Koza, J. R., *Genetic Programming: On the Programming of Computers by Means of Natural Selections*, MIT Press, Cambridge, MA, 1992.

2.   Tunstel, E. and Jamshidi, M., On Genetic Programming of Fuzzy Rule-Based Systems for Intelligent Control, *Intl. J. Intell. Automation Soft Computing*, 2(3), 271, 1996.

3.   Tunstel E., Lippincott, T., and Jamshidi, M., Behavior Hierarchy for Autonomous Mobile Robots: Fuzzy-Behavior Modulation and Evolution, *Intl. J. Intell. Automation Soft Computing*, 3(1), 37, 1997.

4.   Alba, E., Cotta, C., and Troyo, J., Type-Constrained Genetic Programming for Rule-Base Definition in Fuzzy Logic Controllers, *1st Ann. Conf. on Genetic Program.*, 28, Palo Alto, CA, 1996.

5.   Hemami, A., Steering Control Problem Formulation of Low-Speed Tricycle-Model Vehicles, *Intl. J. Control*, 61(4), 783, 1995.

6.   Hemami, A., Mehrabi, M., and Cheng, R., Optimal Kinematic Path Tracking Control of Mobile Robots with Front Steering, *Robotica*, 12(6), 563, 1994.

7.   Gerald, C. and Wheatley, P., *Applied Numerical Analysis*, Addison-Wesley, Reading, MA, 1989.

8.   Lee, C., Fuzzy Logic in Control Systems: Fuzzy Logic Controller, Part I, *IEEE Trans. Syst., Man & Cybern.*, 20(2), 404, 1990.

9.   Battle, D. D., Implementation of Genetic Programming for Mobile Robot Navigation, M.S. Thesis, Department of Electrical Engineering, North Carolina A&T State Univ., Greensboro, NC, 1998.

10.   Homaifar, A. and McCormick,E., Simultaneous Design of Membership Functions and Rule Sets for Fuzzy Controllers using Genetic Algorithms, *IEEE Trans. Fuzzy Systems*, 3(2), 129, 1995.

11.   Homaifar, A., Battle, D., Tunstel, E., and Dozier, G., Genetic Programming Design of Fuzzy Logic Controllers for Mobile Robot Path Tracking, *Intl. J. Knowledge-Based Intell. Eng. Syst.*, 1999.

12.   Tackett, W. and Carmi, A., SGPC: Simple Genetic Programming in C, *Prime Time Freeware for AI*, 1(1), 1993.

13.   Dozier, G., Esterline, A., Homaifar, A., and Bikdash, M., Hybrid Evolutionary Motion Planning Via Visibility-Based Repair, *IEEE Intl. Conf. on Evolutionary Computation*, 1997.

14.   Eshelman, L. J. and Shaffer, J. D., Real-Coded Genetic Algorithms and Interval-Schemata, in *Foundations of Genetic Algorithms II*, Whitley, L. D. (ed.), Morgan Kaufmann, San Francisco, CA, 1993.

15.   Dozier, G., Bowen, J., Homaifar, A., and Esterline, A., Solving Randomly Generated Static and Dynamic Fuzzy Constraint Networks using Microevolutionary Hill-Climbing, *Intl. J. of Intell. Automation Soft Computing*, 3(1), 51, 1997.

16.   O'Reilly, U.M., An Analysis of Genetic Programming, Ph.D. Dissertation, School of Computer Science, Carleton University, Ottawa, Ontario, 1995.

21

PROBLEMS AND MATLAB PROGRAMS

Ali Zilouchian and Mo Jamshidi

21.1 INTRODUCTION

This appendix serves two purposes. First, it provides readers with problems and exercises related to NN and FL and their applications. Second, it presents the MATLAB programs and solutions to problems in the text. The problems are identified based on the subject matter as discussed in the book.

21.2 NEURAL NETWORK PROBLEMS

Chapter 2

1. Consider following the sigmoid function:

$$f(x) = \frac{1}{1+e^{-\alpha x}}$$

(a) What are the upper and lower limit of this function for constant α? Obtain the value of f(x) at x=0

(b) Show that the derivative of f(x) with respect to x is given by:

$$\frac{df}{dx} = f'(x) = \alpha.f(x)[1-f(x)]$$

(c) How would you modify f(x) such that its value at x=0 is equal (i) 0.125; (ii)0.8

(d) What is the value of f'(x) at the origin?

2. Consider the following hyperbolic activation function

$$g(x) = \frac{e^{\alpha x} - e^{-\alpha x}}{e^{\alpha x} + e^{-\alpha x}}$$

(a) What are the upper and lower limits of this function?

(b) Show the derivative of g(x) is given as

$$\frac{dg}{dx} = g'(x) = 2\alpha[1-g^2(x)]$$

455

(c) What is the value of g'(x) at the origin?

3. Consider the activation function f(x) shown below:

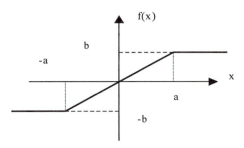

(a) Formulate f(x) as a function of x

(b) Obtain f(x) if either a or b or both are allowed to approach zero.

4. A neuron m receives sensory information from five inputs with the values of 8, -10, 4, -2 and 5. The synaptic weights of neuron m are 0.8, 2.0, 1.0, -0.9 and 0.6. Calculate the output of neuron for the following three situations:

(a) The neuron is a linear model

(b) The neuron is represent by a McCulloch-Pitts model. (Hard limit activation function with negative threshold zero)

(c) The neuron is represented based on a sigmoid function as follows:

$$f(x) = \frac{1}{1 + \exp(-x)}$$

5. Consider the following network:

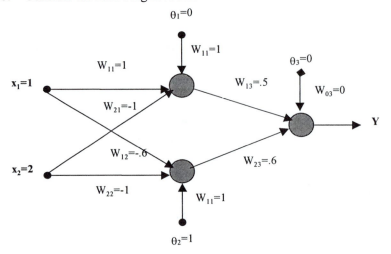

Obtain the output Y for the following cases:
(a) All the neurons are represented by a McCulloch-Pitts model (hard limit activation function with negative threshold zero)
(b) All the neurons are represented based on a sigmoid activation function.

6. Suppose you would like to implement the following logic gates using NN

   (a) OR gate

   (b) AND gate

   (c) XOR gate

   for each case, could you utilize one (or two) hidden layer(s) with linear activation function to achieve your goal?
   If your answer is yes, justify your answer.
   If your answer is no, suggest an alternative solution.

Chapter 3

1. Consider a multi layer feed forward network, all the neurons, which operate in their linear regions. Justify the statement that such a network is equivalent to a single layer feed forward network.

2. Consider the following network with the inputs and outputs as follow:

$$i_1=0.8, \ i_2=1, \ i_3=0.9 \text{ with } d_o=d_1=1$$

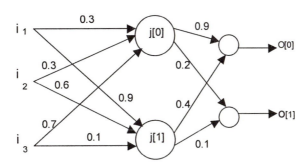

   (a) Derive a step-by-step procedure using back propagation algorithm for two complete iterations of the network with the sigmoid activation functions at the first stage and linear function at the output stage with the initial weights as shown.

   (b) Solve (a) using MATLAB.

   (c) Solve (a) using MATLAB with tangent hyberbolic activation function at the first stage and the linear function at the output stage.

(d) Suppose you select different initial conditions. Are the final weights the same? Justify your answer.

(e) Generate three different random initial weights. Obtain the final weight of the network after training with 0.1 error goal. Compare your results.

3.   Suppose for problem 2, the inputs and the outputs are as follow:

$$i_1 = 12 , i_2 = 3, i_3 = 8 \text{ and } d_o = 9, d_1 = 1$$

Solve the problem with appropriate scaling of the inputs and outputs data sets.

4.   Repeat problem 2 for a radial basis function network.

5.   Compare the radial basis functions neural network and back propagation in term of various aspects such as training phase, recall phase, convergence, and applications.

6.   Consider a two link serial robot manipulator as shown:

(a) Write the forward kinematics equation for the robot.

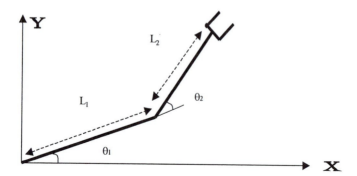

(b) Show how an NN architecture can be used to solve the forward kinematics problem. Is the solution unique?

(c) How do you solve inverse kinematics of the robot using NN. Is the solution unique? Justify your answer,

7.   Consider the two-link robot manipulator of the previous example with $L_1 = 2$(m), and $L_2 = 3$(m).

(a) Generate at least 100 data points for x and y, given the following trajectory:
$\theta 1 = 0.005*t$
$\theta 2 = 0.005*t$
where $\theta 1$ and $\theta 2$ are uniformly distributed in the first quarter and the range of t is from 0 to 500.

(b) Use two different back propagation algorithms (with the adaptive rate and one hidden layer) to solve the forward kinematics problem of the two link robot. The first 80 data points from data set as generated in part (a) should be used in the training phase.

(c) Change the number of hidden layers to 2 and solve the problem as indicated in (c).

(d) Use radial basis network to solve (b).

(e) Compare your results for (b)-(d).

(f) Confirm the fidelity of your trained NN using the remaining 20 data sets for each part (b-e). Draw the error plots for the comparison.

8. Consider a Hopfield network made up of five neurons, which is required to store the following three fundamental memories:

$$\zeta_1 = [+1 \quad +1 \quad +1 \quad +1 \quad +1]^T$$
$$\zeta_2 = [+1 \quad -1 \quad -1 \quad +1 \quad -1]^T$$
$$\zeta_3 = [-1 \quad +1 \quad -1 \quad +1 \quad +1]^T$$

(a) Evaluate the 5-by-5 synaptic weight matrix of the network.

(b) Demonstrate that all three fundamental memories, ξ_1, ξ_2, and ξ_3 satisfy the alignment condition, using asynchronous updating.

(c) Show that

$$\zeta_1 = [-1 \quad -1 \quad -1 \quad -1 \quad -1]^T$$
$$\zeta_2 = [-1 \quad +1 \quad +1 \quad -1 \quad +1]^T$$
$$\zeta_3 = [+1 \quad -1 \quad +1 \quad -1 \quad -1]^T$$

are also fundamental memories of the Hopfield network.

9. Consider a simple Hopfield network made up of two neurons. The synaptic weight matrix of the network is

$$W = \begin{bmatrix} 0 & -1 \\ -1 & 0 \end{bmatrix}$$

The threshold applied to each neuron is zero. The four possible states of the network are

$$S_1 = [+1 \quad +1]^T, \quad S_2 = [-1 \quad +1]^T$$
$$S_3 = [-1 \quad -1]^T \quad S_4 = [+1 \quad -1]^T$$

(a) Using the alignment (stability condition), show that states S2 and S4 are stable.

(b) What are the statuses of S1 and S3?

10. Consider a simple Hopfield network made up of four neurons. The synaptic weight matrix of the network is given as:

$$W = \begin{bmatrix} 0 & 1 & 1 & -1 \\ 1 & 0 & 1 & -1 \\ 1 & 1 & 0 & -1 \\ -1 & -1 & -1 & 0 \end{bmatrix}$$

The threshold applied to each neuron is zero. There are $2^4=16$ possible states of the network.

(a) Using the alignment stability condition, obtain the stable states (fundamental memories).

(b) Show the architectural graph of the Hopfield network.

(c) Suppose the network is initialized at x0=[1, 1, 1, 1]. Show that the network will converge to the nearest fundamental memory (equilibrium state) after iteration.

21.3   FUZZY LOGIC PROBLEMS

Chapter 8

1. Design membership functions to describe the linguistic terms "tall", "average", and "short".

2. Let $A = \dfrac{0.3}{1} + \dfrac{1}{2} + \dfrac{0.2}{3}$ and $B = \dfrac{0.5}{1} + \dfrac{0.6}{2} + \dfrac{0.2}{3}$. Find the following:

   (a) $A \cup B$

   (b) $A \cap B$

   (c) $\overline{A \cup B}$

   (d) $\overline{A \cap B}$

3. Let $A = \dfrac{1}{x_1} + \dfrac{0.9}{x_2} + \dfrac{0.85}{x_3} + \dfrac{0.75}{x_4} + \dfrac{0.5}{x_5} + \dfrac{0.1}{x_6} + \dfrac{0}{x_7}$.

   Find the α-cut sets A_1, $A_{0.8}$, $A_{0.2}$, and A_0.

4. For fuzzy sets A, B, and C defined on the universe
   X={0, 1, 2, 3, 4, 5, 6, 7,8}

   A={0.1/2, 0.7/3, 1/4, 0.3/5, 0.2/6}
   B={0.2/1, 0.3/2, 0.6/3, 1/4, 0.7/5, 0.4/6, 0.1/7}
   C={.4/2, .8/4, 1/5, .6/7, .4/8}

   Answer the following questions:

   (a) Compute the intersections and unions of the fuzzy sets A, B, and C.

   (b) Determine the intersection and union of the complements of fuzzy set B
       and C.

   (c) What are the cardinalities and relative cardinalities of the above fuzzy
       sets?

   (e) Which of the above fuzzy sets are convex and which are not?

5. Consider two fuzzy sets $A_1 = \dfrac{0.1}{x_1} + \dfrac{0.3}{x_2} + \dfrac{0.2}{x_3} + \dfrac{0.5}{x_4}$ and

   $A_2 = \dfrac{0.5}{y_1} + \dfrac{1}{y_2} + \dfrac{0.3}{y_3}$.

   Determine the fuzzy relation among these sets.

6. Consider two fuzzy relations

   $$R = \begin{bmatrix} 1 & 0.1 & 0.3 \\ 0.8 & 0.6 & 0.9 \\ 0 & 0.3 & 0.2 \end{bmatrix} \text{ and } S = \begin{bmatrix} 0.5 & 0.7 & 0.2 \\ 0.2 & 0.4 & 0.9 \\ 0.7 & 0.6 & 0.1 \end{bmatrix}$$

   It is desired to evaluate $R \circ S$ using (a) min-max, and (b) max-product.

7. Let S and R be the matrix representation of fuzzy relations

   $$R = \begin{bmatrix} 0.6 & 0.3 & 0.1 \\ 0.8 & 0.9 & 0.2 \end{bmatrix} \text{ and } S = \begin{bmatrix} 0.5 & 0.8 \\ 0.8 & 0.9 \\ 0.5 & 0.1 \end{bmatrix}$$

   Calculate $R \circ S$, $\overline{R} \circ \overline{S}$ & $(R \cup S) \circ S$ using (a) min-max, and (b)
   max-product.

Chapter 9

1. Prove the truth value of the modus ponens deduction

    $$(A \wedge (A \rightarrow B)) \rightarrow B \qquad \qquad (\textit{Modus Ponens})$$

2. Prove the truth value of the modus tollens inference

    $$(\overline{B} \wedge (A \rightarrow B)) \rightarrow \overline{A} \qquad \qquad (\textit{Modus Tollens})$$

3. Let two universes of discourse be described by X={3,4,5,6} and Y={1,2,3} and define the crisp set A={4,5} on X and B={1,3} on Y. Determine the deductive inference IF A, THEN B.

4. Let three universes of discourse be described by X={5,6,7}, Y={1,2,3,4,5} and define the crisp set A={5,6} on X, B={1,3,5} on Y and C={1,6} on Y. Determine the deductive inference IF A, THEN B, ELSE C.

Chapter 10

1. Given 3 fuzzy sets

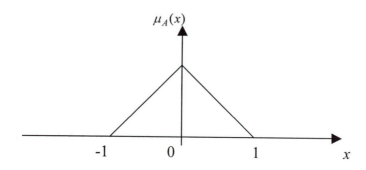

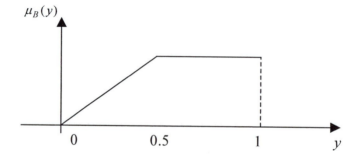

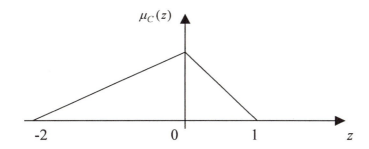

Find the consequent, z, of the following rule

$$\text{IF } x \text{ is } A \cap B \text{ AND } y \text{ is } A \cup B \text{ THEN } z \text{ is } C \cup C$$

2. Design a set of fuzzy rules to control a braking system of a car. Use car speed and distance from an object as the input fuzzy variable, and the strength of braking as the output variable.

3. A fuzzy system is represented by

$$P^1 : \text{IF } x(k-1) \text{ is } A^1 \text{ THEN } x^1(k+1) = 2x(k) - 0.5x(k-1)$$

$$P^2 : \text{IF } x(k-1) \text{ is } A^2 \text{ THEN } x^2(k+1) = -x(k) - 0.5x(k-1)$$

where A^i, i=1,2 are shown in Chapter 10, Figure 10.15. Check if the system is stable by Lyapunov's method. Use x(0)=-1 and x(1)=0 and verify your answer by simulation.

4. Consider a fuzzy feedback control system of the type shown in Chapter 10 Figure 10.10 with the following implications:

$$P^1 : \text{IF } x(k) \text{ is } A^1 \text{ THEN } x^1(k+1) = 1.85x(k) - 0.65x(k-1) + 0.35u(k)$$

$$P^2 : \text{IF } x(k) \text{ is } A^2 \text{ THEN } x^2(k+1) = 2.56x(k) - 0.135x(k-1) + 2.22u(k)$$

$$C^1 : \text{IF } x(k) \text{ is } A^1 \text{ THEN } f^1(k+1) = k_1^1 x(k) - k_2^1 x(k-1)$$

$$C^2 : \text{IF } x(k) \text{ is } A^2 \text{ THEN } f^2(k+1) = k_1^2 x(k) - k_2^2 x(k-1)$$

where A^i, i=1,2 are shown in Chapter 10, Figure 10.17.

Find the closed-loop implications S^{ij}, i=1,2, and j=1,2. Notice that in this case only three rules are needed to describe the closed-loop system. Also find the appropriate feedback gains to stabilize the system.

21.4   APPLICATIONS

1. Consider a nonlinear multivariable dynamic process with two inputs and two outputs. The actuator inputs and the measured outputs data as well as the desired outputs of the plant are provided.

   (a) Draw an identification block diagram to obtain the model of the physical process by NN. What are the advantages/disadvantages to utilized your proposed method in comparison to conventional identification algorithms?

   (b) Suppose you would like to control the given process, explain how you would design a NN controller to achieve your goals. What type of NN algorithm would you utilize for your proposed design and why? Draw the block diagram of your design and various steps to achieve the design goals.

   (c) How would you design a controller using both the NN and conventional methodologies? Draw the block diagrams and explain your design strategy.

2. Consider the D.C. motor given in the figure below. Suppose the transfer function and the tacho-generator of the DC motor are given as:

$$G(s) = \frac{\theta(s)}{V_i(s)} = \frac{K_M}{S+2}$$

$$V_o(t) = K_T \theta(t)$$

$$K_M = 100, \ K_T = 20$$

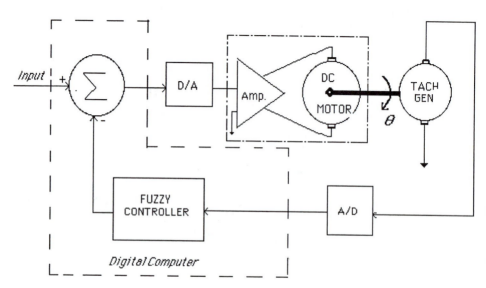

(a) Design a fuzzy controller for angular velocity control of the system.
(b) Please provide step-by-step implementation procedures for the controller design.
(c) Simulate the system using simulink and fuzzy toolbox.
(d) For two various inputs (e.g. step inputs and ramp) simulate your design.
(e) How is the robustness of your proposed fuzzy controller in the present of 10% on DC gain (K), and pole variations?

3. Consider a three-link serial robot manipulator (planar) as shown :

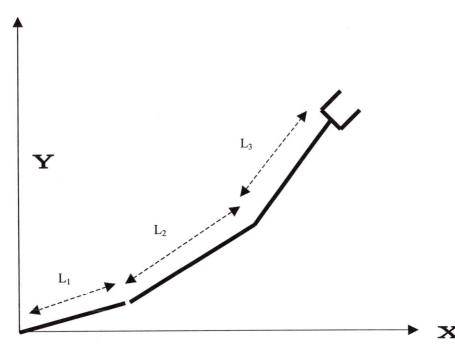

(a) Write the forward kinematics equation for the robot.

(b) How do you solve inverse kinematics of the robot using ANFIS? Is the solution unique?

(c) Simulate the inverse kinematics of the robot for the given values (L1=1 m, L2=1.5 m, L3=1.2 m) using fuzzy logic toolbox.

(d) Draw three dimension plots of joint angles v.s. x and y similar to fuzzy demo plots in fuzzy logic toolbox.

(e) Obtain the errors (both L2 and L*inf*) in comparison with the original values.

4. Consider the controller design for a central air condition system. Suppose you would like to have a steady room temp. around 84° F. The outside temperature can be varied relative to the room temperature depending upon various seasons and weather conditions. The ranges of changes include "very cold", "cold", "medium", "hot", and "very hot" temperature.

   (a) Design a fuzzy logic controller (FLC) for this system. Draw the major block diagrams for the process as well as FLC.

   (b) Show step-by-step your FLC design.

21.5   MATLAB PROGRAMS

Example 3.1:
   The following is the MATLAB code for Example 3.1. A simple feed forward network has been defined and trained to map the input P to output T.

```
P=[1; 0; 1];   % Input Sample
T=[0; 1];       %Desired Output
% Creating the network
net = newff([0 1 ; 0 1 ; 0 1],[2 2],{'tansig' 'purelin'},'traingd');
net.iw{1,1}=[0.1 0.6 0.8;0.2 0.3 0.9]; % Input Weights
net.b{1}=[0; 0];  % Input Threshold Weights = 0
net.lw{2,1}=[0.5 0.7;0.6 0.5]; % Hidden Layer weights
% Printing the Weights
input_layer_weight = net.iw{1,1}
Hidden_layer_weight = net.lw{2,1}
Bias_weights = net.b{1}
Initial_Output=sim(net,P);
% Setting Training Parameters
net.trainParam.lr = 0.5;
net.trainParam.epochs = 50;
net.trainParam.goal = 0.001;
net = train(net,P,T);  % Training the Network
% Evaluating the results
Final_Output = sim(net,P)
input_layer_weight = net.iw{1,1}
Hidden_layer_weight = net.lw{2,1}
Bias_weights = net.b{1}
```

Example 3.2:
   The following MATLAB code is for Example 3.2. It is trained to solve the forward kinematics of a robot manipulator. A path is defined in an excel file for the following trajectory:

$\theta_1 = 0.03$ t
$\theta_2 = 0.03$ t

The network is trained and then the result is compared to the desired value.

```
% Reading the training set from excel file.
A = wk1read('A:\hmwk55.wk1',0,3,'D1..E201');
B = wk1read('A:\hmwk55.wk1',0,6,'G1..H201');
Theta=A';
X=B';
% Creating the Network
net = newff([-pi pi;-pi pi],[15 2],{'tansig' 'purelin'});
% Setting Training Parameters
net.trainParam.goal = 0.001;
net.trainParam.epochs = 50;
net = train(net,Theta,X); % Training the network
Output=sim(net,Theta); % Evaluating the output to the same input data
plot(Output(1,:),Output(2,:),'k+', X(1,:),X(2,:),'ko');
 % Plotting the desired value and output of the Network
```

Example 3.3:
   The following is the MATLAB code for solving Example 3.3. In the example, the problem 3.1 has been solved using a radial basis function network. Then the results have been evaluated for an input similar to the training input.

```
P=[1; 0; 1]  % Input Sample
T=[0; 1] % Desired Output
net = newrb(P,T) % Defining RBFN
Output=sim(net,P) % Evaluating Output
P1=[1.1; -0.3; 0.9]; % A Sample Input other than training set
Output=sim(net,P1) % Output of the network for sample Input
```

Example 3.4:
   This is the MATLAB code for Example 3.4. It solves forward kinematics of robot manipulator of problem 3.2 using an RBFN.

```
% Reading the training set from excel file.
A = wk1read('A:\hmwk55.wk1',0,3,'D1..E201');
B = wk1read('A:\hmwk55.wk1',0,6,'G1..H201');
Theta=A';
X=B';
net = newrb(Theta,X,0.001); % Creating  and Training of the Network
Output=sim(net,Theta); % Calculating the Output
plot(Output(1,:),Output(2,:),'k+',X(1,:),X(2,:),'ko'); % Plotting the Result
```

Example 3.5:
   The following is the MATLAB code for Example 3.5. A Kohonen network has been defined and trained with an input path. It can be seen that the weights of the network are in the form of the training path.

```
angles = -0.5*pi:0.5*pi/99:0.5*pi;
P=[sin(angles);cos(angles)];
figure(1);
plot(P(1,:),P(2,:),'k+');
net=newsom([0 1;0 1],[10]);
net.trainParam.epochs=1200;
net=train(net,P);
figure(2);
plotsom(net.iw{1,1},net.layers{1}.distances)
```

Example 3.6:
   The following is the MATLAB code for Example 3.6. In this example 1000 random input points have been generated. Then a Kohonen network has been trained using this input set. The plot of the weights after training shows that the weights of the network have a uniform distribution.

```
figure(1);
T=rands(2,1000); % Creating 1000 Random Inputs
plot(T(1,:),T(2,:),'k+'); % Plotting the original random input pattern
net=newsom([0 1;0 1],[5 6]);
figure(2);
plotsom(net.iw{1,1},net.layers{1}.distances); % Plotting the Initial Weights
net.trainParam.epochs=1500;
net=train(net,T); % Training the Network
figure(3);
plotsom(net.iw{1,1},net.layers{1}.distances); % Plotting the Final Pattern
```

INDEX